Understanding and Using
Structural Concepts

Second Edition

Understanding and Using
Structural Concepts

Second Edition

Tianjian Ji

The University of Manchester, United Kingdom

Adrian J. Bell

The University of Manchester, United Kingdom

Brian R. Ellis

Former technical director, BRE, United Kingdom

CRC Press
Taylor & Francis Group
Boca Raton   London   New York

CRC Press is an imprint of the
Taylor & Francis Group, an **informa** business

A SPON PRESS BOOK

CRC Press
Taylor & Francis Group
6000 Broken Sound Parkway NW, Suite 300
Boca Raton, FL 33487-2742

© 2016 by Taylor & Francis Group, LLC
CRC Press is an imprint of Taylor & Francis Group, an Informa business

---

Library of Congress Cataloging-in-Publication Data

Ji, Tianjian.
    [Seeing and touching structural concepts]
    Understanding and using structural concepts / Tianjian Ji, Adrian J. Bell and Brian R. Ellis. -- Second edition.
        pages cm
    "The second edition of Seeing and Touching Structural Concepts with a revised title, Understanding and Using Structural Concepts"--Preface.
    Includes bibliographical references and index.
    ISBN 978-1-4987-0729-9
    1. Structural analysis (Engineering) I. Bell, Adrian J. II. Ellis, Brian R. III. Title.

TA645.J53 2016
624.1'71--dc23   2015024987

Visit the Taylor & Francis Web site at
http://www.taylorandfrancis.com

and the CRC Press Web site at
http://www.crcpress.com

Contents

Preface

This is the second edition of *Seeing and Touching Structural Concepts* with a revised title, *Understanding and Using Structural Concepts*. The previous title highlighted ways to aid understanding of structural concepts, whilst the current title emphasises both understanding and application of structural concepts. The objective of understanding structural concepts is to enable their use in a creative way to solve practical engineering problems, leading to safe, economical and elegant designs.

This change of the title reflects the nature and our understanding of structural concepts. Structural concepts are defined in this new edition as follows:

A structural concept is *a qualitative and concise representation of a **mathematical** relationship between **physical** quantities, which captures the essence of the relationship and provides a basis for **practical** applications in structural engineering.*

This definition shows that structural concepts present a sound understanding of theory, bring together mathematics, physical understanding and engineering practice as a whole and indicates a suitable way to learn and to study.

Structural concepts can be learnt from theory in textbooks, in an abstract way, but they can also be learnt, in an intuitive way, through the use of physical models and practical examples. The latter method appears more enlightening, inspiring and interesting, but greater numbers of physical models and appropriate practical examples need to be made available in textbooks. It would be ideal if the two ways of learning could be integrated to complement each other. This book aims to develop such an intuitive way of learning structural concepts in conjunction with conventional learning from theory. This is achieved through:

- Providing a series of simple demonstration models for illustrating structural concepts, or the effect of the concepts, which provides an easy and quick way to comprehend theory.
- Providing associated examples in engineering practice and in everyday life to demonstrate the innovative application of structural concepts, which illustrates the use of structural concepts and helps to bridge the gap between knowledge and practice.
- Converting appropriate research output, which particularly involves structural concepts, into teaching material to improve and update existing course contents.

Physical models are real and simple and can be created to illustrate particular structural concepts. The models in this book are often presented in pairs, one with the involvement of a structural concept, and one without. Thus, the different behaviour of the two can be seen and felt. The models can be demonstrated by lecturers in class or used by students after class. Many photographs are used to illustrate these models and their actions.

Practical examples, in which structural concepts are used innovatively, will show how structural concepts are actually used to produce effective engineering solutions. These examples take students from structural elements to whole structures, and from understanding to application. They also provide an opportunity to appreciate the creativity of engineers and architects in producing solutions to practical problems.

The study of structural concepts brings together practice, research and teaching – each contributing to, and benefiting from, the other two. Research based on emerging engineering problems has led to the identification of new concepts and to the provision of new examples, which effectively enrich the teaching and learning of structural concepts. Some of the contents in several chapters in this book are developed from such research.

In addition to achieving a better understanding, real interest in learning can be generated by seeing many physical models and practical examples. It has been observed in class situations that students show a greater interest in topics which are demonstrated physically than in topics that are explained by words and blackboard/overhead projector/PowerPoint presentations alone. Students show an even greater interest in practical examples which illustrate the use of concepts to solve engineering problems, rather than in coursework examples. Students are motivated by 'hands on' experience and by linking concepts and models to real engineering problems.

The new edition makes the following changes and improvements:

- An introduction chapter is provided to give an overview of structural concepts and a guide on how to learn and to use this book.
- A new part, Synthesis, is added to the existing Statics and Dynamics parts. It contains four new chapters discussing the relationships between static and modal stiffnesses, between static and dynamic problems, between experimental and theoretical studies and between theory and practice.
- A small number of problems are provided at the end of each chapter in the Statics and Dynamics parts.
- Some new models and examples are added.

The Statics part contains 12 chapters and the Dynamics part 7 chapters. The new part, Synthesis, aims to integrate statics and dynamics. Static stiffness and modal stiffness are widely used in engineering practice and they are normally treated independently. Chapter 21 studies such a relationship, provides a theoretical basis for the relationship and explores its applications. The relationships between some static and dynamic problems are examined in Chapter 22, allowing some statics methods to be used to solve problems in dynamics, and some dynamics methods to solve problems in statics, to produce effective solutions to practical problems. The last two chapters present evidence-based studies. Chapter 23 develops relationship models between experimental and theoretical studies, which appear to be research oriented. This is stimulated by the modelling used in economics and illustrated using many real cases. The relationships between theory and practice are explored in Chapter 24, through reviewing the role of structural concepts. The content in this chapter may not yet be mature or complete, but the exploration is supported through the use of many examples.

The contents of this book are an outcome of our many years of teaching, research and engineering practice in which structural concepts are concerned. This book provides numerous demonstrations using physical models and practical examples. A significant amount of material, not found in current textbooks, is included, to enhance the understanding of structural concepts and stimulate interest in learning, creative thinking and design. This book will be of interest to all engineers, from students to consultants. It will be useful to civil and structural engineering students, including postgraduates, in all years of their courses, as well as the more technically minded architecture students and practising engineers.

Accompanying this book, we have created a website, Seeing and Touching Structural Concepts, which can be found at www.structuralconcepts.org. The website contains most of the contents presented in the first, third and fourth sections of each chapter in the Statics and Dynamics parts of the book, and selected student submissions for their coursework on Understanding and Using Structural Concepts.

Tianjian Ji, Adrian J. Bell
University of Manchester, UK

Brian R. Ellis
St. Albans, UK

Acknowledgements

Our work on this topic began in 1999. We started identifying the concepts in textbooks for teaching undergraduates in civil and structural engineering, which can be demonstrated using physical models and the real examples from engineering practice showing the application of the concepts. The work, however, could not be presented in the current form without the input of others.

Several previous undergraduate students at the University of Manchester carried out investigative projects on Seeing and Touching Structural Concepts. They contributed their understanding through personal experience of study in class and made a number of physical models. Among them, Dr. Wai Leng Yip, Mr. Tom Eccles, Dr. Lu Chen and Mr. John Oxhey made significant contributions. Several technical staff in the school gave assistance to these students with making models.

Professor Biaozhong Zhuang, an emeritus professor in engineering mechanics at Zhejiang University, China, contributed his personal experience of solving practical problems using basic theory of mechanics and offered some interesting models used in daily life.

Several individuals and organisations kindly permitted the use of their photographs and they are acknowledged in the captions.

Dr. Tianxin Zheng and Mr. Hector Bobadilla, a previous PhD student and a current PhD student at the University of Manchester, respectively, provided many of the line drawings in the book. Mrs. Linwei Xue, a previous student at the University of Manchester, helped to create the website.

The assistance of Taylor & Francis in the publication of this book is greatly appreciated. We would like to thank Mr. Tony Moore, senior editor, for his encouragement in producing the first and second editions of the book. We are also grateful for the help we received from Catherine Hogan and her team at Deanta Global Publishing Services in preparing the final version of the new edition.

Finally, we would like to acknowledge the financial support provided by The Educational Trust of the Institution of Structural Engineers, The University of Manchester and The Higher Education Academy, for developing and updating the website and for developing the ways of teaching and learning structural concepts.

Authors

Tianjian Ji, MSc, PhD, CEng, FIStructE, is a senior lecturer at the University of Manchester. He worked on the design of structures and on structural investigation with consultants, China Academy of Building Research and Building Research Establishment Ltd., UK, for over 10 years before joining Manchester University in 1996. He has taught courses in structural analysis and structural design at all levels and has carried out research into structural dynamics and structural concepts. Together with Adrian Bell, he received the award for Excellence in Structural Engineering Education from the Institution of Structural Engineers, UK, in 2014.

Adrian J. Bell, BEng, MSc, PhD, is a senior lecturer at the University of Manchester. He has worked with consultants on the design of a wide range of structures including long-span roofs and tower structures. He has taught courses in structural analysis and structural design at all levels for over 30 years and has carried out research into cable, steel and masonry structures. Together with Tianjian Ji, he received the award for Excellence in Structural Engineering Education from the Institution of Structural Engineers, UK, in 2014.

Brian R. Ellis, BSc, PhD, DSc, CEng, MIStructE, was a technical director at the Building Research Establishment Ltd., UK. He worked at the Building Research Establishment for most of his career, where he undertook a wide range of work primarily related to structural dynamics. A significant part of the work involved testing and monitoring various structures in situ. He was also involved in work on national and international standards. Following his retirement, he has undertaken a limited amount of consultancy work.

CHAPTER 1

CONTENTS

Overview of Structural Concepts

1.1 WHAT ARE STRUCTURAL CONCEPTS?

Structural concepts are one of the foundations for study, analysis and design in civil and structural engineering. However, there are several different views on the meaning of structural concepts, as there is no agreed definition. Therefore, there is a need to define structural concepts in the context of this book before they and their uses are examined.

One dictionary definition [1] of a concept is *a truth or belief that is accepted as a base for reasoning or action*. This general definition can be applied to structural concepts if the truth or belief relate to structural engineering. However, a more specific definition of a structural concept is given as:

*A structural concept is a qualitative and concise representation of a **mathematical** relationship between **physical** quantities which captures the essence of the relationship and provides a basis for **practical** applications in structural engineering.*

This definition illustrates the significance and usefulness of understanding structural concepts, as it states the physical essence of mathematical equations for practical application and brings mathematics, physical understanding and practical application together.

This definition of structural concepts can be demonstrated using examples from theory and practice.

1.1.1 A Structural Concept Derived from Theory

The maximum deflection Δ of a simply supported uniform beam subjected to a uniformly distributed load q is:

$$\Delta = \frac{5}{384}\frac{qL^4}{EI} \tag{1.1}$$

Equation 1.1 has a theoretical basis and is derived from the differential equation of equilibrium for the bending of a uniform beam. It shows a relationship between five physical quantities: maximum deflection Δ, uniformly distributed load q, span L, modulus of elasticity E and second moment of area I. A structural concept abstracted from Equation 1.1 can be presented as:

Deflection is proportional to the span to the power of four and to the inverse of the second moment of area of the cross section.

This statement captures the physical essence of the equation and its application can be extended beyond simply supported uniform beams. Applications using this concept include the use of props and cable stays (reducing spans) and I-section beams (increasing the second moment of area). It can even be extended to tall buildings (when treated as cantilever beams) and to floors and roofs that are beyond the model of a beam. This concept has contributed to designing some innovative structures and to solving some challenging practical problems. Several practical examples can be seen in Chapters 4, 8, 10 and 24.

Equation 1.1 also shows that the deflection is proportional to the load q and to the inverse of the modulus of elasticity E, but they are less significant than the span and the second moment of area in practical applications. Thus, they are not included in this presentation of the concept.

1.1.2 A Concept Observed in Practice

Dance-type activities, such as keep-fit exercises and aerobics, are usually held on dance floors and in sports centres, and grandstands that are used for pop concerts may encounter similar activities. Thus, the use and design of these structures should consider the effect of human-induced rhythmic loads. Research has been undertaken to determine the frequencies of human loads on the venues where pop concerts might be held. Site measurements were taken at different venues during pop concerts. It was observed, from dynamic measurements, that the structure under human action responded at the beat frequency of the music and at integer multiples of the beat frequency. This observation can be expressed as

$$f_p = f_m \qquad (1.2)$$

where:

f_p is the first frequency of human loading
f_m is the beat frequency of the music played

This relationship (Equation 1.2) is straightforward and the concept can be easily interpreted as:

The first frequency of human loads at pop concerts is equal to the beat frequency of the music played.

The application of this concept allows the frequency of dance-type loads to be determined from the beat frequency of songs played on these occasions instead of taking vibration measurements on-site. Two undergraduate students, who were music fans, determined the beat frequencies of 210 modern songs that were played at pop concerts from the 1960s to the 1990s [2]. Using the concept, based on Equation 1.2, a significant amount of time and money was saved in gaining a better understanding of how to deal with dance-type loads on structures.

The two examples explain the definitions of structural concepts and also illustrate the effectiveness of understanding and using structural concepts. The first one is general and has a wide range of applications, whilst the second is specific and leads to the solution of a particular problem in a simple and effective manner.

One question related to the present definition of a structural concept is the exclusion of the concepts that are effective and widely used in practice but which have not been linked to a theoretical equation. It is believed that any structural concept, that will be effective in practice, will have a theoretical basis, although it may have not yet been identified. For example, the concept of direct load paths has been used in structural design for many years, without an appropriate theoretical support. However, the theoretical basis for the concept of direct internal force paths has now been established and is presented in Chapter 9.

Structural concepts are different from the conceptual design of structures. The conceptual design of structures is the first design stage, where options and the feasibility of possible structural designs are assessed, covering loading, structure, foundation, material and construction methods and so on. To produce a design that is safe, economical and elegant requires that the designers demonstrate a high level of intuitive understanding of structural behaviour, structural action and structural adequacy. The study and understanding of structural concepts can help the conceptual design of structures.

Understanding structural behaviour is related to, but different from, understanding structural concepts. Structural behaviour describes, mathematically or descriptively, how structures behave under loading. Qualitative analysis, or approximate analysis, is often conducted to help understand the behaviour of a structure, for example the likely deformed shapes and distribution of internal forces, and is a useful stage of the conceptual design process. Structural behaviour and structural concepts emphasise different aspects of structural engineering, but they are interrelated. Structural behaviour and structural concepts are discussed further in Chapter 13.

1.2 WHY STUDY STRUCTURAL CONCEPTS?

Structural concepts provide a basis for study, analysis and design in civil and structural engineering, and therefore their use is ideal for educating civil and structural engineering students. There are many reasons to emphasise the importance of the understanding and use of structural concepts in education and in practice.

1. In the past, our understanding of structural concepts was developed through working with hand calculations, many of which have now been replaced by computer-based calculations. Indeed, understanding structural concepts is fundamental to the sound and innovative design of structures and is of increasing importance due to the wide use of computers and the, often unquestioning, reliance placed on the results of computer analyses. Computed results will be flawed if they are based on incorrect assumptions and modelling, and engineers must understand this fact.
2. Analysis of the structural response to loads is much less demanding today than it was 20 or more years ago. Structural behaviour can be easily and quickly analysed using a personal computer (PC) and finite element (FE) software. It is the input loads and the model of a structure that are critical, as the output from a computer cannot be any better than the input. A good understanding of structural concepts will help to avoid errors when creating computer input.
3. Engineers aim to produce designs that are safe, economical and elegant but these three objectives are often not compatible, though they can, however, sometimes be achieved simultaneously when appropriate structural concepts are used.
4. Buildings become taller, bridges span longer and roofs cover larger areas than before. The effective use of structural concepts can help to deal with these ever-increasing challenges for design.
5. A relatively large civil engineering project is normally completed by different groups of people consisting of a small number of senior engineers who provide ideas and conceptual designs, a greater number of engineers/designers who produce the detailed analysis and designs and many more workers who construct the project on-site. University education should aim to prepare students to be the future senior engineers, for whom the understanding of structural concepts and behaviour will play an essential role.

This, admittedly incomplete list of reasons, shows the significance of structural concepts in learning and in practice. However, there are only a few research papers and books on structural concepts. This is perhaps due to three reasons.

First, it is sometimes thought that *structural concepts are too simple to study as they are taught at university.* People with this view simply miss the opportunity to appreciate the beauty and potential use of structural concepts. It is true that many structural concepts are well established and presented in textbooks. However, they could be presented in a different way, allowing students and engineers to gain a better understanding and appreciation of their application. In addition, new concepts will emerge from future engineering problems.

Second, in contrast to the first reason, it might be thought that *structural concepts are too hard to study as they are too abstract.* Unlike the study of particular structures, where the methodology and techniques have been established, there are no recognised ways to study structural concepts and few clear examples of their use. However, there are opportunities for studying and using structural concepts which can only be beneficial. Some approaches are discussed in Chapter 24.

Third, new structural concepts may have not yet been abstracted from current engineering practice. For example, engineers and architects devise intuitive, creative and ingenious solutions when solving challenging problems and realising innovative designs/solutions. Although some projects may be reported and published, little further work has been undertaken to identify the underlying structural concepts and how they have been used. Therefore, this important source of information may not have been fully explored.

As the understanding and using of structural concepts in education and engineering practice are important, new ways of studying structural concepts should be explored.

1.3 APPROACHES TO LEARNING STRUCTURAL CONCEPTS

The objectives for learning are to understand and recognise the use of structural concepts. Thus, appropriate approaches have been developed to achieve this target [3].

1. Theoretical content should be presented in such a way that structural concepts are highlighted.
2. Physical models should be used, where possible, for demonstrating structural concepts or the effect of the concepts, in order to gain an intuitive understanding.
3. Appropriate practical examples, either from engineering projects or from products in everyday life, should be provided to identify and illustrate the use of structural concepts.
4. Students should be engaged, through coursework of their own choice, on topics relating to structural concepts to stimulate further interest.

1.3.1 Theoretical Content

It would be helpful to students if structural concepts were highlighted in textbooks, such as that given for Equation 1.1 in Section 1.1.1. This would capture the essence of an equation for understanding rather than simply memorising it for an examination. Structural concepts are highlighted in this book in two ways.

First, the definitions and concepts are provided directly using bullet points and presented concisely in a memorable manner without using equations. Thus, readers can easily find related definitions and concepts in this book.

Second, concepts are abstracted and presented immediately after the equations on which they are based, and written in *italics*, as shown for Equation 1.1. Examples are often presented in pairs to demonstrate the effects of concepts in a quantitative manner.

1.3.2 Physical Models

Unlike in many current textbooks, many physical models are used in this book. But why is it so important to use physical models in learning?

When we talked to first-year students some 15 years ago, we asked what particular difficulties they experienced in their studies. Among their comments, the students said that they found it difficult to understand structural theory and structural concepts. They mentioned examples, such as, why does the normal stress due to bending vary linearly and become zero at the neutral axis for a given section, whilst shear stress varies quadrilaterally and becomes a maximum at the neutral axis? They also felt that a lot of the theory and many concepts were abstract since they could not be seen or touched. This triggered the thought that students might gain a better understanding of structural concepts if aspects of the theory could be 'seen' and 'touched'. It is true that stresses cannot be seen or touched easily, but the effects of stresses can be seen and touched. This led to the idea of developing physical demonstration models.

Two questions then had to be answered: How to develop physical models? and which models could be produced? Hence, Year 3 students were encouraged to study 'seeing and touching structural concepts' as Year 3 investigative projects. They reviewed several textbooks used at university level, mainly *Mechanics of Materials* and *Analysis of Structures*, and listed a number of definitions, concepts and methods from the textbooks. These were classified into three categories for possible model making:

1. Concepts that could be easily understood by reading definitions or by examining simple diagrams. For example, loads, displacements, internal forces, strain, redundancy, potential energy and elasticity.
2. Concepts that could not be understood easily from text or diagrams but could be demonstrated by using physical models. For example, force paths, resonance, stress distribution, stability, prestressing, centre of mass, equilibrium and vibration modes.

(a)   (b)

FIGURE 1.1
From a practical case to demonstration models. (a) Cross section of a concrete-filled steel column. (b) Effect of constrained and unconstrained sponge blocks subjected to the same weights.

3. Concepts that could not be understood easily from text or diagrams and were not readily suited to physical models. For example, strain energy, moment area method and Castigliano's theorem.

Students and lecturers examined each concept/definition in the three categories concentrating on, and trying to extend, the contents in Category 2. The models to be developed initially were for demonstration purposes, with the intent of eventually creating finished products that would be portable, durable and capable of reuse. Short assembly times and ease of handling during demonstrations were two desirable factors; however, the key issue was that the models should be able to demonstrate the related concepts effectively.

Other sources for generating ideas for producing demonstration models come from engineering practice and research. Practising engineers have the opportunity to solve challenging practical problems which often reflect intuitive approaches to developing innovative measures for solutions in which structural concepts are embedded. When such examples are identified, physical models can then be developed to demonstrate the effect of such measures. One such example follows.

Concrete-filled steel tube (CFT) column systems have been widely used in practice. Their main advantages are that the buckling capacity of the steel tube is increased by the restraint of the concrete and the load capacity of the concrete is increased by the confining effect of the steel tube. Figure 1.1a shows the cross section of a CFT column. This suggested an idea for producing a model to demonstrate the confining effect. Two similar sponge blocks are placed in two open plastic boxes. One box has gaps between the box walls and the sponge block, allowing unconstrained deformation of the sponge when it is loaded vertically. The other box is a tight fit between the box walls and the sponge block, preventing the sponge from deforming in the horizontal directions. Figure 1.1b shows the significant difference in the deformations of the constrained and the unconstrained sponge blocks when subjected to the same vertical loads. This set of models is used to explain the confined behaviour of the concrete in a CFT column.

Many such physical models have been developed to demonstrate assumptions, principles and concepts and the effects of concepts. For demonstrating the effect of a concept, a pair of similar models is normally produced, such as those shown in Figure 1.1b. This shows the effect of a concept in a qualitative manner and leads to an intuitive understanding.

1.3.3 Practical Examples

The purpose of learning and understanding structural concepts is to use them in analysis and design wherever possible. However, textbooks do not normally provide practical examples to show the applications of structural concepts. Consequently, the use of theory for both teaching and learning is often limited to solving theoretical questions. In parallel to demonstrations using physical models, appropriate real examples should be used to enrich teaching and stimulate interest.

Engineers and architects are creative when dealing with practical projects. Studying their projects can show what particular measures are used in solving challenging problems and what structural concepts are embedded in these measures. Fortunately, many well-known and excellent structures worldwide are well documented in technical papers or in books for the general public. Some professional journals and websites also provide a good source for practical examples in which structural concepts are used creatively.

Valuable lessons can also be learnt from failures, and some failures have been due to misunderstanding or an ignorance of structural concepts. Such examples are provided in Chapters 9 and 11.

Relevant practical examples can also be sought on the basis of particular structural concepts. For example, load paths, or internal force paths, is an important concept used in structural design. Thus, good examples with more direct load paths and other examples with less direct load paths can be sought from well-known structures and from local structures.

Research in structural engineering also helps to identify new structural concepts for understanding and developing new applications based on existing concepts to improve structural efficiency and solve practical problems. This can also be used to further enrich teaching. For example, the contents in Chapters 9, 10, 13, 15, 20 and 21 are developed based on our own research on practical problems.

There are often gaps between theory and practice, but a good understanding of structural concepts and their applications can help to bridge these gaps. Therefore, the need to gain a good understanding of structural concepts is particularly emphasised in this book through the use of dedicated physical models and appropriate practical examples.

1.3.4 Engaging Students

Motivating and engaging students to learn is perhaps as effective as, or more effective than, improving lectures. In addition to providing students with good teaching and learning material, students should be given opportunities for practice.

The Royal Academy of Engineering's publication, *Educating Engineers in the 21st Century* [4], showed that industry's top priorities for engineering graduate skills are *practical application*, *theoretical understanding* and *creativity and innovation*. It summarised the requirements of industry for its graduate entrants as '*A sound understanding* of the relevant engineering fundamentals plus *the ability to apply them* in an *innovative* way to the solution of *practical engineering problems*'. An individual piece of student coursework, containing the three previously mentioned components, has been developed and practised at Manchester for over eight years. This coursework encourages students to use structural concepts and theory for dealing with real problems in a creative way.

Past coursework submissions have been interesting and varied and have included some really creative components. It is hoped that the coursework has encouraged students to consider and explain structural concepts in a simple manner, to develop physical models to demonstrate the effects of structural concepts and to look for examples in everyday life in which structural concepts are innovatively used. These motivate further study and the development of a greater understanding and awareness of structural concepts. This has proved to be the case.

The specification of coursework and the ways to help students to conduct coursework are discussed in Section 1.5.1.

1.4 ORGANISATION OF THE TEXT

This book consists of three parts: Statics, Dynamics and Synthesis, which integrates both statics and dynamics. The Statics part has 12 chapters, the Dynamics part has 7 chapters and the Synthesis part has 4 chapters. The headings of chapters are related to concepts rather than to structures, as the use of the concepts is not limited by the type of structure.

In the first two parts of the book, Statics and Dynamics, each chapter contains four sections.

1. Definitions and concepts: Definitions of the terms used in the chapter are provided. The concepts are presented concisely, in one or two sentences, and in a memorable manner. Key points are also given in some chapters.
2. Theoretical background: If the theory is readily available in textbooks, only a brief summary is presented, together with appropriate references. More details are given when the theory is not readily available elsewhere. Selected examples are provided which aim to show the use of the theory and link it with the demonstration models illustrated in the next section.
3. Model demonstrations: Demonstration models are illustrated with photographs. Normally, two related models are provided to show differences in behaviour, thus illustrating the concept. Small-scale experiments are also included in some chapters.
4. Practical examples: Appropriate engineering examples are given to show how the concepts have been applied in practice. Some examples come from everyday life and will be familiar to most people.

A small number of questions are provided at the end of each chapter for practice. Some of the questions come from our classroom teaching and from our examination questions testing the understanding of structural concepts.

The third part of the book discusses more general topics in structural engineering. These topics are selected as they are fundamental but have not been presented elsewhere. The four chapters consider the relationships between static and modal stiffnesses, static and dynamic problems, experimental and theoretical studies, and theory and practice. All of these relationships are linked to structural concepts. The presentation of the chapters in this part follows the nature of each topic, which is different from the first two parts.

To facilitate learning, bold typeface is used for words that provide definitions. Such definitions are normally presented in the first pages of the chapters in the Statics and Dynamics parts of the book. Italics and bullet points are used to highlight important statements and key sentences in all chapters.

Accompanying this book, we have created a website, Seeing and Touching Structural Concepts, which can be found at www.structuralconcepts.org. The website contains most of the contents presented in the first, third and fourth sections of each chapter in the first two parts of the book. Colour photos can be downloaded from the website and video clips can be played.

1.5 HOW TO USE THIS BOOK

1.5.1 For Students

This book provides useful and interesting information to enhance the understanding of structural concepts and to complement class studies. The level of contents spans from the first year to the fourth year of a typical undergraduate course.

The contents of the book can be used in different ways.

1. You can look at any particular chapter, after you have been introduced to a concept in the classroom, to help you to gain a better understanding. Try to answer the questions at the end of related chapters.
2. You can use the book to revise what you have learnt in the past about structural concepts. In particular, the sections on physical models and practical examples could be useful.
3. You may challenge yourself to produce another physical model to demonstrate one of the structural concepts listed (or indeed one that is not listed) or to identify an existing structure or product in which one structural concept plays an important role.

Many of the models in the book were developed by our students as they knew which concepts were difficult to understand and which concepts could be physically demonstrated. Samples of our student submissions may help you generate similar ideas; these can be found on our website, www.structuralconcepts.org. You may also read the contents in Section 1.5.2 for more detailed information.

1.5.2 For Lecturers

It is hoped that the book provides useful material to supplement the teaching of structural concepts. There are two ways that the contents of the book can assist teaching. The easier way is that you simply download the related photos from our website and insert them into your existing teaching materials. With the aid of related physical models and practical examples, your students may be more attentive in the class and more interested in the contents you deliver.

The other way is to add four one-hour lectures to accompany the study undertaken by students on *Understanding and Using Structural Concepts*. We have conducted this exercise at Manchester since 2006 and improvements have been made year by year. The contents of the lectures are as follows:

1. Lecture 1: *Concepts for designing stiffer structures*. This shows how new structural concepts have been identified from a textbook equation for calculating the displacements of simple trusses and how these concepts can be used effectively to deal with practical engineering problems. It also introduces the route from textbook contents to practical applications and from structural elements to complete structures through the use of examples. The contents of this lecture are covered in Chapters 9 and 10.
2. Lecture 2: *Theory, concepts and practice*. This gives students a global view of studying and understanding structural concepts. It shows, through a number of examples, paths from theory to structural concepts and then to practice, and inversely from practice to structural concepts and then to theory. This helps to bridge the gaps between theory and practice. The contents of this lecture are provided in Chapter 24.
3. Lecture 3 explains and assigns appropriate individual coursework to students and shows many of the good coursework samples produced by our previous students to encourage students to involve themselves in learning about structural concepts and to stimulate their ideas. The coursework specification and samples of student submissions can be found on our website. You may use the examples produced by your own students in subsequent years!
4. Lecture 4 is given at the end of the course unit to summarise the coursework submissions, provide feedback and give prizes to the best three submissions, voted for by the students. This time, you use the submissions produced by your own students.

Students are required to learn from the website, *www.structuralconcepts.org*, at their own pace in conjunction with the assignment of the individual piece of coursework which has the general title, Understanding and Using Structural Concepts. They are asked to design/make/use a physical model to demonstrate one structural concept or to identify an example from engineering practice or everyday life in which a structural concept has been used creatively. As a guide rather than a limitation, students are asked to make a two A4-page submission using a prescribed template which allows them to focus on the quality of the contents. It has been found that students have been stimulated by this coursework. One interesting observation is that the number of coursework submissions has often been larger than the number of students in the class, indicating that some students were sufficiently enthused to make more than one submission!

In Manchester, all the coursework submissions are made through 'Blackboard', an online teaching package. They are slightly edited for consistency of format and compiled into a booklet, effectively written by the students, which is 'published' internally. Thus, students can learn from each other and further improve their understanding of structural concepts. To encourage students to learn from each other, the 'X factor' style of voting is used to select the best three coursework submissions for which small prizes are awarded. Students are also encouraged to design the covers of their booklets for the internal publication.

Figure 1.2a shows the cover of a typical booklet designed by students together with three examples of student submissions. Figure 1.2b shows a string-reinforced ice beam that can carry several

FIGURE 1.2
A cover of a student booklet and three coursework examples. (a) Book cover. (b) String-reinforced ice beam. (c) LEGO model of a footbridge. (d) Torsion of an I-section beam.

weights (an unreinforced ice beam cannot carry a single weight). Figure 1.2c shows a LEGO model of a footbridge and Figure 1.2d shows a sponge model that demonstrates torsion of an I-section beam.

The coursework specification, samples of our student submissions and further information have been provided at our website, www.structuralconcepts.org.

As the learning resource is provided at our website and coursework is specified, students should be able to spend sufficient time learning and conducting the coursework by themselves.

1.5.3 For Engineers

The contents of this book are useful to engineers, in particular recent graduate engineers. Unlike university students, you will have gained practical experience but may have forgotten some structural concepts studied at university. You may find the book useful in four ways.

1. You can quickly and effectively revise many structural concepts.
2. You can read the models and examples sections to develop your intuitive understanding of structures which could eventually be very helpful in your work.

3. You can examine the use of each of the concepts in practice through the examples provided. It is hoped that this may generate your own ideas for applying the concepts, or indeed any other structural concepts, in your work.

4. You may identify, using your own experience, how any of the structural concepts have been used in your work or in the work of your colleagues. Then consider how the application of the concepts helps in enhancing your understanding of structural behaviour and providing more efficient structures.

Engineers aim to achieve safe, economical and elegant structures. A sound understanding of structural concepts will help you to use them productively in your work.

REFERENCES

1. Summers, D. *Longman Dictionary of Contemporary English*, London: Longman Group, 1987.
2. Ginty, D., Derwent, J. M. and Ji, T. The frequency ranges of dance-type loads, *The Journal of Structural Engineer*, 79, 27–31, 2001.
3. Ji, T. and Bell, A. Innovative teaching and learning of structural concepts, *The Structural Engineer*, 92, 10–14, 2014.
4. Royal Academy of Engineering. *Educating Engineers for the 21st Century*, London: Royal Academy of Engineering, 2007.

PART I

STATICS

CHAPTER 2

CONTENTS

Equilibrium

2

2.1 DEFINITIONS AND CONCEPTS

- A body which is not moving is in a state of equilibrium.
- A body is in a state of stable equilibrium when any small movement increases its potential energy so that when released it tends to resume its original position.
- A body is in a state of unstable equilibrium if, following any small movement, it tends to move away from its original position.
- The sum of reaction forces on a body is always equal to the sum of action forces if the body is in a state of equilibrium, regardless of the positions and magnitudes of the action forces.
- If a body is in an equilibrium state, all the moments acting on the body about any point are zero.

2.2 THEORETICAL BACKGROUND

Rigid body: When considering the equilibrium of an object subjected to a set of forces, the object is considered as a rigid body when its deformations do not affect the position of equilibrium or the forces (either directions or amplitudes) or both applied on the body.

Internal forces: Internal forces act between particles within a body and occur in equal and opposite collinear pairs. The sum of the internal forces within a body will therefore be zero and need not be considered when considering the overall equilibrium of the body.

External forces: External forces act on the surfaces of bodies and may be considered to be distributed and applied over contact areas. When the contact areas are very small in comparison with the area of the surface, the distributed forces can be idealised as single concentrated forces acting at points on the surface of the body. A typical example of an external force which is distributed over a surface is wind load acting on the walls of a building. Concentrated external forces may be considered to be applied to a bridge deck by a vehicle whose weight is transmitted to the deck through its wheels which have small contact areas relative to the area of the deck.

Moment: A moment can be either an internal or an external force that tends to cause rotation of a body. A pair of parallel forces with the same magnitude but in opposite directions form a moment or couple. The magnitude of the moment is measured by the product of the magnitude of the forces and the perpendicular distance between the two parallel forces.

Body force: A body force, or the self-weight of a body, is the effect of Earth's gravity on the body. This force is not generated through direct contact like most other external forces. The self-weight of a body can be significant in the design of structures; for example, the self-weight of a concrete floor in a building can be larger than other applied imposed loads such as those due to people or furnishings. Body forces, other than self-weight, can also be generated remotely by magnetic or electric fields (Section 2.3.4).

Reactions: The forces developed in the supports or points of contact between bodies are called *reactions*. Supports are provided to prevent the movement of a body. Different types of support prevent different types of movement. The main types of support which occur in practice in plane structures are as follows:

- **Pin supports:** A pin support restrains translational movements in two perpendicular directions but allows rotation (Figure 2.1a).
- **Roller supports:** A roller support restrains translational movement in one direction but allows translational movements in a perpendicular direction as well as allowing rotation (Figure 2.1b).

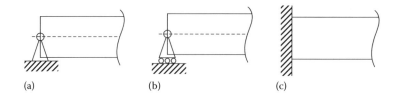

(a)   (b)   (c)

FIGURE 2.1
Examples of support conditions. (a) Pin supports. (b) Roller supports. (c) Fixed supports.

- **Fixed supports:** A fixed support restrains translational movements in two perpendicular directions and prevents rotation (Figure 2.1c).

Conditions of equilibrium of a rigid body: A body is in a state of equilibrium when it does not move. In this state, the effects of all the forces F and moments M applied to the body cancel each other. Considering all the forces and moments applied in the x–y plane, the conditions of equilibrium of the body can be specified using three scalar equations of equilibrium:

$$\sum F_X = 0 \quad \sum F_Y = 0 \quad \sum M_A = 0 \tag{2.1}$$

Equation 2.1 indicates that

- The sums of all the forces in the x and y directions are zero and the sums of the moments induced by these forces about an arbitrary point are zero.
- The three equations of equilibrium can be used to determine three unknowns in an equilibrium problem, such as the reaction forces of a simple body in the x–y plane.

   Free-body diagram: A free-body diagram is a diagrammatic representation of a body or a part of a body showing all forces applied on it by mechanical contact with other bodies or parts of the body itself. Drawing a free-body diagram, which shows clearly and completely all the forces acting on the body, is an important and necessary skill.

EXAMPLE 2.1

Figure 2.2a shows a beam which is supported at the left-hand end on a pin support and at the right-hand end on a roller support. Such a beam is called a *simply supported beam*. The beam carries a concentrated vertical force W acting at a distance x from the left-hand end of the beam. Determine the reaction forces A_x, A_y and B_y at the two supports.

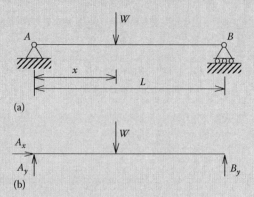

(a)

(b)

FIGURE 2.2
A simply supported beam with (a) a concentrated load and (b) its free-body diagram.

SOLUTION

A free-body diagram of the beam is shown in Figure 2.2b, where the external force and all the support forces are indicated. Using the three formulae in Equation 2.1 gives

$$\sum F_X = 0 \quad A_X = 0$$

$$\sum F_Y = 0 \quad A_Y + B_Y - W = 0$$

$$\sum M_A = 0 \quad B_Y L - Wx = 0$$

Solving the last two equations leads to $A_Y = W - Wx/L$ and $B_Y = Wx/L$. The positive signs of the reaction forces indicate that A_Y and B_Y act in the assumed directions as shown in Figure 2.2b. The second formula shows that *the sum of the reaction forces in the y direction is always equal to the external force W, no matter where it is placed.* A model demonstration of this example is given in Section 2.3.1. Further details can be seen in [1].

2.3 MODEL DEMONSTRATIONS

2.3.1 Action and Reaction Forces

This model demonstrates what has been shown in Example 2.1, namely,

- The sum of reaction forces is always equal to the external force, W, regardless of the location of the external force.
- When the external force, W, is placed above one of the two supports, the reaction force at that support is equal to the external force but acts in the opposite direction and there is no reaction force at the other support.

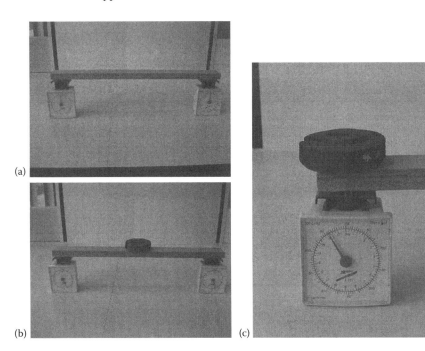

(a)

(b)

(c)

FIGURE 2.3
(a–c) Action and reaction.

Figure 2.3a shows a wooden beam supported by two scales, one at each end. The scales are adjusted to zero when the beam is in place. Locate a weight of 1 lb (454 g) at three different positions on the beam and note the readings on the two scales. It can be seen that

- For the three positions of the weight, the sum of the readings (reactions) from the two scales is always equal to the weight (the external force), see Figure 2.3b.
- The readings on the scales have a linear relation to the distance between the weight and the locations of the supports. The closer the weight is to one of the scales, the larger the reading on the scale. An extreme case occurs when the weight is placed directly over the left-hand scale. The reading on the left-hand scale is the same as the weight and the reading on the other scale is zero (Figure 2.3c).

2.3.2 Stable and Unstable Equilibrium

This model demonstration shows *the difference between stable and unstable equilibrium.*

The equilibrium of a ruler supported on two round pens (or roller supports) can be achieved easily as shown in Figure 2.4a. However, supporting the ruler horizontally on a single round pen alone is very difficult, because the external force (self-weight) from the ruler and the reaction force from the round pen are difficult to align exactly. If the ruler achieves equilibrium on the single round pen, this type of equilibrium is unstable and is not maintained if a slight disturbance is applied to the ruler or to the round pen. The ruler will rotate around the point of contact with the pen until it finds another support (Figure 2.4b).

2.3.3 Wood–Bottle System

This model demonstration shows *how an equilibrium state can be achieved.*

Figure 2.4 shows that the ruler is in stable equilibrium when it has two round pen supports and becomes unstable equilibrium when it has only one round pen support. This, however, does not mean that a body placed on a single support cannot achieve a state of stable equilibrium.

Figure 2.5a shows a bottle of wine and a piece of wood with a hole. The bottle can be supported by the wood when the neck of the bottle is inserted into the hole to the maximum extent, and the two form a single wood–bottle system in equilibrium as shown in Figure 2.5b.

The wood–bottle system, supported on the narrow wooden edge, is and feels very stable, because:

- The two external forces from the weights of the bottle and the wood are equal to the reaction force generated from the table.
- The sum of the moments of the two action forces about the point where the support force acts is equal to zero.

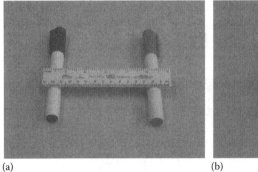 

(a)   (b)

FIGURE 2.4
(a,b) Stable and unstable equilibrium.

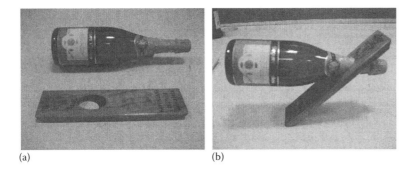

FIGURE 2.5
(a,b) Equilibrium of a bottle and wood system.

Another way of saying this is that *the centre of mass of the wood and bottle system lies over the base of the wood.*

2.3.4 Magnetic 'Float' Model

This model demonstrates *the effect of magnetic force although the force cannot be seen.*

Figure 2.6 shows a model consisting of an axisymmetric body and a base unit. There are two magnetic rings in the axisymmetric body and two magnets in the base unit.

It seems surprising that the axisymmetric body can be in an equilibrium position when no vertical supports are provided. Where is the force that supports the weight of the axisymmetric body? When opposite poles of the magnets in the axisymmetric body and the base unit are the same, the body can be positioned above the base with no visible support (Figure 2.6a and c). When the opposite poles are different, the body rests on the base (Figure 2.6b and d). In both cases, the external force is the weight of the body. In the first case the reaction forces are the magnetic forces that push the body away and up, while in the second case the reaction forces are provided through the points of contact between the unit and the base.

When the free end of the body (Figure 2.6a) is twisted using the thumb and index finger, the body is able to rotate many times before stopping. This is because there is little friction between the rotating body and its lateral glass support.

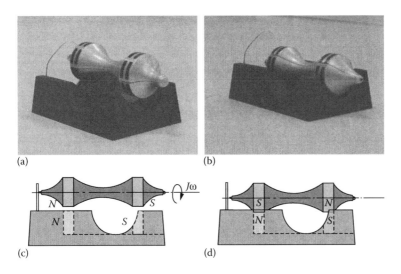

FIGURE 2.6
(a–d) A magnetic float model. (Courtesy of Professor B. Zhuang, Zhejiang University, China.)

FIGURE 2.7
A barrier.

2.4 PRACTICAL EXAMPLES

2.4.1 Barrier

Counterbalanced barriers can be found in many places, such as the one shown in Figure 2.7. The weight fixed to the shorter arm counterbalances much of the weight of the barrier arm, allowing the barrier to be opened easily with a relatively small downward force near the counterbalance weight.

The barrier can be opened when the moment induced by the counterbalance weight and the applied downward force is larger than the moment induced by the weight of the barrier about the support point.

2.4.2 Footbridge

The form of footbridge shown in Figure 2.8 is a development of the simple counterbalanced barrier shown in Figure 2.7.

The moment induced by the weight of the footbridge deck about the supporting points on the wooden frame is slightly larger than that induced by the balance weight, which is placed on and behind a wooden board. The addition of a small force, applied by pulling on a cable, causes the bridge to open.

FIGURE 2.8
A footbridge can be lifted by a single person.

FIGURE 2.9
An equilibrium kitchen scale.

2.4.3 Equilibrium Kitchen Scale

Figure 2.9 shows a kitchen scale that weighs an apple and a banana using the principle of equilibrium.

At the centre of the scale is a pivot and a pair of arms that can rotate about the pivot. The left and right arms are designed to be symmetrical about the central pivot. There are two other pivots at the ends of the left and right arms where two trays are placed. The two arms would be at the same level when an equilibrium state is achieved. The equilibrium equation for moment (the third equation in Equation 2.1) can be used to give

$$\text{Weight of the fruit} \times d_L = \text{standard weights placed} \times d_R$$

where d_L and d_R are the distances between the central pivot and the left- and right-hand points of application of the weights. Due to the symmetry, $d_L = d_R$, the foregoing equation thus states that the weight of the fruits is equal to the standard weights placed on the tray. The operation of the scale is thus based on the principle of equilibrium.

2.4.4 Stage Performance

Equilibrium in the posture of the human body can be important in sport and dance.

The two people shown in Figure 2.10 have positioned themselves horizontally such that the centre of mass of their bodies lies over the feet of one of the performers. The balance shows the strength of the performers and the beauty of the posture.

FIGURE 2.10
Equilibrium on stage. (Photograph of Crazeehorse, copyright of Fremantle Media, reproduced by kind permission.)

2.4.5 Magnetic Float Train

Figure 2.11 shows the first commercial, friction-free magnetic 'float' train, which operates between Shanghai City and Pudong Airport. The floating train can reach speeds in excess of 270 mph. The train, with magnets under the cars, 'floats' above a sophisticated electromagnetic track in the same way that the axisymmetric object 'floats' above the base unit in the demonstration in Section 2.3.4. The weight of the train (external force) is balanced by magnetic forces (reaction forces) generated between the magnets in the cars and the track. Since little friction is generated between the external and reaction forces, much smaller tractive forces are required to propel the train than would normally be the case.

In the United Kingdom, float trains were also run on a prototype route between Birmingham Airport and a nearby train station, but they were abandoned in 1995 because of reliability problems.

FIGURE 2.11
A magnetic float train.

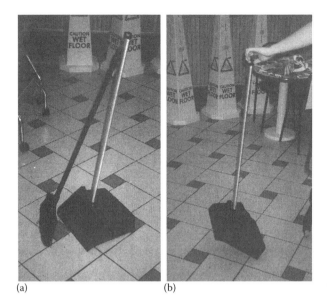

(a)   (b)

FIGURE 2.12
(a,b) An application of inequilibrium.

2.4.6 Dust Tray

Inequilibrium may also be found in several common items used in daily life. Figure 2.12a shows a dust tray which is often used in fast-food restaurants to collect rubbish from floors. When moving the dust tray between locations, one lifts the vertical handle. The tray then rotates through an angle such that the rubbish will be kept at the end of the tray as shown in Figure 2.12b. The rotation occurs because there is a pivot on the tray and the pivot is purposely placed at the position which will cause the tray to turn. Because the centre of mass of the dust tray in its horizontal position is not in line with the handle, the tray rotates until this is the case. If the larger portion of the tray lies behind the handle, the direction of rotation is such that rubbish remains trapped during movement.

PROBLEMS

1. Draw free-body diagrams for the bottle, the wood member and the wood–bottle system shown in Figure 2.5b.
2. Draw free-body diagrams for the woman, the man and the man–woman system shown in Figure 2.10.
3. Identify a practical example in which equilibrium has been used in a clever or an innovative manner.

REFERENCE

1. Hibbeler, R. C. *Mechanics of Materials*, 6th edn, Singapore: Prentice-Hall, 2005.

CHAPTER 3

CONTENTS

Centre of Mass

3

3.1 DEFINITIONS AND CONCEPTS

The **centre of gravity of a body** is the point about which the body is balanced, or the point through which the weight of the body acts.

The location of the centre of gravity of a body coincides with the **centre of mass of the body** when the dimensions of the body are much smaller than those of the earth.

When the density of a body is uniform throughout, the centre of mass and the **centroid of the body** are at the same point.

- If the centre of mass of a body is not positioned above the area over which it is supported, the body will topple over.
- The lower the centre of mass of a body, the more stable the body.

3.2 THEORETICAL BACKGROUND

Centre of gravity: The centre of gravity of a body is a point through which the weight of the body acts or is a point about which the body can be balanced. The location of the centre of gravity of a body can be determined mathematically using the moment equilibrium condition for a parallel force system. Physically, a body is composed of an infinite number of particles and if a typical particle has a weight of dW and is located at (x, y, z), as shown in Figure 3.1, the position of the centre of gravity, G, at $(\bar{x}, \bar{y}, \bar{z})$ can be determined using the following equations:

$$\bar{x} = \frac{\int x\,dW}{\int dW} \quad \bar{y} = \frac{\int y\,dW}{\int dW} \quad \bar{z} = \frac{\int z\,dW}{\int dW} \tag{3.1}$$

Centre of mass: To study problems relating to the motion of a body subject to dynamic loads, it is necessary to determine the centre of mass of the body. For a particle with weight dW and mass dm, the relationship between mass and weight is given by $dW = g\,dm$, where g is the acceleration due to gravity. Substituting this relationship into Equation 3.1 and cancelling g from both the numerator and denominator leads to

$$\bar{x} = \frac{\int x\,dm}{\int dm} \quad \bar{y} = \frac{\int y\,dm}{\int dm} \quad \bar{z} = \frac{\int z\,dm}{\int dm} \tag{3.2}$$

Equation 3.2 contains no reference to gravitational effects (g) and therefore it defines a unique point which is a function solely of the distribution of mass. This point is called the centre of mass of the body. For civil engineering structures, whose dimensions are very small compared with the dimensions of the earth, g may be considered to be uniform and hence the centre of mass of a body coincides with the centre of gravity of the body. In practice, no distinction is made between the centre of mass and the centre of gravity of a body.

The mass, dm, of a particle in a body relates to its volume, dV, and can be expressed as $dm = \rho(x, y, z)\,dV$, where $\rho(x, y, z)$ is the density of the body. Substituting this relationship into Equation 3.2 gives

$$\bar{x} = \frac{\int x\rho(x, y, z)dV}{\int \rho(x, y, z)dV} \quad \bar{y} = \frac{\int y\rho(x, y, z)dV}{\int \rho(x, y, z)dV} \quad \bar{z} = \frac{\int z\rho(x, y, z)dV}{\int \rho(x, y, z)dV} \tag{3.3}$$

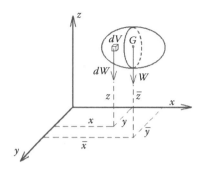

FIGURE 3.1
Centre of gravity.

Centroid: The centroid C of a body is a point which defines the geometrical centre of the body. If a body is composed of one material, or the density of the material ρ is constant, it can be removed from Equation 3.3, and then the centroid of the volume of the body can be defined by

$$\bar{x} = \frac{\int x\,dV}{\int dV} \quad \bar{y} = \frac{\int y\,dV}{\int dV} \quad \bar{z} = \frac{\int z\,dV}{\int dV} \tag{3.4}$$

Equation 3.4 contains no reference to the mass of the body and therefore defines a point, the centroid, which is solely a function of geometry.

If the density of the material of a body is uniform, the locations of the centroid and the centre of mass of the body coincide. However, if the density varies throughout a body, the centroid and the centre of mass will not be the same in general (Section 3.3.2).

When the centroid of a body, such as a uniform thin flat plate, needs to be determined, the volume of a small section of the body, dV, can be expressed as $dV = t\,dA$, where dA is the surface area of the volume dV perpendicular to the axis through the thickness, t, of the volume. Substituting the expression for dV into Equation 3.4 leads to

$$\bar{x} = \frac{\int x\,dA}{\int dA} \quad \bar{y} = \frac{\int y\,dA}{\int dA} \quad \bar{z} = \frac{\int y\,dA}{\int dA} \tag{3.5}$$

Terms such as $x\,dA$ in the numerators of the integrals represent the moments of the area of the particle, dA, about the centroid. The integrals represent the summations of these moments of areas for the whole body and are usually called the **first moments of area** of the body about the axes in question. The integral in the denominators in Equation 3.5 is the area of the body.

When the centroid of a line, effectively a body with one dimension, needs to be determined, Equation 3.5 can be simplified to

$$\bar{x} = \frac{\int x\,dL}{\int dL} \quad \bar{y} = \frac{\int y\,dL}{\int dL} \quad \bar{z} = \frac{\int z\,dL}{\int dL} \tag{3.6}$$

where dL is the length of a short element or part of the line.

Equations 3.4 through 3.6 are based on a coordinate system that may be chosen to suit a particular situation and in general such a coordinate system should be chosen to simplify, as much as possible, the integrals. When the geometric shape of an object is simple, the integrals in the equations can be replaced by simple algebraic calculations.

EXAMPLE 3.1

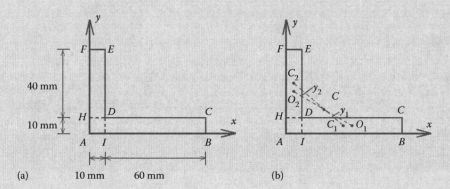

FIGURE 3.2
(a,b) Centroid of an L-shaped object.

Determine the centroid of the uniform L-shaped object shown in Figure 3.2a.

SOLUTION 1

As the L-shaped object consists of two rectangles, *ABCH* and *HDEF*, these areas and the moments of the areas about the axes can be easily calculated without the need for integration. The areas and their centroids for the coordinate system chosen are

$$ABCH: A_1 = 70 \times 10 = 700 \text{ mm}^2, x_{C1} = 35 \text{ mm}, y_{C1} = 5 \text{ mm}$$

$$HDEF: A_2 = 40 \times 10 = 400 \text{ mm}^2, x_{C2} = 5 \text{ mm}, y_{C2} = 30 \text{ mm}$$

Using the first two formulae in Equation 3.5 gives

$$\bar{x} = \frac{A_1 x_{C1} + A_2 x_{C2}}{A_1 + A_2} = \frac{700 \times 35 + 400 \times 5}{700 + 400} = \frac{26,500}{1,100} = 24.09 \text{ mm}$$

$$\bar{y} = \frac{A_1 y_{C1} + A_2 y_{C2}}{A_1 + A_2} = \frac{700 \times 5 + 400 \times 30}{700 + 400} = \frac{15,500}{1,100} = 14.09 \text{ mm}$$

SOLUTION 2

The centroid of the L-shaped object can also be determined using a geometrical method. The procedure can be described as follows:

- Divide the L-shape into two rectangles, *ABCH* and *HDEF*, as shown in Figure 3.2b. Find the centroids of the two rectangles, C_1 and C_2, and then draw a line linking C_1 and C_2. The centroid of the object must lie on the line $C_1 - C_2$. The equation of this line is $y_1 - 5 = -5(x - 35)/6$.
- Divide the shape into two other rectangles, *IBCD* and *AIEF*, as shown in Figure 3.2b. Find the centres of the two rectangles, O_1 and O_2, then draw a line joining their centres. The centroid of the object must also lie on the line $O_1 - O_2$, which is $y_2 - 5 = -4(x - 40)/7$.

- As the centroid of the object lies on both lines $C_1 - C_2$ and $O_1 - O_2$, it must be at their intersection, $\bar{C}$, as shown in Figure 3.2b. Solving the two simultaneous equations for x and y leads to $\bar{x} = 24.09$ cm and $\bar{y} = 14.09$ cm.

It can be seen from Figure 3.2b that the centroid, $\bar{C}$, is located outside the boundaries of the object. As the density of the L-shaped object is uniform, the centre of gravity, the centre of mass and the centroid coincide, so all lie outside the boundaries of the object. From the definitions of centre of mass and centre of gravity given in Section 3.1, the L-shaped object can be supported by a force applied at the centre of mass, perpendicular to the plane of the object. In practice, this may be difficult as the centre of mass lies outside the object. Section 3.3.3 provides a model demonstration that shows how an L-shaped body can be lifted at its centre of mass.

EXAMPLE 3.2

Determine the centre of mass of a solid, uniform pyramid that has a square base of $a \times a$ and a height of h. The base of the pyramid lies in the x–y plane and the apex of the pyramid is aligned with the z-axis as shown in Figure 3.3a.

SOLUTION

Due to symmetry, the centroid must lie on the z-axis, for example

$$\bar{x} = \bar{y} = 0$$

The location of the centroid of the pyramid on the z-axis can be obtained by using the third formula in Equation 3.4 where the integral is taken over the volume of the pyramid. Consider a slice of thickness dz of the pyramid, which is parallel to the base with a distance z from the base, as shown in Figure 3.3b. The length of the side of the square slice is $a(1 - z/h)$, and hence the volume of the slice is $dV = a^2(1 - z/h)^2 dz$. Substituting these quantities into the third formula in Equation 3.4 and integrating with respect to z through the height of the pyramid gives

$$z_c = \frac{\int_0^h za^2\left(1 - \frac{z}{h}\right)^2 dz}{\int_0^h a^2\left(1 - \frac{z}{h}\right)^2 dz} = \frac{\dfrac{a^2h^2}{12}}{\dfrac{a^2h}{3}} = \frac{h}{4}$$

Thus, the centre of mass of a uniform square-based pyramid is located on its vertical axis of symmetry at a quarter of its height measured from the base.

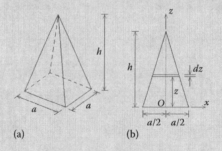

(a) (b)

FIGURE 3.3
(a,b) Centroid of a pyramid.

EXAMPLE 3.3

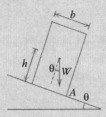

FIGURE 3.4
Centre of mass and stability.

The location of the centre of mass of a uniform block is at a height of h from its base that has a side length of b. Place the block on an inclined surface as shown in Figure 3.4 and gradually increase the slope of the surface until the block topples over. Determine the maximum slope for which the block does not topple (assuming that there is no sliding between the base of the block and the supporting surface).

SOLUTION

It would normally be assumed that the block would not topple over if the action line of its weight remains over the area of the base of the block. Thus, the critical position for the block is when the centre of gravity of the block and the corner of the base lie on the same vertical line. Therefore, the critical angle of tilt for the block is

$$\theta_c = \tan^{-1}\frac{b/2}{h} = \tan^{-1}\frac{b}{2h} \tag{3.7}$$

If the assumption is correct, Equation 3.7 indicates that if $\theta < \theta_c$ the block will not topple. The assumption can be checked by conducting the simple model tests in Section 3.3.5. Further details can be seen in [1].

3.3 MODEL DEMONSTRATIONS

3.3.1 Centre of Mass of a Piece of Cardboard of Arbitrary Shape

This demonstration shows how *the centre of mass of a body with an arbitrary shape can be determined experimentally.*

Take a piece of cardboard of any size and shape, such as the one shown in Figure 3.5, and drill three small holes at arbitrary locations along its edge. Now suspend the cardboard using a drawing pin through one of the holes and hang from the drawing pin a length of string supporting a weight. As the diameter of the hole is larger than the diameter of the pin, the cardboard can rotate about the pin and will be in equilibrium under the action of the self-weight of the cardboard. This means that the resultant of the gravitational force on the cardboard must pass through the pin. In other words, the action force (gravitational force) and the reaction force from the pin must be in a vertical line, which is shown by the string. Now mark a point on the cardboard under the string and draw a straight line between this point and the hole from which the cardboard is supported. Repeat the procedure hanging the cardboard from each of the other two holes in turn, and in each case draw a line between a point on the string and the support hole. It can be seen that the three lines (or their extensions) are concurrent at a single point, which is the centre of mass of the piece of cardboard and of course is also the centre of gravity and the centroid of the cardboard.

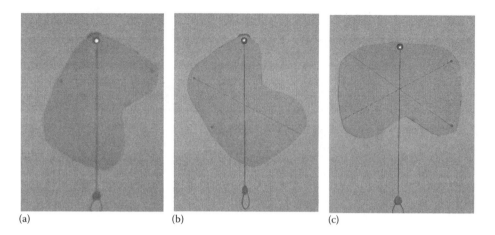

(a)   (b)   (c)

FIGURE 3.5
(a–c) Locating the centre of mass of a piece of cardboard of arbitrary shape.

3.3.2 Centre of Mass and Centroid of a Body

This model is designed to show that *the location of the centre of mass of a body can be different from that of the centroid of the body.*

A 300 mm tall pendulum model is made of wood. The model consists of a supporting base, a mast fixed at the base and a pendulum pinned to the other end of the mast, as shown in Figure 3.6. A piece of metal is inserted into the right arm of the pendulum. The centre of mass of the pendulum must lie on a vertical line passing through the pinned point. The centroid of the pendulum lies on the intersection of the vertical and horizontal members. Due to the different densities of wood and metal, the centroid of the pendulum does not coincide with the centre of mass of the body.

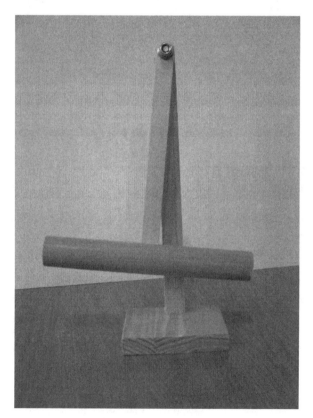

FIGURE 3.6
Centre of mass and centroid.

3.3.3 Centre of Mass of a Body in a Horizontal Plane

This demonstration shows *how to locate the centre of mass of a horizontal L-shaped body.*

Figure 3.2b shows an L-shaped area whose centre of mass, *C*, is outside the body. The following simple experiment, using a fork, a spoon, a toothpick and a wine glass, can be carried out to locate the centre of mass of an L-shaped body made up from a spoon and a fork.

1. Take a spoon and a fork and insert the spoon into the prongs of the fork, to form an L-shape, as shown in Figure 3.7a.
2. Then take a toothpick and wedge the toothpick between two prongs of the fork and rest the head of the toothpick on the spoon, as shown in Figure 3.7b. Make sure the end of the toothpick is firmly in contact with the spoon.
3. The spoon–fork can then be lifted using the toothpick. The toothpick is subjected to an upward force from the spoon, a downward force from the prong of the fork and forces at the point at which it is lifted, as shown in Figure 3.7c.

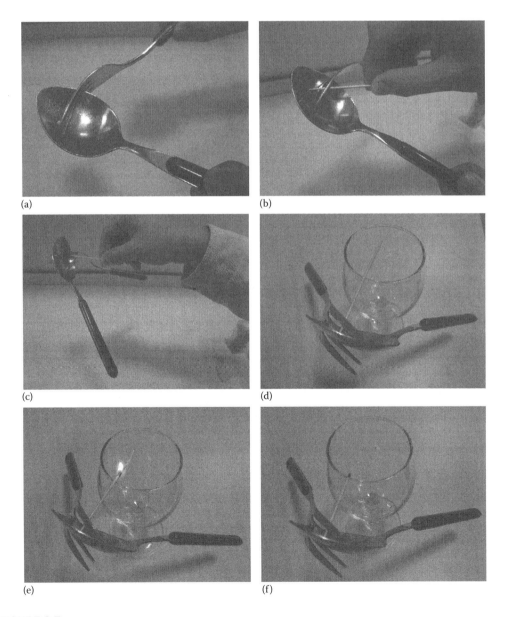

(a) (b) (c) (d) (e) (f)

FIGURE 3.7
(a–f) Centre of mass of an L-shaped body in a horizontal plane.

4. Place the toothpick on the edge of a wine glass, adjusting its position until the spoon–fork–toothpick is balanced at the edge of the glass, as shown in Figure 3.7d. According to the definition of the centre of mass, the contact point between the edge of the glass and the toothpick is the centre of mass of the spoon–fork–toothpick system. The mass of the toothpick is negligible in comparison with that of the spoon–fork, so the contact point can be considered to be the centre of mass of the L-shaped spoon–fork system.
5. To reinforce the demonstration, set alight the free end of the toothpick (Figure 3.7e). The flame goes out when it reaches the edge of the glass as the glass absorbs heat. As shown in Figure 3.7f, the spoon–fork just balances on the edge of the glass.

If the spoon and fork are made using the same material, the centre of mass and the centroid of the spoon–fork system will coincide.

3.3.4 Centre of Mass of a Body in a Vertical Plane

This demonstration shows that *the lower the centre of mass of a body, the more stable the body.*

A cork, two forks, a toothpick, a coin and a wine bottle are used in the demonstration shown in Figure 3.8a. The procedure for the demonstration is as follows:

1. Push a toothpick into one end of the cork to make a toothpick–cork system as shown in Figure 3.8b.
2. Try to make the system stand up with the toothpick in contact with the surface of a table. In practice, this is not possible as the centre of mass of the system is high and easily falls outside the area of the support point. The weight of the toothpick–cork and the reaction from the surface of the table that supports the toothpick–cork are not in the same vertical line and this causes overturning of the system.
3. Push two forks into opposite sides of the cork and place the coin on the top of the bottle.
4. Place the toothpick–cork–fork system on the top of the coin on which the toothpick will stand in equilibrium as shown in Figure 3.8c.

The addition of the two forks significantly lowers the centre of mass of the toothpick–cork system in the vertical plane below the contact point of the toothpick on the coin. When the toothpick–cork–fork system is placed on the coin at the top of the bottle, the system rotates until the action and reaction forces at the contact point of the toothpick are in the same line, producing an equilibrium configuration.

(a)

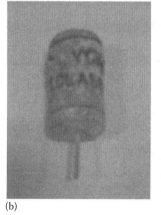
(b)

(c)

FIGURE 3.8
(a–c) Centre of mass of a body in a vertical plane.

3.3.5 Centre of Mass and Stability

This demonstration shows how *the stability of a body relates to the location of its centre of mass and the size of its base.*

Figure 3.9a shows three aluminium blocks with the same height of 150 mm. The square-sectioned block and the smaller pyramid have the same base area of 29×29 mm. The larger pyramid has a base area of 50×50 mm but has the same volume as that of the square-sectioned block. The three blocks are placed on a board with metal stoppers provided to prevent the blocks from sliding when the board is inclined. As the board is inclined, its angle of inclination can be measured by the simple equipment shown in Figure 3.9b. Basic data for the three blocks and the theoretical critical angles calculated using Equation 3.7 are given in Table 3.1. Theory predicts that the largest critical angle occurs with the large pyramid and the smallest critical angle occurs with the square-sectioned block.

The demonstration is as follows:

1. The blocks are placed on the board, as shown in Figure 3.9, in the order of increasing predicted critical angle.
2. The left-hand end of the board is gradually lifted and the square-sectioned block is the first to become unstable and topple over (Figure 3.9b). The angle at which the block topples is noted.
3. The board is inclined further and the pyramid with the smaller base is the next to topple (Figure 3.9c). Although the height of the centre of mass of the two pyramids is the same, the smaller pyramid has a smaller base and the line of action of its weight lies outside the base at a lower inclination than is the case for the larger pyramid. The angle at which the smaller pyramid topples over is noted.
4. When the board is inclined further, the larger pyramid will eventually topple, but its improved stability over the other two blocks becomes apparent (Figure 3.9d). Once again, the angle at which the block topples is noted.

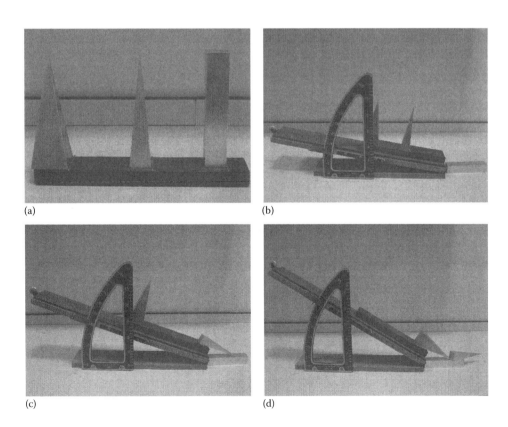

(a)   (b)

(c)   (d)

FIGURE 3.9
(a–d) Centre of mass and stability of three aluminium blocks.

TABLE 3.1   Comparison of the Calculated and Measured Critical Angles

Model	Cuboid	Small Pyramid	Large Pyramid
Height of the body (mm)	150	150	150
Height of the centre of mass (mm)	75	37.5	37.5
Width of the base (mm)	29	29	50
Volume (mm³)	1.26×10^5	0.420×10^5	1.25×10^5
Theoretical max. inclination (°)	10.9	21.9	33.7
Measured max. inclination (°)	10	19	31

The angles at which the three blocks toppled are shown in Table 3.1. The results of the demonstration, as given in Table 3.1, show that

- The order in which the blocks topple is as predicted by Equation 3.7 in terms of the measured inclinations. This confirms that the larger the base, or the lower the centre of mass of a block, the larger the critical angle that is needed to cause the block to topple.
- All the measured critical angles are slightly smaller than those predicted by Equation 3.7.

Repeating the experiment several times confirms the measurements and it can be observed that the bases of the blocks just leave the supporting surface immediately before they topple, which makes the centres of the masses move outwards. In the theory, the bases of the blocks remain in contact with the support surface until they topple.

This demonstration shows that *the larger the base or the lower the centre of gravity of a body* or both, *the larger will be the critical angle needed to cause the body to topple and that this angle is slightly less than that predicted by theory.*

3.3.6 Centre of Mass and Motion

This demonstration shows that *a body can appear to move up a slope unaided.*

Take two support rails which incline in both the vertical and horizontal planes as shown in Figure 3.10a. When a doubly conical solid body is placed on the lower ends of the rails, it can be observed that the body rotates and travels up to the higher end of the rails (Figure 3.10b).

It appears that the body moves against gravity, though, in fact, it moves with gravity. When the locations of the centre of mass of the body, *C*, are measured at the lowest and highest ends of

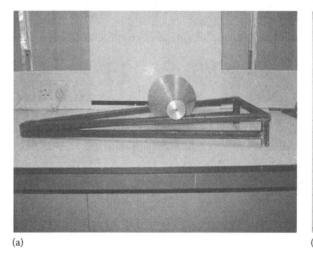
(a)

(b)

FIGURE 3.10
(a,b) Centre of mass and motion.

the rails, it is found that the centre of mass of the body at the lower end is actually higher than that when the body is at the higher end of the rails. It is gravity that makes the body rotate and appear to move up the slope.

The reason for this lies in the design parameters of the rail supports and the conical body. The control condition is that the slope of the conical solid body should be larger than the ratio of the increased height to a half of the increased width of the rails between the two ends.

3.4 PRACTICAL EXAMPLES

3.4.1 Cranes on Construction Sites

Tower cranes are common sights on construction sites. Such cranes normally have large weights placed on and around their bases; these weights ensure that the centre of mass of the crane and its applied loading lie over its base area and that the centre of mass of the crane is low, increasing its stability. Figure 3.11a shows a typical tower crane on a construction site. Concrete blocks are purposely placed on the base of the crane to lower its centre of mass to keep the crane stable, as shown in Figure 3.11b.

3.4.2 Eiffel Tower

Building structures are often purposely designed to be larger at their bases and smaller at their upper parts so that the distribution of the mass of the structure reduces with height. This lowers the centre of mass of the structure and the greater base dimensions reduce the tendency of the structure to overturn when subjected to lateral loads, such as wind. A good example of such a structure is the Eiffel Tower in Paris, as shown in Figure 3.12. The form of the tower is reassuring and appears to be stable and safe.

3.4.3 Display Unit

Figure 3.13 shows an inclined display unit. The centre of mass of the unit is outside the base of the unit, as shown in Figure 3.13a. To make the unit stable and prevent it from overturning, an additional support is fixed to the lower part of the display unit (Figure 3.13b). The added base area effectively increases the size of the base of the unit to ensure that the centre of mass of the display unit will be above the area of the new base of the unit. Thus, the inclined display unit becomes stable. Other safety measures are also applied to ensure that the display unit is stable.

(a)

(b)

FIGURE 3.11
(a,b) Cranes on construction sites.

FIGURE 3.12
The Eiffel Tower.

(a)

(b)

FIGURE 3.13
(a,b) Inclined display unit.

FIGURE 3.14
One of the two Kio Towers.

3.4.4 Kio Towers

Figure 3.14 shows one of the two 26-storey, 114 m high buildings of the Kio Towers in Madrid, which are also known as Puerta de Europa (Gateway to Europe). The Kio Towers actually lean towards each other, each inclined at 15° from the vertical.

The inclinations move the centres of mass of the buildings sideways, tending to cause toppling effects on the buildings. One of the measures used to reduce these toppling effects was to add massive concrete counterweights to the basements of the buildings. This measure not only lowered the centres of the mass of each building but also moved the centre of the mass towards a position above the centre of the base of the building.

PROBLEMS

1. Reproduce the demonstration shown in Sections 3.3.3 and 3.3.4.
2. Think of one example in everyday life or engineering practice in which the centre of mass of a body affects the behaviour of the body.
3. The geometry and parameters of the rail support and the conical body shown in Section 3.4.1 are given in Figure 3.15. Show that the necessary condition that the conical body moves from the lower end to the higher end of the rail support is

$$\frac{r}{d} > \frac{h}{b} \text{ or } \tan\theta > \frac{\tan\gamma}{\tan\phi}$$

Identify four other constraint conditions for designing a feasible rail support and conical body system.

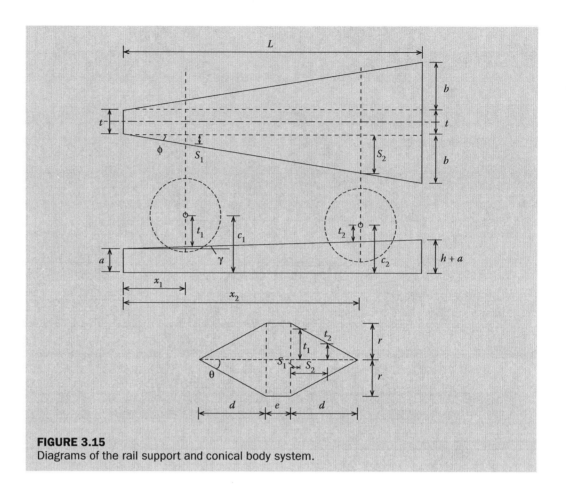

FIGURE 3.15
Diagrams of the rail support and conical body system.

REFERENCE

1. Hibbeler, R. C. *Mechanics of Materials*, 6th edn, Singapore: Prentice-Hall, 2005.

CHAPTER 4

CONTENTS

Effect of Different Cross Sections

4

4.1 DEFINITIONS AND CONCEPTS

Second moment of area (sometimes incorrectly called moment of inertia) is the geometrical property of a plane cross section which is based on its area and on the distribution of the area.

- The further the material of the section is away from the neutral axis of the section, the larger the second moment of area of the section. Therefore, the shape of a cross section will significantly affect the value of the second moment of area.
- The stiffness of a beam is proportional to the second moment of area of the cross section of the beam.

4.2 THEORETICAL BACKGROUND

Second moment of area: The second moment of area of a section can be determined about its x- and y-axes from the following integrals:

$$I_x = \int y^2 dA \qquad I_y = \int x^2 dA \tag{4.1}$$

where x and y are the coordinates of the elemental area dA and are measured from the neutral axes of the section.

The use of this equation can be illustrated by the simple example of the rectangular cross section shown in Figure 4.1, where the x- and y-axes have their origin at the centroid C, and b and h are the width and height of the section. Consider a strip of width b and thickness dy which is parallel to the x-axis and a distance y from the x-axis. The elemental area of the strip is $dA = bdy$. Using the first expression in Equation 4.1 gives the second moment of area of the section about the x-axis as

$$I_x = \int_{-h/2}^{h/2} y^2 bdy = \frac{bh^3}{12} \tag{4.2}$$

It can be seen from Equation 4.2 that I_x *is proportional to the width b and to the height h to the power of three – that is, increasing the height h is a much more effective way of increasing the second moment of area, I_x, than increasing the width b.*

Similarly, consider a strip of width h and thickness dx which is parallel to the y-axis with a distance x from the y-axis. The elemental area of the strip is $dA = hdx$. Using the second expression in Equation 4.1 gives the second moment of area of the section about the y-axis as

$$I_y = \int_{-b/2}^{b/2} x^2 hdx = \frac{hb^3}{12} \tag{4.3}$$

Parallel-axis theorem: To determine the second moment of area of a cross section about an axis which does not pass through the centroid of the section, the parallel-axis theorem may be used. Referring to Figure 4.1, the theorem may be stated in the following manner for the determination of the second moment of area of the section about the p-axis:

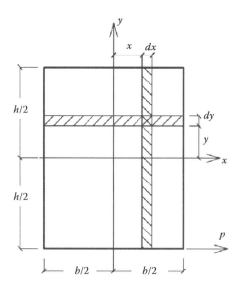

FIGURE 4.1
Second moment of area of a rectangular section.

$$I_p = I_x + Ad^2 \tag{4.4}$$

where:
I_x is the second moment of area of the section about the x-axis
d is the perpendicular distance between the x-axis and the p-axis
A is the area of the section

Equation 4.4 states that *the second moment of area of the cross section with respect to the p-axis is equal to the sum of the second moment of area about a parallel axis through the centroid and the product of the area and the square of the perpendicular distance between the two axes.*

For example, the second moment area of the section in Figure 4.1 about the p-axis can be determined using Equation 4.1, with y being measured from the p-axis and the range of the integration being between 0 and h, that is:

$$I_p = \int_0^h y^2 b \, dy = \frac{bh^3}{3} \tag{4.5}$$

Alternatively, the parallel-axis theorem embodied in Equation 4.4 can be used to determine the same second moment of area about the p-axis:

$$I_p = I_x + Ad^2 = \frac{bh^3}{12} + bh\left(\frac{h}{2}\right)^2 = \frac{bh^3}{3} \tag{4.6}$$

Relationship between deflection and second moment of area of a beam: The second moment of area of a cross section is required to calculate the deflection of a beam. Consider a simply supported uniform beam with a span L, a modulus of elasticity E and a second moment of area I which carries a uniformly distributed load q. The deflection at the midspan of the beam is

$$\Delta = \frac{5qL^4}{384EI} \tag{4.7}$$

Equation 4.7 indicates that the deflection is proportional to the inverse of I. Therefore, to reduce the deflection of the beam it is desirable to have the largest I possible for a given amount of material.

Relationship between normal stress and second moment of area of a beam: When a beam is loaded in bending, its longitudinal axis deforms into a curve, causing stresses that are normal to the cross section of the beam. The normal stress, σ_x, in the cross section at a distance y from the neutral axis can be expressed as

$$\sigma = \frac{My}{I} \tag{4.8}$$

Equation 4.8 indicates that *the normal stress is proportional to the bending moment M and inversely proportional to the second moment of area I of the cross section.* In addition, *the stress varies linearly with the distance y from the neutral axis.*

EXAMPLE 4.1

Figure 4.2 shows three different cross section shapes, A, B and C, two of which are rectangular sections and one is an I-section. They are made up of three identical rectangular plates of width b and thickness t. The three sections have the same areas. Determine and compare the second moments of areas of the three sections with respect to the horizontal x-axis passing through their centroids. For $b=3t$, $6t$, $9t$, $12t$ and $15t$, compare the relative values of the second moments of area of the sections.

SOLUTION

Section A has a width of b and height of $3t$ (Figure 4.2a), thus,

$$I_A = \frac{b(3t)^3}{12} = \frac{27bt^3}{12}$$

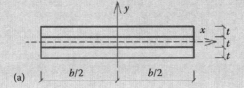

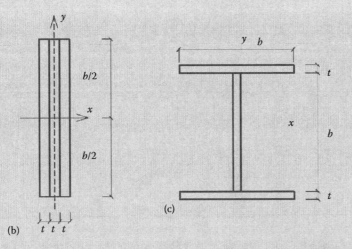

FIGURE 4.2
Different cross sections assembled by three identical strips. (a) Section A. (b) Section B. (c) Section C.

TABLE 4.1   Comparison of the Relative I Values of the Three Cross Sections

	$b = 3t$	$b = 6t$	$b = 9t$	$b = 12t$	$b = 15t$
Section A	1	1	1	1	1
Section B	1	4	9	16	25
Section C	4	12	25	43	65

Section B has a width of $3t$ and height of b (Figure 4.2b), therefore,

$$I_B = \frac{3tb^3}{12} = \frac{tb^3}{4}$$

Section C has an I-shape (Figure 4.2c). Its second moment of area can be calculated using the parallel-axis theorem stated in Equation 4.4:

$$I_C = \frac{tb^3}{12} + 2\frac{bt^3}{12} + 2bt\left(\frac{b}{2} + \frac{t}{2}\right)^2 = \frac{7tb^3}{12} + \frac{2bt^3}{3} + b^2t^2$$

The relative values of second moments of area, about the horizontal x-axis, with respect to the second moment of area of section A, are

$$\frac{I_B}{I_A} = \frac{3tb^3}{12}\frac{12}{27bt^3} = \frac{b^2}{9t^2}$$

and

$$\frac{I_C}{I_A} = \left(\frac{7tb^3}{12} + \frac{8bt^3}{12} + \frac{12b^2t^2}{12}\right)\frac{12}{27bt^3} = \frac{7b^2}{27t^2} + \frac{8}{27} + \frac{12b}{27t}$$

The relative I values are given in Table 4.1 for different ratios of b to t. It can be observed from Table 4.1 that

- When $b = 3t$, sections A and B have the same width and height and they also have the same second moments of area.
- The ratio of the second moments of area of section B to section A increases quadratically with the ratio of b/t.
- Section C (the I-shaped section) has the largest second moment of area. When $b = 9t$, its second moment of area is 25 times that of section A and almost 2.8 times that of section B, demonstrating the beneficial use of material in I-sectioned steel members, which are extensively used in buildings.
- As demonstrated by section C, the farther the areas of the material in a section are from its centroid, the larger will be the second moment of area.

EXAMPLE 4.2

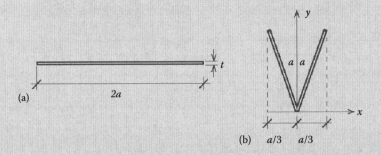

FIGURE 4.3
Two sections with the same area. (a) A flat section. (b) A V-section.

Figure 4.3 shows two cross sections which have the same area. One is a flat section with width $2a$ and thickness t where $2a \gg t$ and the other is in a folded V-form with its dimensions shown in Figure 4.3b. Calculate the second moments of areas of the two cross sections about their neutral axes.

SOLUTION

For the flat form:

$$I_f = \frac{(2a)t^3}{12} = \frac{at^3}{6}$$

For the V-form, the right half of the V-shape can be described by $y = \sqrt{8}x$ (Figure 4.3); the neutral axis can be determined using Equation 3.5:

$$c = \frac{2\int_0^{a/3} y\sqrt{1+(y')^2}\,t\,dx}{2\int_0^{a/3}\sqrt{1+(y')^2}\,t\,dx} = \frac{2\sqrt{2}a^2 \cdot t/3}{2at} = \frac{\sqrt{2}}{3}a$$

The second moment of area of the V-section with respect to the x-axis is

$$I = 2\int_0^{a/3} y^2\sqrt{1+(y')^2}\,t\,dx = \frac{16}{27}a^3t$$

Using the parallel-axis theorem (Equation 4.4), the second moment of area of the V-section about its neutral axis is

$$I_v = I - Ac^2 = \frac{16a^3t}{27} - 2at\left(\frac{\sqrt{2}a}{3}\right)^2 = \frac{4a^3t}{27}$$

The ratio of the second moments of areas of the V-section to the flat section is

$$\frac{I_v}{I_f} = \frac{4a^3t}{27} \cdot \frac{6}{at^3} = \frac{8a^2}{9t^2}$$

When $a = 30$ mm and $t = 0.4$ mm, the ratio becomes 5000! This will be demonstrated in Section 4.3.2.

Further details can be seen in [1].

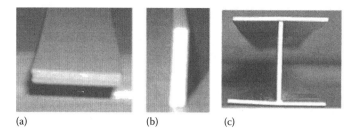

(a)   (b)   (c)

FIGURE 4.4
Three beams with different cross sections. (a) Beam 1. (b) Beam 2. (c) Beam 3.

4.3 MODEL DEMONSTRATIONS

4.3.1 Two Rectangular Beams and an I-Section Beam

This demonstration shows *how the shapes of cross sections affect the stiffness of a beam.*

The three beams shown in Figure 4.4 can be made using plastic strips, each with the same amount of material, for example three 1 mm thick by 15 mm wide strips. Figure 4.4a shows the section of the first beam in which the three strips are glued together to form a section 15 mm wide and 3 mm deep. The beam shown in Figure 4.4b is the same as the first beam but its section is turned through 90°. In the beam shown in Figure 4.4c, the three 1 mm thick strips are arranged and glued together to form an I-section. The second moments of area of the three sections about their horizontal neutral axes are in the ratios: 1:25:65 (Table 4.1). The stiffnesses of the three beams can be demonstrated and felt through simple experiments as follows:

1. Support Beam 1 at its two ends, as shown in Figure 4.5a, and press down at the midspan of the beam. Notice and feel that the beam sustains a relatively large deflection under the applied load. This beam has a small value of second moment of area because the material of the cross section is close to its neutral axis.

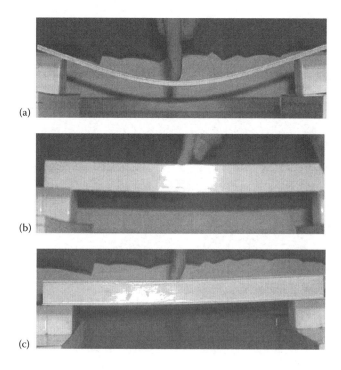

(a)

(b)

(c)

FIGURE 4.5
The deflections of three beams with the same cross section area but different second moments of area. (a) A rectangular section beam with the larger dimension horizontal. (b) A rectangular section with the larger dimension vertical. (c) An I-section.

2. Replace Beam 1 with Beam 2. Press down at the midspan of the beam whilst also supporting one end of the beam to prevent its tendency to twist (Figure 4.5b). Note that Beam 2 deflects much less than Beam 1 and that its stiffness feels noticeably larger. The material of the cross section in Beam 2 has been distributed farther away from its neutral axis than was the case for Beam 1, significantly increasing its second moment of area.
3. Replace Beam 2 with Beam 3 and again press down at the midspan as shown in Figure 4.5c. The beam is stable and feels even stiffer than the rectangular beam shown in Figure 4.5b. The increased stiffness is because two-thirds of the material of the cross section is placed in the flanges, that is, as far away as possible from the neutral axis of the section.

This simple demonstration shows that the I-section Beam 3 is significantly stiffer than the other two beams and that Beam 2 is much stiffer than Beam 1, although all use the same amount of material.

4.3.2 Lifting a Book Using a Bookmark

This demonstration shows again *how the shapes of cross sections affect the stiffness of a member.*

Figure 4.6a shows two bookmarks which are formed from identical strips of card with length 210 mm, width 6 mm and approximate thickness 0.4 mm. The lower bookmark in Figure 4.6a is in its original flat form and the upper one is folded and secured with rubber bands to form the V-shape shown.

If one tries to lift a book using the bookmark in its flat form, the bookmark bends and changes its shape significantly and the book cannot be lifted. This is because the bookmark is very thin and has a small second moment of area around the axis passing through the midthickness of the bookmark, parallel to its width, resulting in a stiffness which is too small to resist the loading.

It is, however, easy to lift the book using the bookmark in its folded form, as shown in Figure 4.6b. This is because the folded bookmark has a much larger second moment area since the material is distributed farther away from its neutral axis than in the flat form, providing sufficient stiffness to resist the deformation induced by the book. As calculated in Example 4.2, the member with the V-shaped section has a stiffness of about 5000 times that of the member of the flat section.

4.4 PRACTICAL EXAMPLES

4.4.1 Steel-Framed Building

I-section members are the most commonly used structural elements in steel building frames. Figure 4.7 shows a typical steel-framed building using I-shaped sections for both beams and columns.

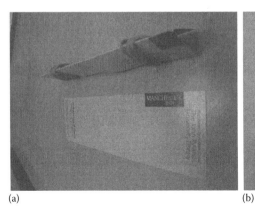
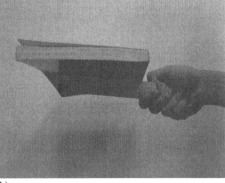

(a)   (b)

FIGURE 4.6
Effect of the shape of bookmarks. (a) Two bookmarks with different shapes. (b) Lifting a book using a folded bookmark.

FIGURE 4.7
A steel-framed building.

FIGURE 4.8
A railway bridge.

4.4.2 Railway Bridge

Figure 4.8 shows a railway bridge. Both the longitudinal and transverse beams are steel I-sections and are used to support the precast concrete slabs of the deck.

4.4.3 I-Section Members with Holes (Cellular Beams and Columns)

The material close to the neutral axis of a section does not contribute efficiently to the second moment of area and hence to the stiffness of a member; thus, this material can be removed, making the member lighter and saving material. In addition, the normal stress due to bending about the neutral axis is small (Equation 4.8), so the removal of the material around the neutral axis does not significantly reduce the loading capacity of the member. It is quite common to see I-section beams and columns with holes along their neutral axes.

Figure 4.9 shows a car showroom in which the external columns support the roof. The external columns are all I-sections with the material around their neutral axes removed to create a lighter structure of more elegant appearance. Cellular columns are most effective in cases where axial loads are small. Figure 4.10 shows cellular beams used in an airport terminal.

FIGURE 4.9
Cellular columns.

FIGURE 4.10
Cellular beams.

PROBLEMS

1. For a member subjected to bending, it is efficient to increase the second moment of area of the member. Identify examples around you in which (1) the second moment of area was purposely enlarged; and (2) the materials are cleverly arranged to increase the second moment of area.

2. Consider the I-section beam in Example 4.1 (Figure 4.2c). If another similar rectangular plate is glued to the bottom of the I-section, what is the second moment of area of the new cross section? What is the increase of the second moment of area relative to the original I-section when $b = 3t$, $6t$, $9t$, $12t$ and $15t$?

REFERENCE

1. Gere, J. M. *Mechanics of Materials*, Belmont, CA: Thomson Books/Cole, 2004.

CHAPTER 5

CONTENTS

Stress Distribution

5.1 CONCEPTS

For a given external or internal force, the smaller the area of the member resisting the force, the higher the stress.

   Saint-Venant's principle: The stresses and strains in a body at points that are sufficiently remote from the points of application of loads depend only on the static resultant of the loads and not the distribution of the load.

5.2 THEORETICAL BACKGROUND

Normal stress is defined as the intensity of force or the force per unit area acting normal to the area of a member. It is expressed as

$$\sigma = \lim_{\Delta A \to 0} \frac{\Delta P}{\Delta A} \tag{5.1}$$

where:

   ΔA is an element of area
   ΔP is the force acting normal to the area

   When the normal stress tends to 'pull' on the area it is called **tensile stress**, and if it tends to 'push' on the area it is termed **compressive stress**. When ΔA tends infinitely towards zero, the stress is actually at a point instead of over an area.

   Average normal stress often needs to be considered in engineering practice in such members as truss members, cables and hangers. It is expressed as

$$\sigma = \frac{P}{A} \tag{5.2}$$

where:

   P is the resultant normal force
   A is the external normal force acting on the area

   Equation 5.2 assumes the following:

- After deformation, the area A should still be normal to the force P.
- The material of the member is homogeneous and isotropic. Homogeneous material has the same physical and mechanical properties at different points in its volume, and isotropic material has these same properties in all directions at any point in its volume. A typical example of a homogeneous and isotropic material is steel.
- The force P is applied at a sufficient distance from the position where the stress is evaluated.

   Equation 5.2 states that, *for a given force P, the smaller the area resisting the force, the larger the normal stress.*

   The third assumption avoids stress concentration at the loading position, where stress distribution can be complicated. This assumption, which will be observed to be true from a model demonstration in Section 5.3.2, is referred to as Saint-Venant's principle [1]. The significance of the principle can be stated as follows: *the stresses and strains in a body at points that are sufficiently remote from the points of application of loads depend only on the static resultant of the loads and not the distribution of the load.*

Further details can be found in [1,2].

5.3 MODEL DEMONSTRATIONS

5.3.1 Balloons on Nails

This demonstration shows *the effect of stress distribution.*

Place a balloon on a single nail and hold a thin wooden plate above the balloon and position the balloon as shown in Figure 5.1a. Gradually transfer the weight of the plate onto the balloon. Before the full weight of the plate rests on the balloon, it will burst. This happens because the balloon is in contact with the very small area of the single nail, resulting in a very high stress and causing the balloon to burst.

Now place another balloon on a bed of nails instead of a single nail. Put the thin wooden plate on the balloon and gradually add weights to the plate as shown in Figure 5.1b. It will be seen that the balloon can carry a significant weight before it bursts. It is observed that the shape of the balloon changes but it does not burst. Due to its changed shape, the balloon and hence the weight on the balloon are supported by many nails. As the load is distributed over many nails, the stress level caused is not high enough to cause the balloon to burst.

A similar example was observed at a science museum, as shown in Figure 5.2. A 6000-nail bed is controlled electronically, allowing the nails to move up and down. When the nails move down below the smooth surface of the bed, a young person lies on the bed. Then the 6000 nails move up slowly and uniformly and lift the human body to the position shown in Figure 5.2. If the body has a total uniform mass of 80 kg and about two-thirds of the nails support the body, each loaded nail only carries a force of 0.2 N or a mass of 20 g. Thus, the person is not hurt by the nails.

A similar but simpler demonstration can be conducted following the observation of the nail bed. Figure 5.3a shows 49 plastic cups which are placed upside down and side by side. Place two thin wooden boards on them and invite a person to stand on the boards. The boards spread the weight of the person, say 650 N, over the 49 cups, each cup carrying about 13 N, which is less than the 19 N capacity of a cup.

(a)   (b)

FIGURE 5.1
(a,b) Balloon on nails.

FIGURE 5.2
A person lying on a nail bed.

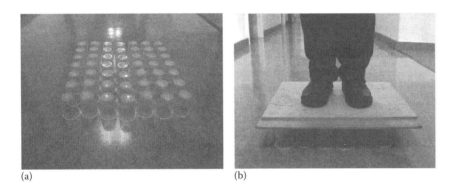

(a)   (b)

FIGURE 5.3
Uniform force distribution. (a) Forty-nine plastic cups placed upside down. (b) A person standing on the cups.

5.3.2 Uniform and Nonuniform Stress Distributions

This demonstration shows *the effect of nonuniform stress distribution.*

Place a wooden block on a long sponge and apply a concentrated force at the centre of the block, as shown in Figure 5.4a. The sponge under the block deforms uniformly. If the concentrated force is placed at one end of the block, as shown in Figure 5.4b, the wooden block rotates with part of the

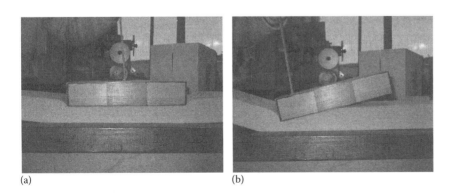

(a)   (b)

FIGURE 5.4
Uniform and nonuniform stress distributions. (a) Uniform stress distribution. (b) Nonuniform stress distribution. (The model demonstration was provided by Mr. P. Palmer, University of Brighton.)

block remaining in contact with the sponge and part of the block separating from the sponge. It can be observed that the sponge deforms nonuniformly, indicating a nonuniform stress distribution in the sponge.

5.3.3 Stress Concentration

This demonstration shows *the stress distributions of a sponge block model due to a compressive distributed load and a compressive concentrated load, and illustrates Saint-Venant's principle.*

The model comprises a piece of sponge on which a number of horizontal lines are marked; a T-shaped plate, by which a distributed load or a concentrated load can be applied; and a base plate, with two vertical metal tubes to ensure that the load is applied vertically. Figure 5.5 shows the components of the model and the sponge block before deformation.

When a distributed load is applied on the sponge, as shown in Figure 5.6a, it is observed that all the horizontal lines of the block remain horizontal. In this case, the stresses in the cross section of the sponge are uniform and can be determined using *P/A*, where *P* is the total load and *A* is the area of the cross section.

When a concentrated load is applied, as shown in Figure 5.6b, it can be observed that

- Near to the top end of the block, where the concentrated load is applied, the lines form a pattern that radiates out from the loading point, indicating a radial stress configuration.
- At a short distance from the loading point, the lines remain parallel to the base and appear to be the same as they were with the distributed load.

The second observation is an illustration of Saint-Venant's principle.

FIGURE 5.5
The model.

(a)   (b)

FIGURE 5.6
Deformations of the sponge due to uniformly distributed and concentrated loads. (a) Applying a uniformly distributed load. (b) Applying a concentrated load.

5.3.4 Core of a Section

These models demonstrate that *if a vertical load is applied within the core of a section the whole section will keep contact with the supporting sponge. Otherwise, there will be separations between the section and the sponge.*

Figure 5.7 shows one rectangular and one circular wooden section, with the core areas raised on top of the pieces. The demonstrations show the effect when a vertical compressive load is applied within or outside the core area of the two wooden pieces. To illustrate the effect, they are placed on a sponge block.

When an eccentric compressive load is applied perpendicular to the core area of the rectangle, as shown in Figure 5.8a, it will cause a uniform stress distribution plus a linearly distributed stress distribution over the rectangle. The uniform stress distribution is due to the action of the compressive load if it is applied at the centre of the rectangle, while the linearly distributed stress is induced by a bending moment that is the product of the load and the distance between the point of the load and the centre of the rectangle. In this case, the maximum stress induced by the moment is smaller than the uniform stress. Thus, all the areas of the wooden piece are subjected to compressive stress. In other words, all the area of the rectangle is in contact with the sponge and there are no gaps between the wooden piece and the sponge.

When the load is applied outside the core area, the linearly distributed stress will increase proportionally with the distance between the load point and the centre of the rectangle. In this case

FIGURE 5.7
A rectangular and a circular section with their core areas.

(a)   (b)

FIGURE 5.8
A rectangular section subject to a vertical load. (a) Load applied in the core area. (b) Load applied outside the core area.

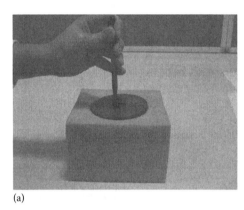

(a)   (b)

FIGURE 5.9
A circular section subject to a vertical load. (a) Load applied in the core area. (b) Load applied outside the core area.

(Figure 5.8b), the maximum stress due to the moment is larger than the uniformly distributed stress and this causes a separation between the wooden piece and the sponge, and a stress redistribution.

Similarly, when the experiment is applied to the circular wooden section, the same phenomena can be observed:

- When the load is applied within the core area, all stress applied by the wooden section is compressive, which means that the whole of the section maintains contact with the sponge (Figure 5.9a).
- When the load is applied outside the core area, it causes separation between part of the area of the wooden section and the sponge and the contact area experiences larger nonuniform stresses (Figure 5.9b).

5.4   PRACTICAL EXAMPLES

5.4.1   Flat Shoes versus High-Heel Shoes

A woman wearing high-heel shoes will exert the same force on a floor as when she is wearing flat shoes, but with significantly different stress levels. Figure 5.10 shows a woman, who weighs 50 kg, wearing (a) flat shoes and (b) high-heel shoes. The stresses exerted by the shoes can be estimated as follows.

Flat shoes: The contact area of one of the flat shoes (Figure 5.10a) is $12 \times 10^{-3}\,\mathrm{m}^2$. Assuming the body weight is uniformly distributed over the contact area, the average stress is

(a)

(b)

FIGURE 5.10
Stresses exerted from flat and high-heel shoes. (a) A flat shoe. (b) A high-heel shoe. (Courtesy of Miss C. Patel.)

$$\sigma_{flat} = \frac{50 \times 9.81}{12 \times 10^{-3} \times 2} = 20.4 \text{ kN m}^{-2}$$

High-heel shoes: Assuming that half of the body weight is carried by the high heels (Figure 5.10b), each high heel carries a quarter of the body weight. The area of the high heel is $1.0 \times 10^{-4} \text{ m}^2$. Thus, the average stress under each high heel is

$$\sigma_{heel} = \frac{50 \times 9.81}{1 \times 10^{-4} \times 2 \times 2} = 1226 \text{ kN m}^{-2}$$

This shows that the stress under the heels of the high-heel shoes is about 60 times that under the flat shoes.

A typical adult African elephant weighs about 5000 kg and the area of each foot is about 0.08 m². Thus, the average stress exerted by an elephant's foot is

$$\sigma_{elephant} = \frac{5000 \times 9.81}{0.08 \times 4} = 153.3 \text{ kN m}^{-2}$$

Elephants have large feet to distribute their body weight. The stress exerted on the ground by the foot of an elephant is much less than that under the high-heel shoes.

5.4.2 Leaning Tower of Pisa

The tilt of one of the world's most famous towers, the Leaning Tower of Pisa, developed because of uneven stress distribution on the soil supporting the upper structure. The eight-storey tower weighs 14,500 t and its masonry foundations are 19.6 m in diameter. If the stress was uniformly distributed, this would lead to an average stress of 470 kN m^{-2}, from Equation 5.2. As the underlying ground consists of about 10 m of variable soft silty deposits (Layer A) and then 40 m of very soft and sensitive marine clays (Layer B), the tower shows that the surface of Layer B is dish-shaped due to the weight of the tower above it [3]. Thus, the uneven stresses on the soft soil under the tower caused the foundation to settle unevenly, making the tower lean.

To redress the problem, many large blocks of lead were placed on the ground on the side of the tower where the settlement was least, as shown in Figure 5.11. The new stress in the ground caused the firmer soil on this side of the tower to compress, in turn preventing further leaning of the tower and returning the tower back towards the vertical.

(a) (b)

FIGURE 5.11
(a,b) Lead blocks used to reduce the uneven stress distribution on the foundation.

PROBLEMS

1. Determine the positions of the core areas for the rectangular and circular wooden sections shown in Figure 5.7. Assume that the length and width of the rectangular section are *a* and *b*, respectively, and the circular section has a radius of *r*.

2. Figure 5.12 shows a foam cuboid encased in a Perspex box with the four vertical surfaces of the foam restrained against deformations perpendicular to their surfaces. Assume that uniform compressive stress, σ_z, in the vertical direction is applied on the top surface of the foam and that σ_x and σ_y in the two perpendicular horizontal directions have the same magnitude and are uniform on the four vertical surfaces. Ignore the friction stresses between the vertical surfaces and the foam. The relationships between strain and stress in the three perpendicular directions are

$$\varepsilon_x = [\sigma_x - \nu(\sigma_y + \sigma_z)]/E$$

$$\varepsilon_y = [\sigma_y - \nu(\sigma_z + \sigma_x)]/E$$

$$\varepsilon_z = [\sigma_z - \nu(\sigma_x + \sigma_y)]/E$$

where:
  ν is Poisson's ratio
  E is Young's modulus

Show that the strain in the *z* direction is

$$\varepsilon_z = \frac{\sigma_z}{E}\frac{1 - \nu - 2\nu^2}{1 - \nu}$$

FIGURE 5.12
A constrained foam cuboid subjected to vertical loads.

REFERENCES

1. Craig, R. *Mechanics of Materials*, New York: John Wiley, 1996.
2. Hibbeler, R. C. *Mechanics of Materials*, 6th edn, Singapore: Prentice-Hall, 2005.
3. Burland, J. B. The Leaning Tower of Pisa, in N. Parkyn ed., *The Seventy Architectural Wonders of Our World*, 34–38, London: Thames & Hudson, 2002.

CHAPTER 6

CONTENTS

Bending

6.1 DEFINITIONS AND CONCEPTS

For a beam subjected to bending:

- Elongation occurs on one surface and shortening on the opposite surface of the beam. There is a (neutral) plane through the beam which does not change in length during bending.
- Plane cross sections of the beam remain plane and perpendicular to the neutral axis of the beam.
- Any deformation of a cross section of the beam within its own plane is neglected.
- The normal stress on a cross section of the beam is distributed linearly with the maximum normal stresses occurring on surfaces farthest from the neutral plane.

6.2 THEORETICAL BACKGROUND

Beams: Normally the length of a beam is significantly greater than the dimensions of its cross section. Many beams are straight and have a constant cross-sectional area.

Shear force and bending moment diagrams: When loaded by transverse loads, a beam develops internal shear forces and bending moments that vary from position to position along the length of the beam. To design beams, it is necessary to know the variations of the shear force and the bending moment, and the maximum values of these quantities. The shear forces and bending moments can be expressed as functions of their position x along the length of the beam, which can be plotted along the length of the beam to produce shear force and bending moment diagrams.

Relationships between shear force (V), bending moment (M) and loading (q) can be represented by the following equations:

$$\frac{dV}{dx} = -q(x) \tag{6.1}$$

$$\frac{dM}{dx} = V \tag{6.2}$$

These two equations can be obtained from the free-body diagram for a segment, dx, of a loaded beam (as illustrated in Figure 6.1), making use of vertical force and moment equilibrium. Equations 6.1 and 6.2 are particularly useful for drawing and checking the shear force and bending moment diagrams of a beam as they state, respectively,

- The slope of the shear force diagram at any point along the length of the beam is equal to minus the intensity of the distributed load at the point.
- The slope of the bending moment diagram at any point along the length of the beam is equal to the shear force at the point.

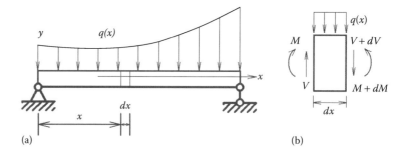

(a)   (b)

FIGURE 6.1
Diagrams for deriving Equations 6.1 and 6.2. (a) A loaded beam. (b) Free-body diagram for element dx.

EXAMPLE 6.1

Draw the shear force and bending moment diagrams for a simply supported beam subject to a uniformly distributed load q, as shown in Figure 6.2a, and determine the maximum bending moment.

SOLUTION

Taking the free-body diagram of the beam (Figure 6.2b) and using Equation 2.1, it is possible to determine the reaction forces from the two supports as indicated in Figure 6.2b. In order to determine the shear force (V) and the bending moment (M) at a cross section, the equilibrium of the free-body diagram shown in Figure 6.2c can be considered. Applying the two equations of equilibrium gives

$$\sum F_y = 0 \quad \frac{qL}{2} - qx - V = 0 \quad V = q\left(\frac{L}{2} - x\right) \tag{6.3}$$

$$\sum M = 0 \quad -\frac{qL}{2}x + qx\frac{x}{2} + M = 0 \quad M = \frac{q}{2}\left(Lx - x^2\right) \tag{6.4}$$

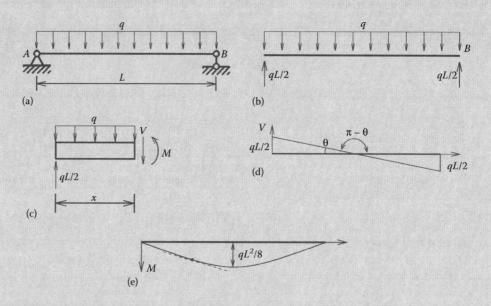

FIGURE 6.2
(a–c) A simply supported beam. (d) Shear force diagram. (e) Bending moment diagram.

The shear force and bending moment diagrams for the beam can be obtained by plotting Equations 6.3 and 6.4, as shown in Figure 6.2d and e. Equations 6.1 and 6.2 can now be used to check the correctness of the shear force and bending moment diagrams. As the distributed load is a constant along the length of the beam (Figure 6.2a), the slope of the shear force diagram is $-[qL/2 - (-qL/2)]L = -q$. According to Equation 6.2 and the shear force diagram in Figure 6.2d, the shear force changes from positive to negative linearly so the bending diagram should be a curve. The slope of the bending moment diagram, that is, the shear force, decreases continuously from $qL/2$ at $x = 0$ to zero at $x = L/2$, then further decreases from zero to $-qL/2$ at $x = L$. The maximum moment occurs at the centre of the beam where the shear force is zero. Thus, from Equation 6.4:

$$M_C = \frac{q}{2}\left[L \times \left(\frac{L}{2}\right) - \left(\frac{L}{2}\right)^2\right] = \frac{qL^2}{8} = 0.125qL^2$$

Equation 6.4 also shows that the shape of the bending moment diagram is a parabola. In order to achieve a safe and economical design, designers often seek ways to reduce the maximum bending moment and the associated stress by reducing spans, or creating negative bending moments at supports to offset part of the positive bending moment.

EXAMPLE 6.2

If the two supports of the beam in Figure 6.2a move inward symmetrically by distances of μL, the beam becomes an overhanging beam with its ends freely extending over the supports, as shown in Figure 6.3a. Determine the value of μ at which the maximum negative and positive bending moments in the beam are the same and determine the corresponding bending moment.

SOLUTION

Figure 6.3b shows the free-body diagram for the beam. The bending moments at supports B and D and at midspan C can be shown to be

$$M_B = M_D = -\tfrac{1}{2}q\mu^2 L^2$$

$$M_C = \frac{1}{2}qL\left(\frac{1}{2} - \mu\right)L - \frac{1}{2}q\left(\frac{L}{2}\right)^2 = \frac{1}{8}qL^2 - \frac{1}{2}q\mu L^2$$

Figure 6.3c and d show the shear force and bending moment diagrams for the beam. When the magnitudes of M_B and M_C are the same, it leads to

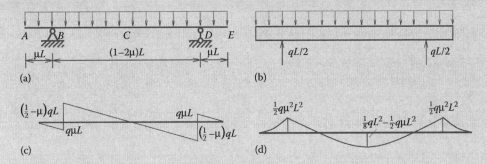

FIGURE 6.3
(a,b) A simply supported beam with overhangs (c) Shear force diagram. (d) Bending moment diagram.

$$\tfrac{1}{2}q\mu^2 L^2 = \tfrac{1}{8}qL^2 - \tfrac{1}{2}q\mu L^2 \quad \text{or} \quad 4\mu^2 + 4\mu - 1 = 0$$

The solution of this quadratic equation is $\mu = 0.207$. Substituting $\mu = 0.207$ into the expression for M_B or M_C gives

$$M_B = -\tfrac{1}{2}q(0.207)^2 L^2 = 0.0214qL^2$$

$$M_C = \frac{1}{8}qL^2 - \frac{0.207}{2}qL^2 = 0.0214qL^2$$

Comparing the maximum bending moments in the simply supported beam ($0.125qL^2$) in Figure 6.2a and the overhanging beam ($0.0214qL^2$) in Figure 6.3a, it shows that the former is nearly six times the latter. The reasons for this are as follows.

- The reduced distance between the two supports. The bending moment is proportional to the span squared, and the shortened span effectively reduces the bending moment.
- The compensation of bending moment. The negative bending moments over the supports, due to the use of the overhangs, compensate for part of the positive bending moment.

In engineering practice, $\mu = 0.2$ is used instead of the exact solution of $\mu = 0.207$ for an optimum simply supported overhanging beam. Further theoretical background and examples, including the calculation of stresses in beams, can be found in many textbooks, including [1–3].

6.3 MODEL DEMONSTRATIONS

6.3.1 Assumptions in Beam Bending

This demonstration examines *some of the basic assumptions used in the theory of beam bending.*

A symmetric sponge beam model is made which can be bent and twisted easily (Figure 6.4a). Horizontal lines on the two vertical sides of the beam are drawn at mid-depth, indicating the neutral plane, and vertical lines are made at equal intervals along the length of the sponge, indicating the different cross sections of the beam.

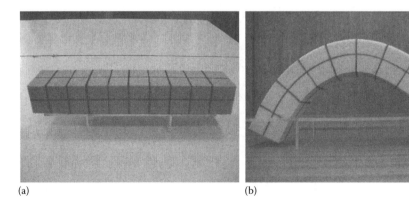

(a)   (b)

FIGURE 6.4
(a,b) Examination of beam-bending assumptions.

FIGURE 6.5
Thin and thick sponge beams.

By bending the beam, as shown in Figure 6.4b, it can be observed that

- All of the vertical lines, which indicate what is happening to the cross sections of the beam, remain straight.
- The angles between the vertical lines and the centroidal line (neutral axis) remain at 90°.
- The upper surface of the beam extends and the bottom surface shortens.
- The length of the centroidal (neutral) axis of the beam does not change.

6.3.2 Thin Beam and Thick Beam

This demonstration shows that *the assumptions used to study the bending of a thin beam do not hold for a thick beam as significant shear deformations occur in the thick beam.*

Figure 6.5 shows two sponge beams that have square cross sections but have different length to width ratios. The beams are set up as vertical cantilevers to be acted on by transverse loads. The loads will cause bending and shear stresses and deformations throughout the beams. To make these phenomena visible, each beam is marked with several equally spaced horizontal lines indicating the transverse cross sections and a vertical line indicating the neutral axis.

The thin beam has a length to width ratio of about 6. Figure 6.6a shows the deformation when a transverse load is applied at the free end of the beam. It can be seen that plane sections remain plane and normal to the neutral axis, even with the large deformations.

The thick beam has a length to width ratio of about 1.5. In this case, the effect of shear deformation cannot be neglected. Applying the transverse force to the top of the beam causes warping of the horizontal lines that are no longer straight and perpendicular to the neutral axis (Figure 6.6b). This phenomenon indicates that the stresses are no longer linearly distributed along the cross sections of the beam.

Therefore, it can be seen that the behaviour of thin beams and thick beams in bending is different, and this indicates that the equations derived for thin beams do not apply to thick beams.

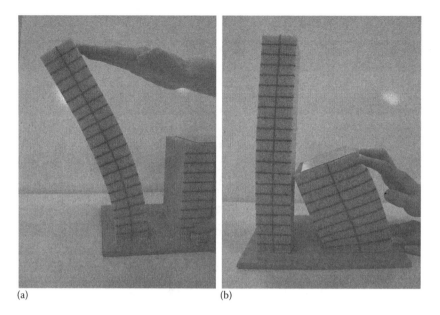

(a)   (b)

FIGURE 6.6
Deformations of the thin and thick beams. (a) Deformation of the thin beam. (b) Deformation of the thick beam.

6.4 PRACTICAL EXAMPLES

6.4.1 Profiles of Girders

Large curtain or window walls are often seen at airport terminals. Figure 6.7 shows three such walls. These large window walls are supported by a series of plane girders. The wind loads applied on the window walls are transmitted to the girders and through the girders to their supports.

The girders act like vertical simply supported 'beams'. The bending moments in the 'beams' induced by wind loads are maxima at or close to their centres and minima at their ends. If the wind load is uniformly distributed, the diagrams of bending moments along the 'beams' will be parabolas. Thus, it is reasonable to design the girders to have their largest depths at their centres and their smallest depths at their ends, with the profiles of the girders being parabolas. The girders shown in Figure 6.7c reflect this and appear more elegant than those shown in Figure 6.7a and b.

(a)   (b)   (c)

FIGURE 6.7
(a–c) A series of girders supporting windows at airport terminals.

FIGURE 6.8
Overhangs used to reduce bending moments and deflections. (Courtesy of Mr. J. Calverley.)

6.4.2 Reducing Bending Moments Using Overhangs

Figure 6.8 shows a steel-framed, multistorey car park where cellular beams are used. The vertical loads from floors are transmitted to the beams and then from the beams through bending to the supporting columns. Overhangs are used in the structure which can reduce the bending moments and deflections of the beams. Examining the first overhang, two steel wires are placed to link the free end of the overhang and the concrete support. A downward force on the free end of the overhang is provided by tensions induced in the pair of steel wires. This force will generate a negative bending moment over the column support which will partly offset the positive moments induced in the loaded beam by the applied loading.

The overhang is subjected to concentrated forces at its free end and therefore the bending moment varies linearly along the overhang with the minimum at its free end and the maximum at its fixed end. The cross section of the overhang is designed accordingly to vary linearly, and the overhang appears lighter and more elegant than it would be if a constant cross section were used throughout its length.

6.4.3 Failure due to Bending

Figure 6.9a shows a bench which consists of wooden strips, serving as seating and backing, and a pair of concrete frames supporting the seating and backing. Figure 6.9b shows cracks at the end of the cantilever of one of the concrete frames. The two main cracks in the upper part of the cantilever (Figure 6.9b) are a consequence of the normal stresses induced by bending exceeding the limit of tensile stress of concrete.

(a)   (b)

FIGURE 6.9
A bench with cracks. (a) A bench. (b) Cracks at the end of the cantilever.

6.4.4 Deformation of a Staple due to Bending

Staplers are common in offices. Figure 6.10 shows a typical stapler which creates closed staples through a process of bending and plastic deformation. In this case, it can be seen that the two symmetric indents or grooves in the front part of the base of the stapler have bowed shapes.

Figure 6.11 shows the process of the deformation of a staple by a stapler. Initially, a staple looks like a frame structure without supports at the ends of the two vertical elements. When the stapler is

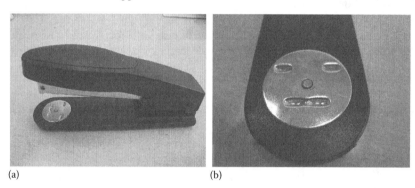

(a)   (b)

FIGURE 6.10
(a,b) A typical stapler.

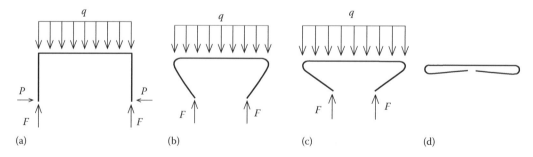

(a)   (b)   (c)   (d)

FIGURE 6.11
(a–d) The process of the deformation of a staple.

depressed, distributed loads are applied on the horizontal element (the middle part of the frame) and these loads push the staple through the multiple sheets of paper being joined, to meet the metal base of the stapler. When the staple reaches and enters the grooves in the base of the stapler, both vertical and lateral forces are applied to the ends of the staple as shown in Figure 6.11a. The lateral forces, which are induced due to the shape of the grooves, make the two vertical elements of the staple bend and move inwards. When the staple reaches the position shown in Figure 6.11b, the ends of the staple leave the grooves and no lateral forces act. In this position, the vertical forces are sufficient to continue to bend the two vertical elements further (Figure 6.11c). Finally, the shape of the staple reaches that shown in Figure 6.11d. The large deformations induced are plastic and permanent.

PROBLEMS

1. The practical example in Section 6.4.2 can be simplified and represented as an overhanging beam, shown in Figure 6.12, with a uniformly distributed load q and the cable force F acting with an angle of θ to a vertical line. Determine the maximum cable force F as a function of the span L, load q and angle θ, which leads to an economical design.

2. Figure 6.13a shows a splice joint consisting of four plates tightly assembled using bolts. It is assumed that there is no relative movement at the interface of the plates when loaded. The idealised analysis model of the joint and its dimensions are shown in Figure 6.13b. A uniformly distributed load of q (N mm^{-2}) is applied at the two ends of the joint. The self-weight of the joint can be neglected and the specimen only experiences small deformation. Answer the following questions.

   a. Due to symmetry, only the upper right quadrant of the joint needs to be modelled and analysed. Draw the free-body diagram for the quadrant.

   b. Draw the anticipated deformed shape of the quadrant. Draw the stress distribution along the height of the cross section at the vertical axis of symmetry and estimate the stresses at the top and the bottom.

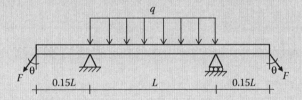

FIGURE 6.12
An overhanging beam subjected to loads and designed forces.

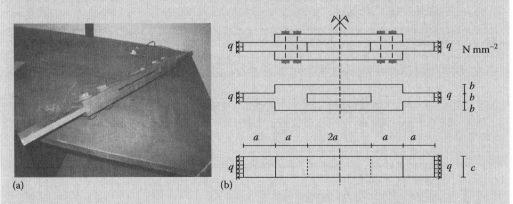

(a)

(b)

FIGURE 6.13
A splice joint and its analysis model. (a) A splice joint. (b) A simplified model of the joint for analysis.

REFERENCES

1. Hibbeler, R. C. *Mechanics of Materials*, 6th edn, Singapore: Prentice-Hall, 2005.
2. Williams, M. S. and Todd, J. D. *Structures: Theory and Analysis*, London: Macmillan, 2000.
3. Gere, J. M. *Mechanics of Materials*, Belmont, CA: Thomson Books/Cole, 2004.

CHAPTER 7

CONTENTS

Shear and Torsion

7.1 DEFINITIONS AND CONCEPTS

Shear: A force (or stress) which tends to slide the material on one side of a surface relative to the material on the other side of the surface in directions parallel to the surface is termed a *shear force* (or *shear stress*). The maximum shear stress in a solid rectangular section occurs at the neutral axis of its cross section and is 1.5 times the average value of the shear stress across the section.

Torsion: A moment that is applied about the longitudinal axis of a member which tends to twist the member about this axis is called a *torque* and is said to cause torsion of the member.

For a circular shaft or a closed circular section member subjected to torsion:

* Plane circular cross sections remain plane and the cross sections at the ends of the member remain flat.
* The length and the radius of the member remain unchanged.
* Plane circular cross sections remain perpendicular to the longitudinal axis.

For a noncircular section member or an open section member subjected to torsion:

* Plane cross sections of the member do not remain plane and the cross sections distort in a manner which is called **warping**; in other words, the fibres in the longitudinal direction deform unequally.

Shear centre is related to beams having thin-wall open sections that have only one or no axis of symmetry and if a force is applied at this point, no twisting occurs. The position of the shear centre is a function only of the geometry of the beam sections.

7.2 THEORETICAL BACKGROUND

7.2.1 Shear Stresses due to Bending

When a beam bends, it is also subjected to shear forces as shown in Section 6.2, which cause shear stresses in the beam. Considering a vertical cross section of a loaded, uniform, rectangular section beam, the shear force and shear stresses in the cross section are as shown in Figure 7.1a. It is reasonable to assume that

* Shear stresses τ act parallel to the shear force V, that is, parallel to the vertical cross section.
* The distribution of the shear stresses is uniform across the width (b) of the beam.

Consider a small element between two adjacent vertical cross sections and between two planes that are parallel to the neutral surface, as shown in Figure 7.1b. The existence of the shear stresses on the top and bottom surfaces of the element is due to the requirement for equilibrium of the element, which will be demonstrated in Section 7.3.2. The shear stresses on the four planes have the same magnitude and have directions such that both stresses point towards or both stresses point away from the line of intersection of the surfaces (Figure 7.1c) [1]. Therefore, for determining shear stresses, either the vertical or the horizontal planes may be considered.

Figure 7.2a and b show the element *EFCD* which lies between the adjacent cross sections *AC* and *BD* separated by a distance dx (Figure 7.1a). The shear stresses on *EF* can be evaluated using equilibrium in the direction in which the shear stresses act as [1]

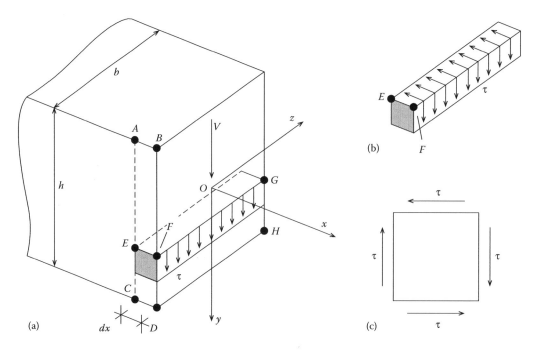

FIGURE 7.1
(a–c) Shear stress on a section. (From Gere, J. M., *Mechanics of Materials*, Thomson Books/Cole, Belmont, CA, 2005.)

$$\tau = \frac{V\int ydA}{bI} = \frac{VA\bar{y}}{bI} \tag{7.1}$$

where:

  V is the shear force at the section where the shear stress τ is to be calculated
  $\int ydA$ is the first moment of the area $DFGH$ about the neutral axis when τ is to be determined across the surface defined by the line EF

For regular sections, the integration can be replaced by the product $A\bar{y}$ in which A is the area of $DFGH$ and $\bar{y}$ is the distance between the neutral axis of the cross section and the centroid of $DFGH$; b is the width of the beam; and I is the second moment of area of the cross section of the beam about the neutral axis.

EXAMPLE 7.1

Determine the distribution of shear stress and the maximum shear stress due to a shear force V in a solid rectangular cross section, shown in Figure 7.2, which has a width of b and a depth of h.

SOLUTION

To determine the distribution of the shear stress across the depth of the section, consider a horizontal plane defined by the line FG at a distance y_1 from the neutral axis and calculate the shear stress along this plane (Figure 7.2b). For the area $DFGH$, the distance between the centroid of the area and the neutral axis of the section is $\bar{y}$, where

$$\bar{y} = y_1 + \frac{1}{2}\left(\frac{h}{2} - y_1\right)$$

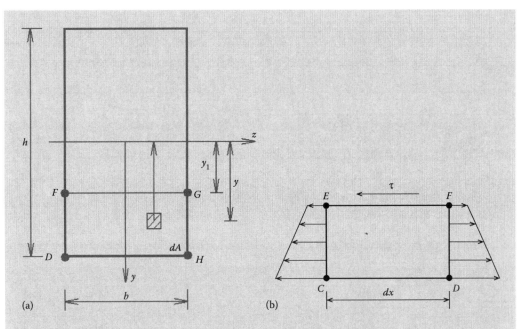

FIGURE 7.2
Example 7.1. (a) A cross section. (b) *EFDC* in the plane parallel to the *x–y* plane.

Thus, the first moment of the area about the neutral axis of the section is

$$A\bar{y} = b\left(\frac{h}{2} - y_1\right) \times \left[y_1 + \frac{1}{2}\left(\frac{h}{2} - y_1\right)\right] = \frac{b}{2}\left[\left(\frac{h}{2}\right)^2 - y_1^2\right]$$

Substituting the expression for $A\bar{y}$ into Equation 7.1 gives the shear stress at a depth y_1 from the neutral axis as

$$\tau = \frac{V}{bI} \times \frac{b}{2}\left[\left(\frac{h}{2}\right)^2 - y_1^2\right] = \frac{V}{2I}\left[\left(\frac{h}{2}\right)^2 - y_1^2\right]$$

This result shows that

- Shear stress varies quadratically along the height of the cross section.
- The shear stress is zero at the outer fibres of the section where $y_1 = \pm h/2$.
- The maximum shear stress occurs at the neutral axis of the cross section where $y_1 = 0$, and this maximum stress is

$$\tau_{max} = \frac{V}{2I}\left(\frac{h}{2}\right)^2 = \frac{V}{2(bh^3/12)}\left(\frac{h}{2}\right)^2 = \frac{3}{2}\frac{V}{bh}$$

The maximum shear stress in a solid rectangular section is 1.5 times the average value of the shear stress across the section.

More detailed information about shear stresses and further examples can be found in [1].

7.2.2 Shear Stresses due to Torsion

Consider a uniform, straight member which is subjected to a constant torque along its length. If the member has a solid circular section or a closed circular cross section, the relationship between torque, shear stress and the angle of twist is [1,2]

$$\frac{T}{J} = \frac{\tau}{r} = \frac{G\theta}{L} \qquad (7.2)$$

where:

T is torque (Nm)
J is the polar second moment of area (m⁴)
τ is the shear stress (N m⁻²) at radius r (m)
G is the shear modulus (N m²)
θ is the angle of twist over length L (m)

Equation 7.2 is derived using the equations of equilibrium, the compatibility of deformation and the stress–strain relationship for the material of the member.

For a solid circular member with radius r:

$$J = \frac{\pi r^4}{2} \qquad (7.3)$$

For a hollow circular shaft with inner and outer radii r_1 and r_2, respectively,

$$J = \frac{\pi}{2}(r_2^4 - r_1^4) \qquad (7.4)$$

For a thin-walled member with thickness t and mean radius r:

$$J = 2\pi r^3 t \qquad (7.5)$$

For thin-walled, noncircular sections, the torque–twist relationship in Equation 7.2 becomes

$$\frac{T}{J} = \frac{G\theta}{L} \qquad (7.6)$$

$$J = \frac{4A_e^2}{\int ds/t} \qquad (7.7)$$

where:

A_e is the area enclosed by the mean perimeter of the section
t is the thickness of the section

When the section has a constant thickness t around its perimeter s, Equation 7.7 can be written as

$$J = \frac{4A_e^2 t}{s} \qquad (7.8)$$

The shear stress across the section is given by

$$\tau = \frac{T}{2A_e t} \qquad (7.9)$$

For open sections made with thin plates:

$$J = \sum \frac{1}{3}bt^3 \qquad (7.10)$$

where:

b is the width
t is the thickness of each length of plate which makes up the cross section

EXAMPLE 7.2

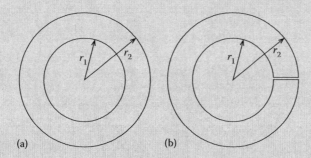

FIGURE 7.3
Hollow circular sections. (a) Without a slit. (b) With a slit.

There are two members with the same length L, one has a hollow circular section with inner and outer radii r_1 and r_2, respectively (Figure 7.3a) and the other has a similar circular section which has a thin slit cut along its full length (Figure 7.3b). Compare the torsional stiffness of the two members and determine the ratio of the stiffnesses when $r_1 = 12.5$ mm and $r_2 = 40$ mm.

SOLUTION

For the closed hollow circular section, Equation 7.4 should be used and the stiffness is

$$\frac{T}{\theta} = \frac{GJ}{L} = \frac{G\pi}{2L}(r_2^4 - r_1^4)$$

The circular section with a slit should be treated as a thin rectangular plate using Equation 7.10 to determine the stiffness as

$$\frac{T}{\theta} = \frac{GJ}{L} = \frac{G}{L}\frac{2\pi}{3}\left[r_1 + \frac{(r_2 - r_1)}{2}\right](r_2 - r_1)^3 = \frac{G\pi}{3L}(r_2 + r_1)(r_2 - r_1)^3$$

Thus, the ratio of the torsional stiffness of the closed section to that of the open section is

$$\frac{3}{2}\frac{(r_2^4 - r_1^4)}{(r_2 + r_1)(r_2 - r_1)^3} = \frac{3}{2}\frac{(r_2^2 + r_1^2)}{(r_2 - r_1)^2}$$

It can be seen that this ratio is independent of the material used for the members of the hollow circular sections. Substituting $r_1 = 12.5$ mm and $r_2 = 40$ mm into the preceding formula gives the ratio to be 3.5. This example will be demonstrated in Section 7.3.4.

EXAMPLE 7.3

Figure 7.4 shows a hollow square section and an I-section each of which has the same cross-sectional area, second moment of area about the x-axis and length L. Compare the torsional stiffnesses of the two members when $b = 16$ mm and $t = 1$ mm.

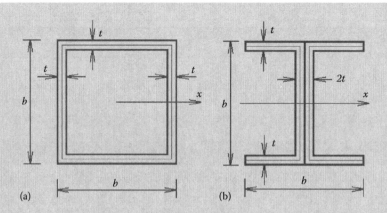

FIGURE 7.4
Two sections with the same area. (a) Closed section. (b) I-section.

SOLUTION

For the square hollow section: This is a thin-walled, noncircular hollow section and Equation 7.7 or Equation 7.8 should be used to determine J and hence the torsional stiffness as

$$\frac{T}{\theta} = \frac{GJ}{L} = \frac{G}{L}\frac{4A_e^2 t}{s} = \frac{4G}{L}\frac{(b-t)^4 t}{4(b-t)} = \frac{Gt(b-t)^3}{L}$$

For the I-section: This may be considered to be an assembly of three thin plates and Equation 7.10 should be used to determine the torsional stiffness as

$$\frac{T}{\theta} = \frac{GJ}{L} = \frac{G}{L}\sum\frac{bt^3}{3} = \frac{G}{3L}[2bt^3 + (b-2t)(2t)^3] = \frac{2Gt^3(5b-8t)}{3L}$$

Thus, the ratio of the torsional stiffnesses of the closed section to the open section is

$$\frac{3}{2}\frac{(b-t)^3}{t^2(5b-8t)}$$

When $b=16$ mm and $t=1$ mm, the ratio is 70. This example will be demonstrated in Section 7.3.5.

Examples 7.2 and 7.3 show that a closed section has a much larger torsional stiffness than an open section although the two cross sections have the same cross-sectional area and second moment of area. This explains why box girder beams rather than I-section beams are used in practical situations where significant torsional forces are present. A practical example will be given in Section 7.4.4. Further theoretical details and examples can be found in [1–4].

7.2.3 Shear Centre

In practice, the thickness of the material of some beams is small compared with the overall geometry and there is only one or no axis of symmetry, such as in a beam with a channel section as shown in Figure 7.5.

Due to the lack of symmetry of such a section, when a force is applied through the centroid of the section both bending and twisting of the beam occur. To avoid twisting and cause only bending, the force needs to act at a particular point that is called the *shear centre*. This position can be

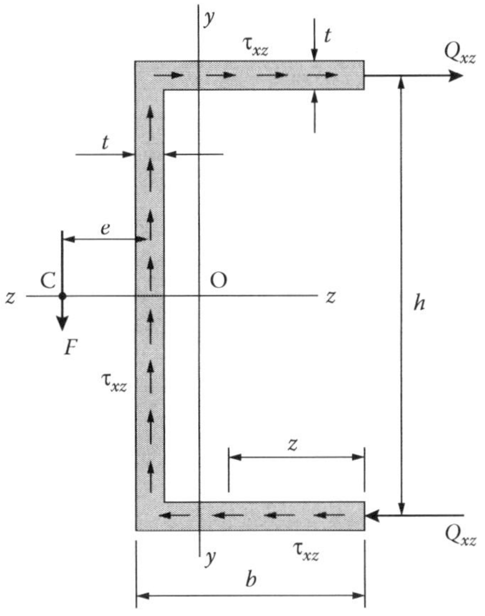

FIGURE 7.5
Location of shear centre of a channel section.

illustrated using the channel section shown in Figure 7.5 which is loaded by a vertical force F. The y- and z-axes are the principal axes of the section.

For the shear stress in the flanges, the analysis is similar to that for the I-section and Equation 7.1 can be used [2]:

$$\tau_{xz} = \frac{V}{tI_z} \int_A y \, dA = \frac{V}{tI_z} \int_A \frac{h}{2} t \, dz = \frac{V}{tI_z} \times \frac{htz}{2} = \frac{Vh \times z}{2I_z}$$

where V is the vertical shear force on the section induced by the applied force F and $V = F$. The shear stress varies linearly with z from zero at the left to a maximum at the centre line of the web:

$$\tau_{xz,\max} = \frac{Vhb}{2I_z}$$

The average shear stress is $Vhb/4I_z$ and therefore the horizontal shear force in the top and bottom flange is:

$$Q_{xz} = \frac{Vhb^2 t}{4I_z}$$

The couple, about the x-axis, of these shear forces that would cause twisting of the section is:

$$Q_{xz}d = \frac{Vh^2 b^2 t}{4I_z}$$

If there is an opposing couple of the same magnitude, twisting can be avoided. Considering the vertical force F acting through a point C, the shear centre, at a distance e from the middle of the web (Figure 7.5), the equilibrium condition in the rotational direction is:

$$Ve = Q_{xz}h = \frac{Vh^2 b^2 t}{4I_z}$$

This leads to

$$e = \frac{h^2 b^2 t}{4I_z} \tag{7.11}$$

Equation 7.11 gives the location of the shear centre of a channel section and indicates that the position of the shear centre is only a function of the geometrical properties of the section. If the vertical force F is applied at the left of the shear centre, the section will be twisted in the anticlockwise direction; if F is applied at the right of the shear centre, the section will be twisted in the clockwise direction. This intuitive understanding will be demonstrated in Section 7.3.6.

7.3 MODEL DEMONSTRATIONS

7.3.1 Effect of Torsion

This demonstration shows *the effect of shear stress in a noncircular section member induced by torsion.*

Take a length of sponge of rectangular cross section and mark longitudinal lines down the centre of each face. Then, add perpendicular lines at regular intervals along the length of the sponge. Restrain one end of the sponge in a plastic frame as shown in Figure 7.6 and twist the other end. It can be observed that

- The lines defining the cross sections of the beam are no longer straight
- The angles between the horizontal longitudinal lines, which define the neutral axis, and the vertical lines are no longer 90°

These observations are different from those of the beam in bending – see Section 6.3.1 (Figure 6.6).

7.3.2 Effect of Shear Stress

This demonstration shows *the existence of shear stress in bending and how shear resistance/ stresses between beams/plates/sheets can significantly increase the bending stiffness of a beam.*

Take two identical, thick catalogues and drill holes through one of them; then put bolts through the holes and tighten the bolts, as shown in Figure 7.7a. Place the two catalogues on a wooden board up against two wood strips, say a quarter of the thickness of a catalogue, which are secured to the board, and apply a horizontal force at the top right edge of each of the two catalogues, as shown in Figure 7.7b. It is apparent that the thin pages slide over each other in the unbolted catalogue while in the bolted catalogue there is no movement of the pages. This is because the bolts and the friction between the pages provide horizontal shear resistance and prevent the pages sliding between each other.

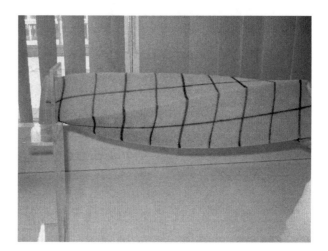

FIGURE 7.6
Effect of torsion.

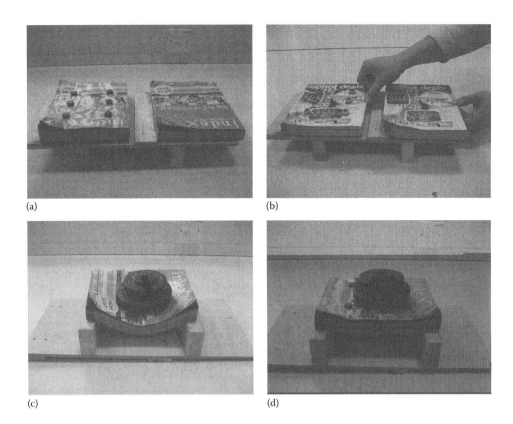

FIGURE 7.7
(a–d) Effect of shear stress in catalogues.

Support the two catalogues at their ends on two wooden blocks on the board as shown in Figure 7.7c and d, and place the same weight at the midspan of each of the two catalogues. Figure 7.7c and d show the bending deflections of the two catalogues with the unbolted catalogue experiencing large deformations, while the bolted catalogue experiences only small deformations. The bolts and the friction between the pages provide horizontal shear resistance between the pages of the bolted catalogue and make this act as a single member, a 'thick' beam or plate, whereas in the unbolted catalogue the pages act as a series of very 'thin' beams or plates.

A similar but simpler demonstration of the effect of shear stress can be provided using beams made up of two strips of plastic, as shown in Figure 7.8a. For one beam, the two plastic strips are loosely bound with elastic bands; for the other beam, the two plastic strips are securely held together with four bolts that act as shear connectors and, because of the tension in the tightened bolts, provide compressive forces on the two strips. Suitable sizes for the strips would be 300 mm long, 25 mm wide and 5 mm thick.

By applying similar horizontal forces to the ends of the beams, one can observe and feel the effect of the shear connection. When bending the beam without shear connectors, one can see the two strips rub against each other and slightly move relative to one another. This becomes noticeable at the ends of the strips as there is little shear resistance between the two strips. When bending the beam with shear connectors, one can feel that the beam is much stiffer than the beam without shear connectors and it is possible to see that there is no relative movement between the strips.

7.3.3 Effect of Shear Force

This demonstration shows *how lateral forces can be resisted and transmitted in frame structures through the use of shear elements, such as shear walls.*

Figure 7.9a shows a three-dimensional frame system in which the columns and beams are made of steel springs with wooden joints linking the beams and columns. Applying a horizontal force to

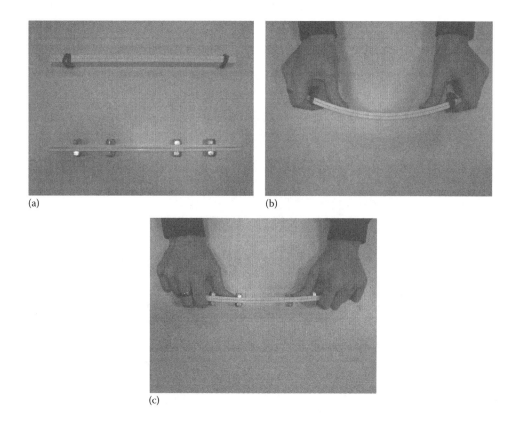

FIGURE 7.8
(a–c) Effect of shear resistance in beams.

FIGURE 7.9
(a–d) Effect of shear walls. (The demonstration model was provided by Mr. P. Palmer, University of Brighton.)

the top corner of the frame causes the frame to move in the direction of loading. It can be seen from Figure 7.9a that the angles between the beams and columns are no longer 90° as was the case in the unloaded frame.

If a wooden board is fitted into a lower vertical panel of the frame as shown in Figure 7.9b and the same force is applied as was the case in the last demonstration, it is observed that the upper storey experiences horizontal or shear deformations while the lower storey is almost unmoved. The wood panel acts as a shear wall which has a large in-plane stiffness enabling it to transmit horizontal loads in the lower storey directly to the frame supports.

If the wooden board is replaced in a vertical plane in the lower part of one of the end frames as shown in Figure 7.9c, and a force is applied at the joint which is one bay away from the wooden board, it is observed that the frame to which the force is applied deforms significantly and the frame where the wooden board is placed has little movement.

If a further wooden board is placed in a horizontal plane, at the first-storey level, next to the vertical board as shown in Figure 7.9d, and a force is applied at the joint at the corner of the horizontal panel, it can be seen that there is little movement at the point where the force is applied. This is because the horizontal and vertical panels have large stiffnesses in their planes and the shear force is transmitted directly through the wooden boards to the supports.

7.3.4 Open and Closed Sections Subject to Torsion with Warping

This demonstration *shows the difference in the torsional stiffness of two circular members: one with a closed section and the other with an open section where warping can be observed.*

Figure 7.10 shows two foam pipes that are used for insulation and each of the two pipes has a length of 450 mm. The sections have machined slits and in one of the pipes the slit is sealed using tape and glue. One pipe thus effectively has an open circular section and one has a closed circular section. The sections have been analysed in Example 7.2 where it was predicted that the torsional stiffness of the closed section was 3.5 times that of the open section. By twisting the two foam pipes, it is possible to feel the significant difference in their torsional stiffnesses.

When twisting the two foam pipes with a similar effort it can be seen from Figure 7.10 that

* It is much easier to twist the open section than the closed section
* The effect of warping can be observed (Figure 7.10a) at the right-hand end of the pipe which has a slit along its length
* There is little warping effect on the pipe with the closed section (Figure 7.10b)

(a)

(b)

FIGURE 7.10
(a,b) Open and closed sections subject to torsion with warping.

7.3.5 Open and Closed Sections Subject to Torsion without Warping

This demonstration shows *the difference in the torsional stiffness of two noncircular members: one with a closed section and the other with an open section where warping is restrained.*

Figure 7.11a shows two 500 mm long steel bars, one with a square hollow section and the other with an I-section which is made by cutting a square hollow section into two halves along its length and welding the resultant channel sections back to back. Handles are welded to the ends of the bars to allow end torques to be easily applied. The sections have been analysed in Example 7.3. Due to the addition of the handles, the warping that occurs in open sections, as shown in Section 7.3.4, is now restrained. In other words, the model with the I-section would be stiffer than that analysed in Example 7.3.

By applying torques at the ends of the two bars, it is readily felt that the bar with a closed section is much stiffer than the bar which has an open section.

7.3.6 Shear Centre of a Thin-Walled Open Section

This demonstration shows *the location of the shear centre of a thin-walled open section. When a vertical force is applied through the shear centre, no twisting or torsion occurs.*

Figure 7.12 shows three identical plastic beams with channel cross sections. The shear centre of a cross section can be identified from the definition of the shear centre. A different loading position for each beam is selected. From left to right, the loading positions are to the left of the shear centre, at the shear centre and to the right of the shear centre. Loads are applied by pulling on wires attached to the loading points.

Figure 7.13 shows the responses of three cantilevers to the applied loads. It can be observed from Figure 7.13 that

- The beam twists in the anticlockwise direction when the load is applied to the left of the shear centre
- The beam deforms vertically, without twisting, when the load is applied through the shear centre
- The beam twists in the clockwise direction when the load is applied to the right of the shear centre

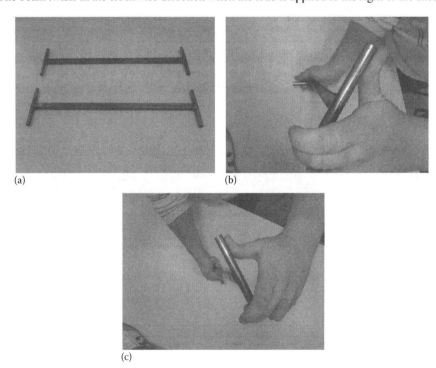

(a) (b)

(c)

FIGURE 7.11
(a–c) Open and closed sections subjected to torsion without warping.

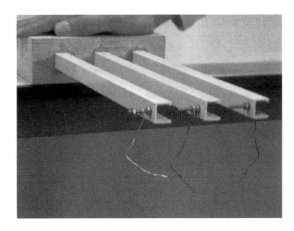

FIGURE 7.12
Different loading positions for three identical cantilever beams with open sections.

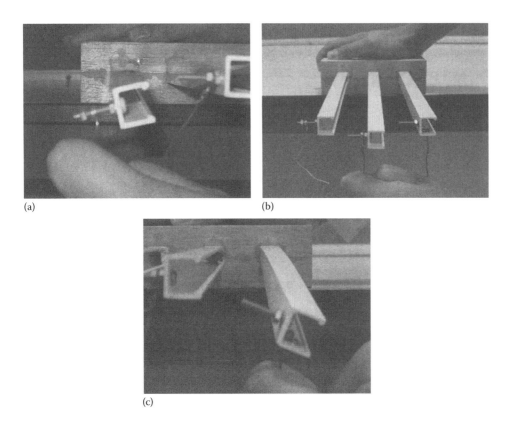

(a)　　　　　　　　　　　　　(b)

(c)

FIGURE 7.13
(a–c) Deformation of cantilevers with a channel section due to vertical loads.

7.4 PRACTICAL EXAMPLES

7.4.1 Composite Section of a Beam

Figure 7.14 shows a number of steel plates or thin beams which are bolted together to form a thick beam. As demonstrated in Section 7.3.2, due to the action of the bolts, there are no relative sliding movements between the thin plates/beams when the beam is loaded. Due to the shear resistance of the connecting bolts, the thin plates/beams act together as a single member that is many times stiffer than a member which would result from the plates/beams acting independently.

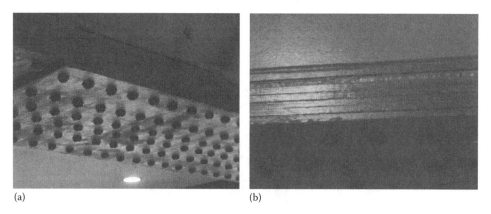

(a)   (b)

FIGURE 7.14
(a,b) Composite sections.

7.4.2 Shear Walls in a Building

Shear walls and bracing members are often present in buildings to provide lateral stiffness and to transmit lateral loads, such as wind loads, to the foundations of the buildings. Figure 7.15 shows one end of a steel-framed building where masonry walls and bracing members are used from the bottom to the top of the building to increase the lateral stiffness of the frame structure.

The dynamic behaviour of this building has been examined, experimentally and numerically, at five distinct construction stages, including the building with and without the walls [5]. The walls and bracing members contributed significant stiffness to the building, which was reflected in the

FIGURE 7.15
Shear walls in a steel-framed building.

natural frequencies and their associated mode shapes. The walls and bracing members in the ends of the building increased the fundamental transverse natural frequency from 0.72 Hz for the bare frame structure to 1.95 Hz for the braced structure. The walls in the other two sides of the building were only one-quarter storey high and they increased the fundamental natural frequency from 0.71 to 0.89 Hz.

7.4.3 Opening a Drink Bottle

When opening the lid of a common plastic drink bottle, a torque T applied to the cap is gradually increased until the plastic connectors between the cap and the bottle experience shear failure.

Figure 7.16 shows the shear failure of the connectors that keep the bottle sealed before opening. If there are n connectors, each with an area A, uniformly distributed along the circumference of the cap which has a radius of r and the failure shear stress of the plastic connectors is τ_f, the equation of torsional equilibrium for the cap before being opened is $T = nA\tau r$. To open the bottle, the failure shear stress τ_f has to be reached; and the design of the cap and its connection to the bottle, in terms of the size of the cap and the number and size of the connectors, need to be such that this can be achieved with a modest effort.

7.4.4 Box Girder Highway Bridge

Figure 7.17 shows a box girder bridge in which the main beams are girders in the shape of a hollow box. The box girder shown in Figure 7.17 is trapezoidal in cross section but rectangular girders are also common.

One of the main advantages of box girders over I-section beams is that they have a much better resistance to torsion. As shown in Section 7.3.5, an enclosed section has a significantly higher torsional stiffness than an open section with the same shape and area. As a bridge is often subjected to nonsymmetric loads, such as traffic on one side of the bridge, this leads to both bending and torsion of the bridge deck. Torsion effects are also more significant when a bridge curves in plan. In addition, the presence of two webs allows wider, and hence stronger, flanges to be used, allowing wider spans.

FIGURE 7.16
Shear failure of the connectors in a drinks bottle.

(a)   (b)

FIGURE 7.17
A closed box girder for a highway bridge. (a) Cross section of a box girder. (b) Construction of a box girder bridge. (Courtesy of Mr. B. Duguid, Mott MacDonald.)

PROBLEMS

1. Consider the example in Section 7.4.1, and assume that the beam consists of seven similar plates and each plate has a width of b and a thickness of t. Compare the second moment of area of the beam when the plates are bolted, as shown in Figure 7.14, and when the plates act independently without the bolts.
2. Obtain an unopened plastic bottle of soft drink, similar to that shown in Figure 7.16, and study the following:
   a. Estimate the area of the shear connectors in the cap and the shear failure stress.
   b. Estimate the torque that is needed to open the cap – that is, all shear connectors on the cap experience shear failure.
   c. Examine possible improvements of the design of the cap.

REFERENCES

1. Gere, J. M. *Mechanics of Materials*, Belmont, CA: Thomson Books/Cole, 2004.
2. Benham, P. P., Crawford, R. J. and Armstrong, C. G. *Mechanics of Engineering Materials*, Harlow: Addison Wesley Longman, 1998.
3. Williams, M. S. and Todd, J. D. *Structures: Theory and Analysis*, London: Macmillan, 2000.
4. Millais, M. *Building Structures: From Concepts to Design*, Abingdon: Spon, 2005.
5. Ellis, B. R. and Ji, T. Dynamic testing and numerical modelling of the Cardington steel framed building from construction to completion, *The Structural Engineer*, 74, 186–192, 1996.

CHAPTER 8

CONTENTS

Span and Deflection

8

8.1 CONCEPTS

- For distributed loads, the deflection of a beam is proportional to its span to the power of four, and for a concentrated load the deflection is proportional to its span to the power of three.
- To reduce deflections, an increase in the depth of a beam is more effective than an increase in its width (for a rectangular cross section).

8.2 THEORETICAL BACKGROUND

The relationships between displacement v, slope θ, bending moment M, shear force V and load q for a uniform beam are as follows:

$$\frac{dv}{dx} = \theta \tag{8.1}$$

$$EI\frac{d^2v}{dx^2} = -M \tag{8.2}$$

$$EI\frac{d^3v}{dx^3} = -V \tag{8.3}$$

$$EI\frac{d^4v}{dx^4} = q \tag{8.4}$$

Equation 8.2 indicates that the displacement function, v, for a loaded beam, can be obtained by integrating the bending moment function, M, twice with respect to the coordinate x. Equation 8.4 shows that the displacement function v, for a loaded beam, can be obtained by integrating the loading function q four times with respect to x. When carrying out the integrations, the integration constants can be uniquely determined using the available boundary conditions of the beam.

Sign conventions differ in different textbooks [1–4]. The sign conventions used in the derivation of the formulae in Equations 8.1 through 8.4 are defined as shown in Figure 8.1a and b.

The directions of the forces and deflections shown in Figure 8.1 are all positive.

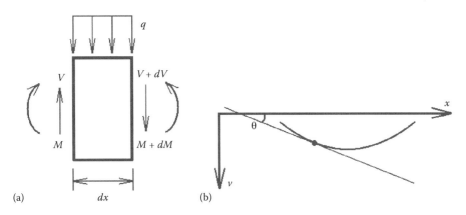

(a)   (b)

FIGURE 8.1
Positive sign convention. (a) Sign convention for bending moment, shear force and load. (b) Sign convention for displacement and slope.

EXAMPLE 8.1

A simply supported uniform beam subject to a uniformly distributed load q is shown in Figure 6.2a. The beam has a length L and a constant EI. Use Equation 8.2 to determine the displacement of the beam and its maximum deflection.

SOLUTION

The bending moment diagram for the simply supported beam is illustrated in Example 6.1 and Figure 6.2e. From Equation 6.4, the bending moment at x is

$$M = \frac{q}{2}(Lx - x^2)$$

Using Equation 8.2 gives

$$EI\frac{d^2v}{dx^2} = -\frac{q}{2}(Lx - x^2)$$

Integrating once with respect to x leads to

$$EI\frac{dv}{dx} = -\frac{q}{4}Lx^2 + \frac{q}{6}x^3 + C_1$$

From symmetry, $dv/dx=0$ at $x=L/2$. Using this condition, the integration constant, C_1, can be determined to be $C_1=qL^3/24$. Thus,

$$\frac{dv}{dx} = -\frac{q}{2EI}\left(\frac{L}{2}x^2 - \frac{1}{3}x^3 - \frac{L^3}{12}\right) \tag{8.5}$$

The slopes at the two ends of the beams can be determined and are $qL^3/24$ at $x=0$ and $-qL^3/24$ at $x=L$. Integrating Equation 8.5 gives

$$v = -\frac{q}{2EI}\left(\frac{L}{6}x^3 - \frac{1}{12}x^4 - \frac{L^3}{12}x\right) + C_2$$

As $v=0$ at $x=0$, $C_2=0$, the deflection becomes

$$v = -\frac{q}{12EI}\left(\frac{1}{2}x^4 - Lx^3 - \frac{L^3}{2}x\right) \tag{8.6}$$

The maximum deflection occurs at the centre of the beam, when $x=L/2$. Substituting $x=L/2$ into Equation 8.6 gives

$$v_{max} = \frac{q}{12EI}\left(\frac{1}{32}L^4 - \frac{L^4}{8} + \frac{L^4}{4}\right) = \frac{5qL^4}{384EI} \tag{8.7}$$

EXAMPLE 8.2

Figure 8.2a shows a uniform beam, fixed at its two ends, carrying a uniformly distributed load q. The beam has a length L and a constant EI. Determine the deflections of the beam using Equation 8.4.

SOLUTION

Equation 8.4 gives

$$EI\frac{d^4v}{dx^4} = q$$

Integrating this equation three times and four times leads to

$$\frac{dv}{dx} = \frac{q}{EI}\left(\frac{x^3}{6} + \frac{C_1}{2}x^2 + C_2x + C_3\right)$$

$$v = \frac{q}{EI}\left(\frac{x^4}{24} + \frac{C_1}{6}x^3 + \frac{C_2}{2}x^2 + C_3x + C_4\right)$$

The four integration constants, C_1 to C_4, can be determined using the boundary conditions of the beam. As $dv/dx = 0$ and $v = 0$ at $x = 0$, it can be shown that $C_3 = 0$ and $C_4 = 0$. The other two integration constants can be determined using two other boundary conditions, $dv/dx = 0$ and $v = 0$ at $x = L$. There is also a symmetry condition available, that is, $dv/dx = 0$ at $x = L/2$. Using the two slope conditions gives

$$\frac{qL^3}{6} + \frac{L^2}{2}C_1 + LC_2 = 0$$

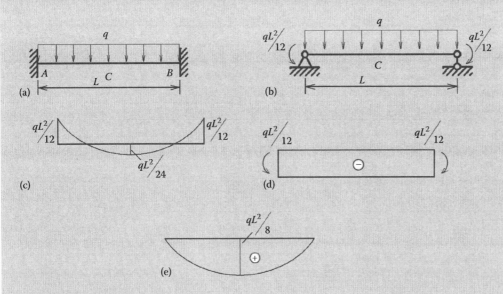

FIGURE 8.2
(a–e) A beam with two fixed ends subjected to uniformly distributed loads.

$$\frac{qL^3}{48} + \frac{L^2}{8}C_1 + \frac{L}{2}C_2 = 0$$

Solving these simultaneous equations gives $C_1 = -qL/2$ and $C_2 = qL^2/12$. Thus,

$$v = \frac{q}{12EI}\left(\frac{x^4}{2} - Lx^3 + \frac{L^2}{2}x^2\right) \qquad (8.8)$$

Substituting $x = L/2$ into Equation 8.8 gives the maximum deflection of the beam at its centre:

$$v_{max} = \frac{qL^4}{384EI} \qquad (8.9)$$

The bending moments in the beam can be obtained using Equation 8.2:

$$M = -\frac{q}{2}\left(x^2 - Lx + \frac{L^2}{6}\right) \qquad (8.10)$$

Therefore, the bending moment is $-qL^2/12$ at the two fixed ends, and it is $qL^2/24$ at the centre of the beam. The bending moment diagram is shown in Figure 8.2c.

From Figure 8.2c, it can be seen that the difference in the bending moments at the ends of the beam and the centre of the beam is $qL^2/8$, which is the same as the maximum bending moment in the simply supported beam in Example 6.1. By removing the rotational restraints at the ends of the beam in Figure 8.2a and replacing them with two moments, as shown in Figure 8.2b, it can be seen that the two beams are equivalent. This means that for the beam with two fixed ends, the bending moment diagram in Figure 8.2c can be interpreted as the summation of the bending diagram due to the end moments (Figure 8.2d), and the bending moment diagram for a simply supported beam carrying the distributed load (Figure 8.2e).

Table 8.1 summarises the maximum bending moments and the maximum deflections for a uniform beam with various support conditions carrying either a uniformly distributed load or a concentrated load applied at the most unfavourable position. All the beams have a length L and a constant EI.

From Table 8.1, the following can be observed:

- *The maximum deflection of a beam is proportional to its span to the power of four for uniformly distributed loads or to its span to the power of three for a concentrated load.* This conclusion is also applicable for other types of distributed loading and for concentrated loads applied at different locations.
- *Larger maximum deflections correspond to situations where larger maximum bending moments occur.* The most common explanation of this observation would lie with the different boundary conditions. Whilst this is true, the explanation can also be stated in terms of the general concept, *the smaller the internal forces, the stiffer the structure*, and this will be explained and demonstrated in detail in Chapters 9 and 10.

TABLE 8.1 Maximum Bending Moments and Deflections of Single-Span Beams

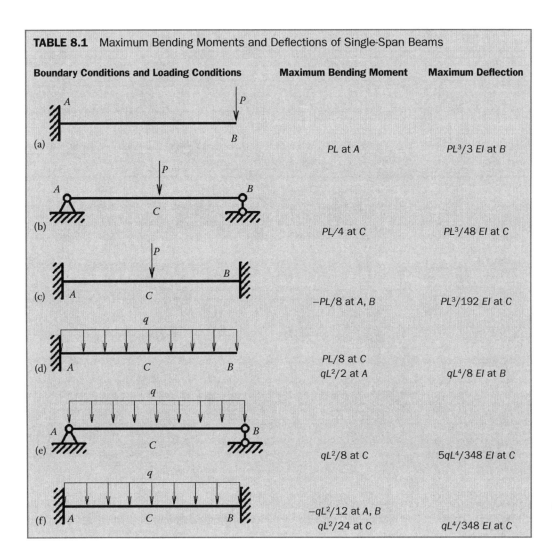

Boundary Conditions and Loading Conditions	Maximum Bending Moment	Maximum Deflection
(a)	PL at A	$PL^3/3\,EI$ at B
(b)	$PL/4$ at C	$PL^3/48\,EI$ at C
(c)	$-PL/8$ at A, B	$PL^3/192\,EI$ at C
(d)	$PL/8$ at C $qL^2/2$ at A	$qL^4/8\,EI$ at B
(e)	$qL^2/8$ at C	$5qL^4/348\,EI$ at C
(f)	$-qL^2/12$ at A, B $qL^2/24$ at C	$qL^4/348\,EI$ at C

8.3 MODEL DEMONSTRATIONS

8.3.1 Effect of Spans

This demonstration shows *the effects of span and second moment of area of a beam on its deflections.*

A 1 m wooden ruler with a cross section of 5×30 mm is used, and a metal block is attached to one of its two ends. One end of the ruler, with the long side of the cross section horizontal, is supported to create a cantilever with the concentrated load at its free end.

1. Observe the displacement at the free end of the cantilever with a span of say 350 mm. It can be seen from Figure 8.3a that there is a small deflection at the free end.
2. Double the span to 700 mm as shown in Figure 8.3b and a much larger end displacement is observed. According to the results presented in Section 8.2, the end deflection for a span of 700 mm should be eight times that for a span of 350 mm when the effect of the self-weight of the ruler is negligible in comparison with the weight of the metal block.
3. Turn the ruler through 90° about its longitudinal axis as shown in Figure 8.3c and repeat the tests when a much reduced end displacement will be seen. The formula in Table 8.1 shows that the deflection is proportional to the inverse of the second moment of area, which for a rectangular section is given by $I = bh^3/12$. For the current test, the second moment of area of the section about the horizontal axis is 36 times that of the section used in the last test, resulting in maximum deflections of about one thirty-sixth of those in the second test (Figure 8.3b).

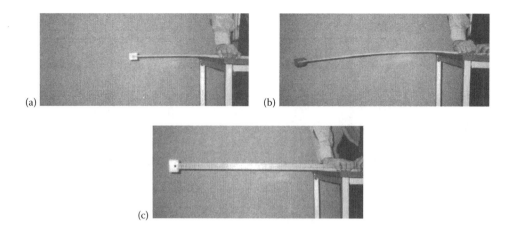

FIGURE 8.3
(a–c) Deflections of a cantilever beam subjected to a concentrated load.

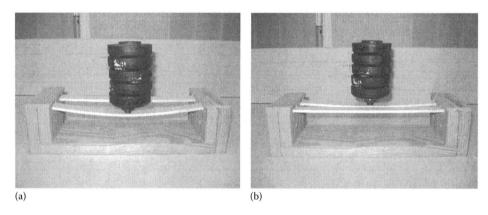

FIGURE 8.4
Effect of boundary conditions. (a) Deflections of a simply supported beam. (b) Deflections of a fixed beam.

8.3.2 Effect of Boundary Conditions

This demonstration shows *the effects of the boundary conditions, or supports, on the deflections of a uniform beam*, and it shows that *fixed boundary conditions produce a stiffer beam than do pinned boundary conditions.*

Figure 8.4 shows the demonstration model which comprises a wooden frame and two plastic strips with the same length and cross section. For the fixed beam, a plastic strip is securely attached to the frame with screws and glue at each end, and for the simply supported beam a plastic strip is encased at its ends which are free to rotate.

A qualitative demonstration can be quickly conducted. By pressing down at the centre of each of the two beams, it is possible to qualitatively feel the difference in the stiffnesses of the two beams. Based on the results in Table 8.1, the fixed beam is four times as stiff as the simply supported beam.

The loads applied and the deflections produced can be quantified. For a particular set of plastic strips, it is found that a concentrated load of 22.3 N produces a measured maximum deflection under the load of 3.5 mm for the beam with the fixed ends, whereas for the simply supported beam, the maximum deflection is 13 mm. The ratio of the measured displacements is close to the theoretical ratio.

8.3.3 Bending Moment at One Fixed End of a Beam

This demonstration shows *how the end moment in a fixed beam can be measured.*

At the fixed end of a beam, both the displacement and the rotation, or slope, are zero. Using the condition that the slope at a fixed end equals zero, a fixed-end condition can be created and the moment associated with this condition can be determined.

Figure 8.5a shows a simply supported beam with a supporting frame. A hanger is placed at the centre of the beam so loads can be added. The ends of two vertical arms at the supports are attached to displacement gauges. Readings from the gauges divided by the lengths of the arms are the end rotations of the beam. If weights are placed on two end hangers, they will induce rotations in the opposite directions to those induced by the load applied at the centre of the beam. When the readings from the gauges are reduced to zero, a beam with two fixed ends has been created and the fixed-end moments are the products of the weights on the end hangers and the horizontal distances between the ends of the hangers and the supports.

Figure 8.5b shows a mass of 5 kg placed on the hanger at the centre of the beam and Figure 8.5c shows the rotation of the beam at the left support (and the reading of 2.99 mm). By adding a mass of 3 kg to each of the two end hangers (Figure 8.5d), the gauge at the left end shows a reading of 0.01 mm, indicating that a fixed boundary condition has been created. The associated fixed-end moment is $3 \times 9.81 \times 0.125$ (the distance between the end hanger and the support) $= 3.68$ Nm in this case.

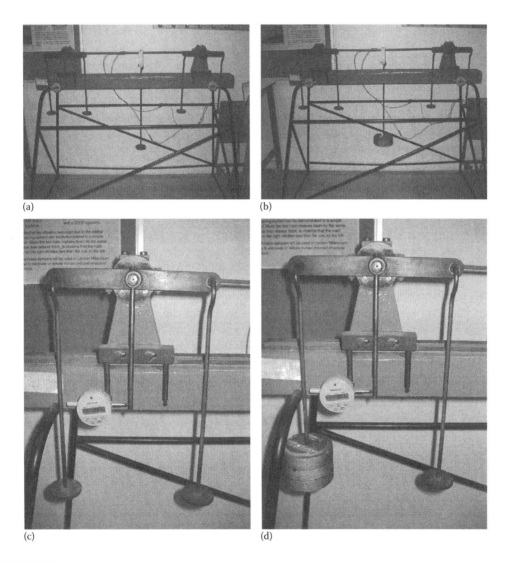

(a)   (b)

(c)   (d)

FIGURE 8.5
Bending moments at the fixed ends of a uniform beam. (a) Before loading. (b) Adding a weight at the centre of the beam. (c) Rotation at the end and the reading. (d) Adding loads to remove the rotation.

8.3.4 Lateral Stiffnesses of Vertical Members

This model examines *the relative stiffnesses of a free cantilever and a restrained cantilever and the manufacturing errors involved in the models.*

The model shown in Figure 8.6a aims to create a cantilever with the top end completely free and a cantilever with the top end restrained against rotation but free to translate. A lateral force can be applied at a distance L from the base of each cantilever (Figure 8.6b). A support frame comprises two pairs of vertical aluminium members rigidly connected to a base member. Two identical steel strips, which are much more flexible than the support frame, are placed vertically and clamped at their bases. The left-hand one is a cantilever as indicated in the left of Figure 8.6b. To prevent rotation at the top of the right column, three small rollers are placed as illustrated in the right of Figure 8.6b, to form a cantilever with the top end restrained against rotation. A horizontal member located between and near the top of the two supporting frames can move horizontally with guide ball bearings to apply a force, at a distance L from the bases of the two vertical members. A mechanical force gauge is used to apply and measure the horizontal force on the right end of the horizontal member.

For similar displacements of the horizontal member, Figure 8.7 shows gauge readings indicating the forces applied to the top of the cantilevers. For the free cantilever the gauge reading is approximately $4 \times 0.25 \times 9.81 = 10$ N and for the restrained cantilever the gauge reading is approximately

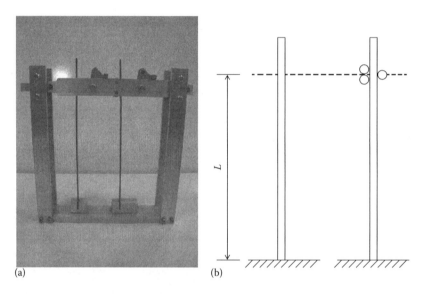

(a)　　　　　　　　　　(b)

FIGURE 8.6
Physical model and idealised models. (a) The physical model. (b) The idealised models.

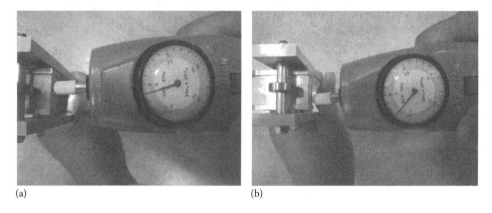

(a)　　　　　　　　　　(b)

FIGURE 8.7
The readings of the loads on the two columns. (a) Loading on the cantilever. (b) Loading on the restrained cantilever.

(a)   (b)

FIGURE 8.8
Deformations of the two columns with different support conditions. (a) Deformation of the cantilever. (b) Deformation of the restrained cantilever.

$12 \times 0.25 \times 9.81 = 30$ N. In other words, the lateral stiffness of the restrained cantilever is about three times that of the free cantilever. Beam theory, however, predicts a value of 4. So what is wrong?

Figure 8.8 shows the deformed shapes of the two cantilevers when subjected to lateral loading. It can be observed that the idealised rigid boundary condition at the top of the restrained cantilever is effectively not fully realised as some rotation does occur at the top end of the member as shown in Figure 8.8b. This is due to the design of the device as indicated on the right of Figure 8.6b, with the three roller fixings not actually creating a fully fixed boundary condition. This results in a lower stiffness than that of a cantilever with the top end fully restrained.

8.4 PRACTICAL EXAMPLES

8.4.1 Column Supports

Floors which extend outside normal building lines can seldom act as cantilevers due to the effect of the relationship between span, deflection and loading, and thus need additional column supports as shown in Figure 8.9.

FIGURE 8.9
Column supports.

FIGURE 8.10
Prop roots.

8.4.2 Phenomenon of Prop Roots

In rain forests, plants such as *Ficus* have prop roots (Figure 8.10). In humid and shaded conditions, when the large branches reach a certain length, aerial roots grow downwards from the branches. When these aerial roots reach the ground, they are similar to stems that support the upper branches, forming the unique phenomenon of prop roots. The prop roots provide the necessary and additional vertical supports to the branches of the tree, allowing them to extend their spans further.

8.4.3 Metal Props Used in Structures

The most effective ways to increase the stiffnesses or reduce the deflections of a structure are to reduce spans or to add supports, as shown in the last example obtained from nature.

The Cardiff Millennium Stadium was selected to hold the Eve of the Millennium concert on 31 December 1999. However, the stadium had been designed for sports events rather than for pop concerts. During pop concerts, spectators bounce and jump in time to the music beat and produce dynamic loading on the structure that is larger than the loading due to their static weight. If one of the music beat frequencies occurs at, or is close to, one of the natural frequencies of the cantilever structure, resonance or excessive vibration may occur, which affects both the safety and the serviceability of the structure.

To enable the Eve of the Millennium concert to take place, the cantilever structure of the Cardiff Millennium Stadium had to be reinforced with temporary metal props, which are similar to the prop roots in Section 8.4.2. In this way, the spans of the cantilevers were effectively reduced and consequently their stiffnesses and natural frequencies were increased above the range where any unacceptable resonance induced by the spectators was possible.

Figure 8.11 shows a steel prop used to support the deck of a footbridge, which is a critical structural member to the bridge.

 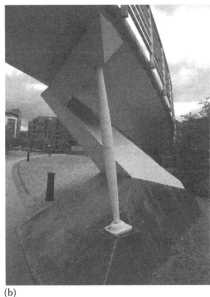

(a)   (b)

FIGURE 8.11
(a,b) Props used to support a footbridge.

PROBLEMS

1. Figure 8.12a shows a uniform cantilever that has a length of 10 m and a rigidity of
   EI and is subjected to a triangularly distributed load $q(y) = q_0 y/10$ (N m^{-1}) along its
   height. The maximum displacement of the cantilever is 0.1 m. In order to reduce
   the displacement, the height of the cantilever is reduced to 9.0 m (Figure 8.12b).
   If the properties of the cantilever and the loading remain the same, what is the
   maximum displacement of the modified cantilever? (The effect of self-weight of
   the cantilever can be neglected.)
   Hint: The solution is based on the information given in the 10 m column and there
   is no need to use the formula for a cantilever subjected to triangularly distributed
   loads.

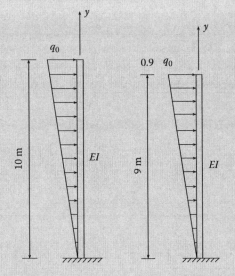

FIGURE 8.12
(a,b) Two cantilevers with different lengths.

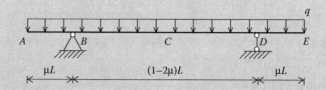

FIGURE 8.13
A beam with overhangs.

2. Figure 8.13 shows a beam with symmetric overhangs that has a constant rigidity, *EI*, and is subjected to a uniformly distributed load *q*. Answer the following questions.
   a. Determine the value of μ that leads to the minimum displacement of the beam – that is, when the displacements at points *A* and *C* are the same.
   b. Calculate the corresponding displacements.
   c. Compare the μ value determined in Example 6.2 with $\mu = 0.207$ which has been determined based on the equal absolute values of the bending moment at points *B* and *C* in Example 6.2, and explain the finding.

REFERENCES

1. Hibbeler, R. C. *Mechanics of Materials*, 5th edn, Singapore: Prentice-Hall, 2005.
2. Williams, M. S. and Todd, J. D. *Structures: Theory and Analysis*, London: Macmillan, 2000.
3. Gere, J. M. *Mechanics of Materials*, Belmont, CA: Thomson Books/Cole, 2004.
4. Benham, P. P., Crawford, R. J. and Armstrong, C. G. *Mechanics of Engineering Materials*, Harlow: Addison Wesley Longman, 1998.

CHAPTER 9

CONTENTS

Direct Force Paths

9

9.1 DEFINITIONS, CONCEPTS AND CRITERIA

The **stiffness** of a structure is its ability to resist deformation. The stiffness of a structure represents the efficiency of transmitting loads on the structure to its supports.

Internal forces in members are induced when they transmit loads from one part to another part of the structure. The internal forces can be tension, compression, shear, torque or bending moment, or a combination of all or some of them.

There are three interrelated concepts relating to the internal forces in a structure:

- The more direct the internal force paths, the stiffer the structure.
- The more uniform the distribution of internal forces, the stiffer the structure.
- The smaller the internal forces, the stiffer the structure.

Following the first concept, five simple criteria can be adopted for arranging bracing members in frame structures to achieve a direct force path leading to a stiffer structure.

- Bracing members should be provided in each storey from the support (base) to the top of the structure.
- Bracing members in different storeys should be directly linked.
- Bracing members should be linked in a straight line where possible.
- Bracing members in the top storey and in the adjacent bays should be directly linked where possible. (Suitable for temporary grandstands and scaffolding structures where the number of bays is larger than the number of storeys.)
- If extra bracing members are required, they should be arranged following the previous four criteria.

9.2 THEORETICAL BACKGROUND

9.2.1 Introduction

In recent years, buildings have become taller, floors wider and bridges longer. It is expected that the trend of increasing heights and spans will continue, but how can engineers cope with the ever-increasing heights and spans, and design structures with sufficient stiffness? The basic theory of structures provides the conceptual relationships between span (L), deflection (Δ), stiffness (K_S) and natural frequency (ω) for a single-span beam subject to distributed loads as follows:

$$\Delta = \frac{c_1}{K_S} = c_2 L^4 \tag{9.1}$$

$$\omega = c_3 \sqrt{K_S} = \frac{c_4}{L^2} \tag{9.2}$$

where c_1, c_2, c_3 and c_4 are dimensional coefficients. The two equations state that

- The deflection of the beam is proportional to its span to the fourth power.
- The fundamental natural frequency of the beam is proportional to the inverse of the span squared.

- Both the deflection and fundamental natural frequency are related to the stiffness of the structure.

The limitations on displacements or the fundamental natural frequency of a structure specified in building codes or both, actually imply that the structure must possess sufficient stiffness. Adding supports, reducing spans or increasing the sizes of cross sections of members can effectively increase structural stiffness. However, these measures may not always be possible for practical designs due to aesthetic, structural or service requirements.

The stiffness of a structure is generally understood to be *the ability of the structure to resist deformation*. Structural stiffness describes the capacity of a structure to resist deformations induced by applied loads. Stiffness (K_S) is defined as the ratio of a force (P) acting on a deformable elastic medium to the resulting displacement (Δ), that is [1]:

$$K_S = \frac{P}{\Delta} \tag{9.3}$$

This definition of stiffness provides a means of calculating or estimating the stiffness of a structure, but it does not suggest how to make a structure stiffer. The question of how to design a stiffer structure (the form and pattern of a structure) is a fundamental and practical problem. It may even be a problem that is more challenging than how to analyse the structure.

9.2.2 Concepts for Achieving a Stiffer Structure

9.2.2.1 Definition of stiffness

Consider a structure that consists of s members and n joints, with no limitation on the layout of the structure and the arrangement of members. To evaluate its stiffness at a particular point in a required direction, a unit force should be applied to the point in the direction where the resulting deflection is to be calculated.

Point stiffness is defined as *the inverse of a displacement in the load direction of a node where a unit load is applied*. Thus, the point stiffness relates to a unit force which is a function of position and direction. In other words, the point stiffnesses at different positions and in different directions are different.

Define the stiffness of a structure in a given direction as the smallest value among all n point stiffnesses, that is:

$$K_S = \min\left\{k_1, k_2, \ldots, k_j, \ldots, k_n\right\} \tag{9.4}$$

where:
- K_S is the stiffness
- k_j is the point stiffness at the jth node in the given direction
- n is the number of nodes in the structure

An alternative expression of this definition is: *the inverse of the stiffness of a structure in the direction of loading equals the largest value of the nodal displacements induced by a unit force applied at each of the nodal locations in turn*, that is:

$$\frac{1}{K_S} = \max\{u_1, u_2, \ldots, u_j, \ldots, u_n\} \tag{9.5}$$

where u_j is the displacement in the direction of load of the jth node when a unit force is applied at that node. The node location where the maximum displacement occurs is the **critical point**. The critical points of many structures can be easily identified. For a horizontal cantilever, the critical point for a vertical load would be at the free end of the cantilever. For a simply supported rectangular plate, the critical point would be at the centre of the plate for a vertical load. For a plane frame supported at its base, the critical point for horizontal loading would be at the top of the frame. Thus, the static stiffness of a structure in a specified direction can be calculated directly by applying the unit load at the critical point in the specified direction. A more detailed discussion about stiffness is given in Chapter 21.

9.2.2.2 Pin-jointed structures

Consider a pin-jointed structure, such as a truss, containing s bar members and n pinned joints, with a unit load applied at the critical point of the structure. The displacement at the critical point and the internal forces in the members can be obtained by solving the static equilibrium equations and can be expressed in the following form:

$$1 \times \Delta = \sum_{i=1}^{s} \frac{N_i^2 L_i}{E_i A_i} \tag{9.6}$$

where:

N_i is the internal force of the ith member induced by a unit load at the critical point
L_i, E_i and A_i $(i = 1, 2, \ldots, s)$ are the length, Young's modulus and the area of the ith member, respectively

Equation 9.6 provides the basis of a standard method for calculating the deflection of pin-jointed structures, and can be found in many textbooks [2]. According to the definition given by Equation 9.5, the stiffness of the structure is the inverse of the displacement due to a unit load, that is:

$$K_S = \frac{1}{\displaystyle\sum_{i=1}^{s} \frac{N_i^2 L_i}{E_i A_i}} = \frac{1}{\displaystyle\sum_{i=1}^{s} N_i^2 e_i} \tag{9.7}$$

where $e_i = L_i/E_i A_i$ and is known as the flexibility of the ith member. Three concepts embodied in Equation 9.6 or Equation 9.7 can be explored.

The force N_i in Equation 9.7 is a function of the structural form, and for statically indeterminate structures it is also a function of material properties. Therefore, finding the largest stiffness of a pin-jointed structure may be considered as a topology optimisation problem of structures. As Equation 9.7 forms an incompletely defined optimisation problem, optimisation techniques may not be applied directly at this stage.

Maximising K_S is achieved by minimising the summation $\sum_{i=1}^{s} N_i^2 e_i$. The characteristics of typical components of the summation are

1. $e_i > 0$.
2. N_i can be null.
3. $N_1^2 \geq 0$, regardless of whether the member is in tension or compression.

Therefore, to make the summation $\sum_{i=1}^{s} N_i^2 e_i$ as small as possible, three conceptual solutions relating to the internal forces can be developed as follows:

1. As many force components as possible should be zero.
2. No one force component should be significantly larger than the other nonzero forces.
3. The values of all nonzero-force components should be as small as possible.

The three conceptual solutions, which are interrelated and not totally compatible, correspond to three structural concepts.

Direct force path: If many members of a structure subjected to a specific load are in a zero-force state, the load is transmitted to the supports of the structure without passing through these

TABLE 9.1 Comparison of Three Sets of Data

Set	Five Data	Largest Difference	$\sum_{i=1}^{5} a_i$	$\sum_{i=1}^{5} a_i^2$
1	1, 2, 3, 4, 5	4	15	55
2	2, 2, 3, 4, 4	2	15	49
3	3, 3, 3, 3, 3	0	15	45

members, that is, the load travels a shorter distance or follows a more direct force path to the supports. This suggests that *shorter or more direct force paths from the load to the structural supports lead to a larger stiffness for a pin-jointed structure.*

Uniform force distribution: Consider three sets of data, each consisting of five numbers as shown in Table 9.1. The sums of the three sets of data are the same, but the largest differences between the five numbers in the three sets are different. Consequently, the sums of the square of the three sets are different. The larger the difference of the five numbers in each of the three sets, the larger the sum of the squares in the example. The comparison between the sums of squares in Table 9.1 shows the effect of the differences between a set of data, which is a simplified case of Equation 9.6.

If the largest absolute value of the internal force, $|N_i|$, is not significantly bigger than other absolute values of nonzero forces, it means that the absolute values of the internal forces N_i ($i = 1, 2, \ldots, s$) should be similar. In other words, *more uniformly distributed internal forces result in a bigger stiffness of a pin-jointed structure.*

Smaller force components: If the values of N_i^2 ($i = 1, 2, \ldots, s$) are small, it means that the force components, either compression or tension, are small. In other words, *smaller internal forces lead to a bigger stiffness of a pin-jointed structure.*

9.2.2.3 Beam types of structure

For a beam-type structure in which bending dominates, an equation, similar to Equation 9.6, exists as [2]:

$$\Delta = \sum_{i=1}^{s} \int_0^{L_i} \frac{M_i^2(x)}{E_i I_i} dx \tag{9.8}$$

where $M_i(x)$, L_i, E_i and I_i are the bending moment, length, Young's modulus and the second moment of area of the cross section of the ith member, respectively. If $E_i I_i$ is constant for the ith member, then the integral in Equation 9.8 becomes $\int_0^{L_i} M_i^2(x)dx$, which can be expressed by $\bar{M}_i^2 L_i$ with the same value. Thus, Equation 9.8 becomes:

$$\Delta = \sum_{i=1}^{s} \frac{\bar{M}_i^2 L_i}{E_i I_i} \tag{9.9}$$

Equation 9.9 has the same format as Equation 9.6 where the numerator contains the square of the internal force. Thus, the three concepts derived for pin-jointed structures can also be extended to beam types of structure associated with Equation 9.9.

9.2.2.4 Expression of the concepts

As the previous derivation has not been related to any particular material properties, loading conditions or structural form, the three concepts are valid for any bar or beam types of structure and can be used for designing stiffer structures. The three concepts may be summarised in a more concise form as follows [3].

- The more direct the internal force paths, the stiffer the structure.
- The more uniform the distribution of the internal forces, the stiffer the structure.
- The smaller the internal forces, the stiffer the structure.

The concepts are general and valid when Equation 9.6 or Equation 9.8 is applicable.

9.2.3 Implementation

9.2.3.1 Five criteria

Bracing systems may be used for stabilising structures, transmitting loads and increasing lateral structural stiffness. There are many options to arrange bracing members and there are large numbers of possible bracing patterns, as evidenced in tall buildings, scaffolding structures and temporary grandstands. Five criteria, based on the first concept derived in Section 9.2.2, have been suggested for arranging bracing members for temporary grandstands. These criteria are also valid for many types of structure, such as tall buildings and scaffolding structures. The five criteria are [4]:

- Criterion 1: Bracing members should be provided in each storey from the support (base) to the top of the structure.
- Criterion 2: Bracing members in different storeys should be directly linked.
- Criterion 3: Bracing members should be linked in a straight line where possible.
- Criterion 4: Bracing members in the top storey and in the adjacent bays should be directly linked where possible.
- Criterion 5: If extra bracing members are required, they should be arranged following the previous four criteria.

The first criterion is obvious since the critical point for a multistorey structure is at the top of the structure and the load at the top must be transmitted to the supports of the structure. If bracing is not arranged over the height of a structure, its efficiency will be significantly reduced. There are a number of ways to achieve the first criterion, but the second and the third criteria suggest a way using a *shorter force path*. The first three criteria mainly concern the bracing arrangements in different storeys of a structure. For some structures, such as temporary grandstands, the number of bays is usually larger than the number of storeys. To create a shorter force path or more zero-force members in such structures, the fourth criterion gives a means for considering the relationship of bracing members across the bays of the structure. The fifth criterion suggests that when extra bracing members are required, usually to reduce the internal forces of bracing members and distribute the forces more uniformly, they should be arranged using the previous criteria.

EXAMPLE 9.1

Two four-bay and four-storey plane pin-jointed structures with the same dimensions but with different bracing arrangements are shown in Figure 9.1. All the members are made of the same material and have the same cross-sectional area. The vertical and horizontal members have the same length of 1 m. On each structure, the same concentrated loads of 0.5 N are applied at the two corner points in the horizontal direction. (If a unit horizontal load is applied on either the top left or top right node, it is difficult to determine the internal forces of all members of the two frames by hand.) Determine the maximum displacements of the two structures.

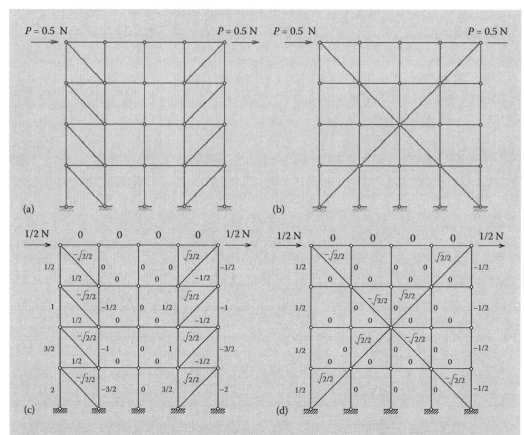

FIGURE 9.1
Two plane frames with different bracing systems. (a) Bracing arrangement following the first criterion (Frame A). (b) Bracing arrangement following the first three criteria (Frame B). (c) Internal forces (N) in Frame A. (d) Internal forces (N) in Frame B.

SOLUTION

The two structures are statically indeterminate. However, they are both symmetric structures subjected to antisymmetric loads. According to the concept that *a symmetric structure subjected to antisymmetric loading will result in only antisymmetric internal forces*, the internal forces in the members of the two frames can be directly calculated using the equilibrium conditions at the pinned joints using the left half of the frames. For example, the internal forces in the horizontal bars in the second and third bays of Frame A must be zero as the forces in the two bays must be antisymmetric and must be in equilibrium at the nodes on the central column. Thus, all the internal forces of the two frames can be easily calculated by hand and are marked directly next to the elements, as shown in Figure 9.1c and d, where the positive values indicate the members in tension and the negative values indicate the members in compression.

The internal forces are summarised in Table 9.2. The second row shows the magnitudes of the internal forces; the third row gives the numbers of members that have the same force magnitude and the fourth row shows the product N^2L of the corresponding members. The sum of N^2L that have the same force magnitude is given by ΣN^2L.

It can be seen from Table 9.2 that

- There are more zero-force members in Frame B than in Frame A.
- The differences between the magnitudes of the internal forces in Frame B are smaller than those in Frame A.
- The magnitudes of the internal forces in Frame B are smaller than those in Frame A.

TABLE 9.2 Summary of the Internal Forces of the Two Frames

		Frame A						Frame B	
Force magnitudes (N)	0	$|1/2|$	$|\sqrt{2}/2|$	$|1|$	$|3/2|$	$|2|$	0	$|1/2|$	$|\sqrt{2}/2|$
No. of elements	16	10	8	4	4	2	28	8	8
N^2L (N²m)	0	1/4	$\sqrt{2}/2$	1	9/4	4	0	1/4	$\sqrt{2}/2$
N^2L (N²m)	0	5/2	$4\sqrt{2}$	4	9	8	0	2	$4\sqrt{2}$

$$\sum_{i=1}^{44} \frac{N_i^2 L_i}{EA} \text{ (Nm)}^a \qquad \frac{23.5+4\sqrt{2}}{EA} = \frac{29.16}{EA} \qquad \frac{2+4\sqrt{2}}{EA} = \frac{7.657}{EA}$$

> [a] The unit in Equation 9.6 is a Newton metre (Nm). As the force on the left side of Equation 9.6 is 1 N, the energy and the displacement have the same value. For this case, the displacement at the top left and top right nodes of the two frames are the same. Thus, $0.5\Delta + 0.5\Delta = 1\Delta$.

As Frame A satisfies the first criterion while Frame B satisfies the first three criteria, according to the concepts given in Section 9.2.2.4, Frame B should be stiffer than Frame A. The maximum displacements of the two frames induced by the same loading are given in the bottom row in Table 9.2. In other words, the lateral stiffness of Frame B is 3.81 times (29.16/7.657) that of Frame A, although the same amount of material is used in the two frames. This demonstrates the effect of using the concepts and criteria defined earlier. Experimental and physical models of the two frames will be provided to demonstrate the difference between the lateral stiffness of the two frames in Section 9.3.

9.2.3.2 Numerical verification

To examine the efficiency of the concepts, consider a pin-jointed plane frame, consisting of four bays and two storeys with six different bracing arrangements as shown in Figure 9.2. All frame members have the same Young's modulus E and cross-sectional area A with EA equal to 1000 N. The vertical and horizontal members have unit lengths (1 m). A concentrated horizontal load of 0.2 N is applied to each of the five top nodes of the frames. The lateral stiffness can be calculated as the inverse of the averaged displacement of the top five nodes in the horizontal direction.

The bracing members in the six frames are arranged in such a way that the efficiency of each criterion given in Section 9.2.3.1 can be identified. The features of the bracing arrangements can be summarised as follows:

- The bracing members in frame (a) satisfy the first criterion.
- The bracing members in frame (b) satisfy the first two criteria.
- The bracing members in frame (c) satisfy the first three criteria.
- The bracing members in frame (d) satisfy the first four criteria.
- Two more bracing members are added to frame (c) to form frame (e), but the added bracing members do not follow the criteria suggested.
- Four more bracing members are added to frame (c) or frame (d) to form frame (f), and the arrangement of bracing members in frame (f) satisfies all of the five criteria.

Table 9.3 lists the total numbers of members, bracing members and zero-force members, the five largest absolute values of member forces and the average horizontal displacements of the five top nodes of the six frames. The relative stiffnesses of the six frames are also given for comparison.

The force paths, which transmit the loads from the tops to the supports of the frames, are indicated by the dashed lines in Figure 9.2. To emphasise the main force paths, forces of less than 3% of the maximum force in each of the first three frames have been neglected in Figure 9.2.

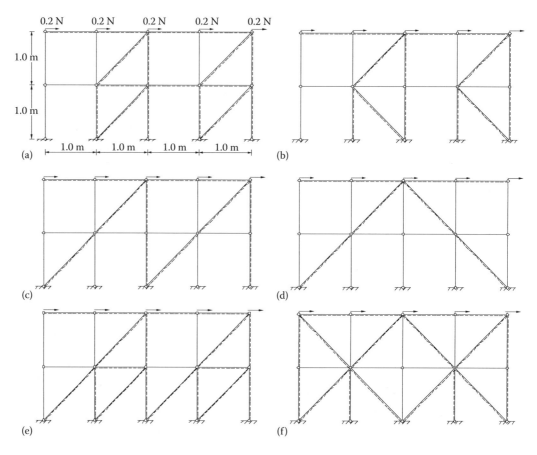

FIGURE 9.2
Frames with different bracing arrangements and force paths (dashed lines). Frame (a). Frame (b). Frame (c). Frame (d). Frame (e). Frame (f).

From the concepts suggested on the basis of Equation 9.7, the following points can be made from Table 9.3 and Figure 9.2:

- Frame (a) has a conventional form of bracing and the loads at the top are transmitted to the base through the bracing and the vertical and horizontal members. There are five members with zero force.

TABLE 9.3 A Summary of the Results for the Six Frames (Figure 9.2)

Frame	(a)	(b)	(c)	(d)	(e)	(f)
No. of elements	22	22	22	22	24	26
No. of bracing elements	4	4	4	4	6	8
No. of zero-force elements	5	8	10	14	6	8
The absolute values of the five largest element forces (N)	1.04 (v)[a]	1.03 (v)	0.75 (b)	0.71 (b)[a]	0.74 (b)	0.40 (b)
	0.96 (v)	0.97 (v)	0.72 (b)	0.71 (b)	0.67 (b)	0.40 (b)
	0.78 (b)	0.74 (b)	0.69 (b)	0.71 (b)	0.64 (v)	0.37 (b)
	0.72 (b)	0.71 (b)	0.67 (b)	0.71 (b)	0.59 (b)	0.37 (b)
	0.69 (b)	0.71 (b)	0.53 (v)	0.40 (v)	0.58 (v)	0.33 (b)
The average horizontal displacement of the five top nodes (mm)	6.60	6.12	4.12	3.23	3.93	1.69
The relative stiffness	1	1.08	1.60	2.04	1.68	3.91
The ratio of stiffness to the total area of bracing members	1	1.08	1.60	2.04	1.12	1.95

[a] (b): bracing members; (v): vertical members.

- In frame (b), the forces in the bracing members in the upper storey are directly transmitted to the bracing and vertical members in the lower storey without passing through the horizontal members that link the bracing members in the two storeys. Thus, frame (b) provides a shorter force path with three more zero-force members and yields a higher stiffness than frame (a).
- In frame (c), a more direct force path is created with two vertical members in the lower storey, which have the largest forces in frame (b), becoming zero-force members. The shorter force path produces an even higher stiffness, as expected.
- To transmit the lateral loads at the top nodes where bracing members are involved, forces in vertical members have to be generated to balance the vertical components of the forces in the bracing members in frame (c). In frame (d), two bracing members with symmetric orientation are connected at the same node, with one in compression and the other in tension. The horizontal components of the forces in these bracing members balance the external loads while the vertical components of the forces are self-balancing. Therefore, all vertical members are in a zero-force state and frame (d) leads to the highest stiffness of frames (a)–(d).
- Two more members are added to frame (c) to form frame (e), but a comparison between frame (d) and frame (e) indicates that bracing members following the criteria set out can lead to a higher stiffness than more bracing members which do not fully follow the criteria.
- Frame (f) shows the effect of the fifth criterion. Four more bracing members are added to frame (d) and arranged according to the first three criteria. Now the lateral loads are distributed between more members, creating a smaller and a more uniform force distribution, which results in an even higher stiffness.

It can be seen from Table 9.3 that the structure is stiffer when the internal forces are smaller and more uniformly distributed although the first four criteria are derived based on the concept of direct force paths. These examples are simple and the variation of bracing arrangements is limited, but they demonstrate the efficiency of the concepts and the criteria.

The lateral stiffnesses of the frames are provided by the bracing members. It is interesting to examine the ratio of the relative lateral stiffness to the total area of bracing members. In this way, frame (d) has the highest ratio.

9.2.4 Discussion

The design of a structure needs to consider several requirements and a stiffness requirement is one of them. The concept of direct force paths, and the criteria which follow from it, may be useful for the design of those structures when increasing stiffness is important.

9.2.4.1 Safety, economy and elegance

A useful definition of structural engineering has been given in the *Journal of the UK Institution of Structural Engineers* [5] as follows:

> *Structural engineering is the science and art of designing and making, with economy and elegance, buildings, bridges, frameworks and other similar structures, so that they can safely resist the forces to which they may be subjected.*

There are three key factors in the statement, *safety*, *economy* and *elegance*. The discipline of structural engineering allows structures to be produced with satisfactory performance at competitive costs. Elegance, which is not particularly related to safety and economy, should also be considered. However, the beauty of the three structural concepts presented in this chapter lies in integrating the safety, economy and elegance of a structure as a whole. This may be demonstrated by the following example.

Figure 9.3a shows the bracing arrangement at the back of part of an actual temporary grandstand, where alternative bays were braced from the bottom to the top and a bracing member was placed at the first-storey level in all the other bays. It can be seen that the bracing system satisfies the

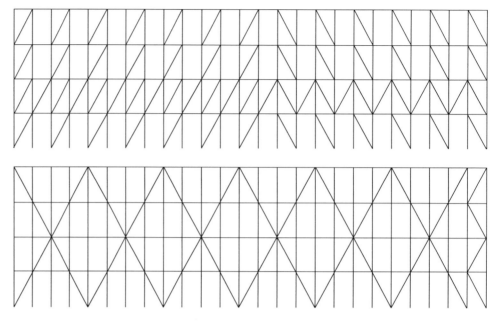

FIGURE 9.3
Bracing arrangements for a temporary grandstand. (a) Original bracing system. (b) Improved bracing system. (From Ji, T. and Ellis, B. R. *The Structural Engineer*, 75, 6, 1997.)

TABLE 9.4	Comparison of the Efficiency of Two Braced Frames (Figure 9.3)	
	Lateral Stiffness	**No. of Bracing Members Used**
Original frame (OF) (Figure 9.3a)	3.16 mN m^{-1}	64
Improved frame (IF) (Figure 9.3b)	8.96 mN m^{-1}	52
(IF)/(OF)	284%	81%

first criterion and partly satisfies the third criterion (Section 9.1). A significant increase in the lateral stiffness of the grandstand can be achieved by using the concept of direct force paths. Without considering the safety, economy and elegance of the structure but following the concept of direct force paths and the first four criteria, the bracing members can be rearranged as shown in Figure 9.3b. The calculated lateral stiffnesses of the two frames are summarised in Table 9.4 [4].

The comparison shows that the lateral stiffness of the improved structure is higher, being 284% of the stiffness of the original structure. The improved structure is also more economical as the number of bracing members is reduced by 19%. Looking at the appearance of the two frames, one would probably feel that the frame with the improved bracing arrangement is more elegant.

Application of the concepts leads to structures with larger stiffness and smaller and more uniform distributions of internal forces, meeting the requirements of safety and economy. It is difficult to show that the concepts also lead to elegant designs. However, the examples of the John Hancock Tower, the Bank of China (Section 9.4.1) and the Raleigh Arena (Section 10.4.1) are all well-known safe, economical and elegant structures in which the concepts presented in Section 9.1 were used.

9.2.4.2 Optimum design and conceptual design

The three concepts presented are derived on the basis of making displacements (Equation 9.6 or Equation 9.9) as small as possible. It is useful to compare the general characteristics of optimum design methods and the use of the concepts. Table 9.5 compares the general characteristics of some optimum design methods and the design methods using the proposed concepts.

Compared with optimum design methods, design based on the concepts presented does not involve an analysis for choosing member cross sections, does not seek the stiffest structure and is not subjected to explicitly applied constraints. Therefore, design using the concepts becomes simple

TABLE 9.5   Comparison of Design Methods

	Design Using Optimum Methods	Design Using the Concepts
Objective	Seek a maximum or a minimum value of a function, such as cost, weight or energy	Seek a stiffer structure rather than the stiffest structure
Constraints	Explicitly applied	Implicitly applied
Solution method	Computer-based mathematical methods	The concepts and derived criteria
Loading	The optimum design depends on loading conditions	The design is independent of loading conditions
Cross-sectional sizes of members	Provided as the solution of the optimum design	To be determined
Design	The optimum design may not be practical	The design is practical
Users	Specialists and researchers	Engineers

and many engineers can make direct use of the concepts. It is useful that the choice of structural form, relating to force paths, and the selection of the sizes of cross sections are conducted separately.

As the concepts are fundamental and they can be applied to a structure globally, designs of bracing arrangement based on the concepts may be more optimal and rational than some designs resulting from optimum design processes. Figure 9.4a shows the optimal topology of a bracing system for a steel building framework with an overall stiffness constraint under multiple lateral loading conditions [6]. The solution was obtained by gradually removing the elements with the lowest strain

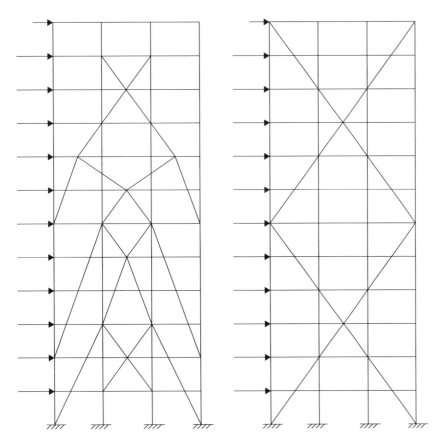

FIGURE 9.4
Comparison of two designs of bracing system. (a) Optimal topology of bracing system. (b) Bracing system following the concept of direct force path. (From Lian, Q., Xie, Y. and Steven, G. *Journal of Structural Engineering*, 126, 7, 2000.)

TABLE 9.6 Displacements of the Two Designs

	Frame without Bracing Members	Frame with Bracing Suggested in [6]	Frame with Bracing Using the First Concept
Horizontal displacement at the top of the frame	630 mm	87.4 mm	78.5 mm
Maximum horizontal displacement and its location	630 mm (top-storey level)	87.4 mm (top-floor level)	84.1 mm (11th-storey level)
Bracing members used	0%	100%	67.5%

energy from a continuum design, which implied creating a direct force path in an iterative manner. Figure 9.4b shows the design of the bracing system using the concept of direct force paths, that is, the first three criteria. This design takes only a few minutes. Using the dimensions, cross-sectional sizes and load conditions given in [6], the maximum displacements of the two frames have been calculated and are listed in Table 9.6.

For this example, it can be seen that the design using the first concept is more practical, economical and is stiffer than a design obtained using an optimisation technique.

9.3 MODEL DEMONSTRATIONS

9.3.1 Experimental Verification

These simple experiments *verify the concept that the more direct the internal force paths is, the stiffer the structure will be, and the first three corresponding criteria.*

Three aluminium frames were constructed with the same overall dimensions of 1025×1025 mm. All members of the frames have the same cross section of 25×3 mm. The only difference between the three frames is the arrangement of the bracing members as shown in Figure 9.5. It can be seen from Figure 9.5 that

1. Frame A is traditionally braced with eight members, which satisfies the first criterion
2. Eight bracing members are again used in Frame B, but are arranged to satisfy the first three criteria

FIGURE 9.5
Aluminium test frames. (Frames A, B and C are placed from left to right.)

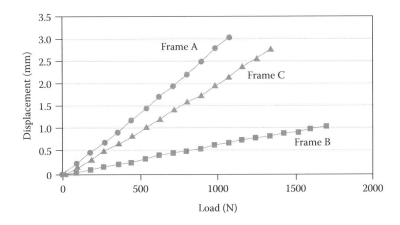

FIGURE 9.6
Load-deflection curves for Frames A, B and C.

   3. A second traditional bracing pattern is used for Frame C, with 16 bracing members arranged satisfying the first two criteria

    The three frames were tested using a simple arrangement. The frames were fixed at their supports and a hydraulic jack was used to apply a horizontal force at the top right-hand joint of the frame. A micrometre gauge was used to measure the horizontal displacement at the top left-hand joint of the frame. A lateral restraint system was provided to prevent out-of-plane deformations [7].

    The horizontal load-deflection characteristics of the three frames are shown in Figure 9.6. It can be seen that the displacements of Frame B, which satisfied the first three criteria, are about one-quarter of those of Frame A for the same load. Frame C, with eight more members but not satisfying the third criterion, is obviously less stiff than Frame B. For example, the displacements corresponding to the load of about 1070 N are 3 mm for Frame A, 0.73 mm for Frame B and 2.2 mm for Frame C, respectively. The experiment results for Frame A and Frame B align with the conclusions obtained in Example 9.1.

9.3.2 Direct and Zigzag Force Paths

This model demonstration *allows one to feel the relative stiffnesses of two similar plastic frames and shows the effect of internal force paths.*

    In order to 'feel' the effect of the force paths, two frames were made of plastic, with the same overall dimensions 400 × 400 mm and member sizes of 25 × 2 mm (see Figure 9.7 [7]). The only difference

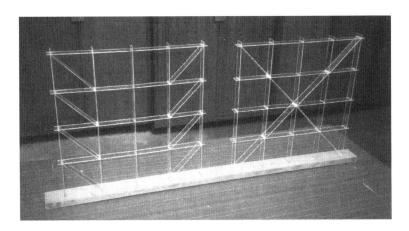

FIGURE 9.7
Braced frame models showing direct and zigzag force paths.

between the two frames is the arrangement of the bracing members. The forms of the two frame models are the same as those in Example 9.1 in Section 9.2.3 and the test Frames A and B in Section 9.3.1. The relative stiffnesses of the two frames can be felt by pushing a top corner joint of each frame horizontally. The frame on the right side feels much stiffer than the one on the left. In fact, the stiffness of the right frame is about four times that of the left frame. The load applied to the right frame is transmitted to its supports through a direct force path while for the frame on the left, the force path is zigzag.

9.4 PRACTICAL EXAMPLES

9.4.1 Bracing Systems of Tall Buildings

The John Hancock Center in Chicago, a 100-storey 344 m tall building, has an exterior-braced frame tube structure. An advance on the steel-framed tube, this design added global cross-bracing to the perimeter frame to increase the stiffness of the structure as shown in Figure 9.8a. Some $15 million was saved on the conventional steelwork by using these huge cross-braces [8]. It was regarded as an extremely economical design which achieved the required stiffness to make the building stable. One of the reasons for the success was, as can be seen from Figure 9.8a, that the required lateral stiffness of the structure was achieved by using cross-braces resulting in direct force paths and smaller internal forces according to the first concept or the first three criteria (Sections 9.1 and 9.2). The Bank of China, Hong Kong (Figure 9.8b), also adopts a similar bracing system.

9.4.2 Bracing Systems of Scaffolding Structures

Scaffolding structures are temporary structures that are an essential part of the construction process. Scaffolding imposes certain design restrictions that can be ignored in the design of other structures. For example, scaffolding structures must be easily assembled and taken apart, and the components should also be relatively light to permit construction workers to handle them. Although scaffolding structures are light and temporary in the majority of cases, their design should be taken seriously. The concept of direct force paths and the five criteria are applicable to scaffolding structures.

(a)   (b)

FIGURE 9.8
Bracing systems used in buildings satisfying the first three criteria. (a) John Hancock Center.
(b) Bank of China in Hong Kong.

FIGURE 9.9
Collapse of a scaffolding structure. (Courtesy of Mr. J. Anderson.)

9.4.2.1 Collapse of a scaffolding structure

The scaffolding structure shown in Figure 9.9 collapsed in 1993 [9], though no specific explanation was given. Using the concept of direct force paths and the understanding gained from the previous examples, the cause of the incident may be suggested. It can be seen that in this scaffolding structure no diagonal (bracing) members were provided, that is, no direct force paths were provided. The scaffolding structure worked as an unbraced frame structure, and the lateral loads, such as wind loads, on the structure were transmitted to its supports through bending of the slender scaffolding members. The structure did not have enough lateral stiffness and collapsed under wind loads only.

9.4.2.2 Some bracing systems used for scaffolding structures

For convenience in erecting the scaffolding structures shown in Figure 9.10, standard units were used. The unit shown in Figure 9.10a consists of two horizontal members, two vertical members

(a)   (b)

FIGURE 9.10
(a,b) Inefficient bracing systems for scaffolding structures.

and two short bracing members. The unit is useful for transmitting the vertical loads applied to the top horizontal member to the vertical members that support the unit at its two ends. The unit is equivalent to a thick beam in the structure and the scaffolding structure becomes a deep beam and a slender column system. The diagonal members used in the structure do not provide the force paths to transmit the lateral loads on the structure from the top to the bottom of the structure and do not follow the basic criteria for arranging bracing members. Therefore, it can be seen that the scaffolding structure has a relatively low lateral stiffness based on the first concept of direct force paths.

Bracing members are also provided in the scaffolding structure shown in Figure 9.10b. However, these bracing members are linked in the horizontal direction but not connected from the top to the bottom of the structure, and do not create direct force paths. Therefore, without any calculation, it can be judged that the scaffolding structure possesses a relatively low lateral stiffness.

PROBLEM

1. Figure 9.11 shows four braced six-bay by three-storey scaffolding structures, A, B, C and D, with the same support and loading conditions. A 1 kN horizontal force is applied at each of the top nodes of the frames. The structures have the same overall dimensions and all the members of the structures have the same cross-sectional properties. Each horizontal and vertical member is 1 m long. The differences between the four scaffolding structures are the arrangements of the bracing members. Buckling does not need to be considered in this problem. Answer the following questions.
   a. For each structure, show the force path that transmits the loads from the top of the structure to the supports.
   b. Judge the relative lateral displacements of the four structures and list them from small to large.
   c. Give reasons to justify your answer in (b).
   d. Rearrange the six bracing members to make the frame even stiffer and show your design and the force path of the design.
   e. Calculate the relative displacement of your design to Frame A.

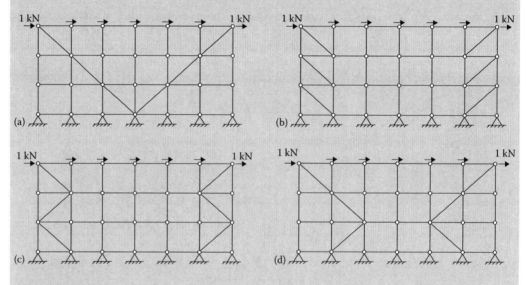

FIGURE 9.11
Frames with different bracing arrangements. (a) Frame A. (b) Frame B. (c) Frame C. (d) Frame D.

REFERENCES

1. Parker, S. P. *Dictionary of Engineering*, 5th edn, New York: McGraw-Hill, 1997.
2. Gere, J. M. *Mechanics of Materials*, Belmont, CA: Thomson Books/Cole, 2004.
3. Ji, T. Concepts for designing stiffer structures, *The Structural Engineer*, 81, 36–42, 2003.
4. Ji, T. and Ellis, B. R. Effective bracing systems for temporary grandstands, *The Structural Engineer*, 75, 95–100, 1997.
5. Ji, T. and Ellis, B. R. Floor vibration induced by dance-type loads: Theory, *The Structural Engineer*, 72(3), 37–44, 1994.
6. Lian, Q., Xie, Y. and Steven, G. Optimal topology design of bracing systems for multi-story steel frames, *Journal of Structural Engineering*, 126, 823–829, 2000.
7. Roohi, R. Analysis, testing and model demonstration of efficiency of different bracing arrangements. Investigative Project Report, UMIST, 1998.
8. Bennett, D. *Skyscrapers: Form and Function*, New York: Simon & Schuster, 1995.
9. Anderson, J. Teaching health and safety at university, Proceedings of the Institution of Civil Engineers, *Journal of Civil Engineering*, 114, 98–99, 1996.

CHAPTER 10

CONTENTS

Smaller Internal Forces

10

10.1 CONCEPTS AND A CRITERION

- The more direct the internal force paths, the stiffer the structure.
- The more uniform the distribution of internal forces, the stiffer the structure.
- The smaller the internal forces, the stiffer the structure.

Smaller internal forces in a structure, or a stiffer structure, can be achieved by

- Providing additional supports to the structure.
- Reducing the spans of the structure.
- Making a self-balanced system of forces in the structure before the forces are transmitted to the supports of the structure.

The first two measures are obvious. The contents of this chapter are related to the third measure.

Criterion: If members can be added into a structure in a way that offsets some of the effects of the external loads, or balances some of the internal forces before the forces are transmitted to the supports of the structure, then the internal forces in the structure will be smaller and the structure will be stiffer.

10.2 THEORETICAL BACKGROUND

10.2.1 Introduction

Chapter 9 discussed the ways to achieve direct force paths between the load at the critical point of a structure and its supports, leading to a stiffer and more economical design. In this chapter, the ways to achieve smaller internal forces in a structure will be considered, which will also lead to a stiffer and more economical design.

Increasing the sizes of the cross sections of members in a structure will effectively reduce their stress levels but not necessarily the internal forces in the members. This measure usually increases the amount of material used and therefore the weight of the structure. Reducing spans is a very effective way of reducing the magnitudes of internal forces and increasing the stiffness of the structure. However, in many practical cases, this measure may not be feasible. Following the third concept derived in Chapter 9, if the internal forces can be partly balanced by introducing new structural members, smaller internal forces will be created and the structure will be stiffer. The theoretical background for this concept and for the criterion given in Section 10.1 is provided in this section.

Consider a beam type of structure with *s* members with any given loading applied to the structure. The maximum displacement of the structure in the direction of loading can be determined using the principle of virtual work [1,2]:

$$v_1 = \sum_{i=1}^{s} \int_0^{L_i} \frac{M_i^P(x)\bar{M}_i(x)}{E_i I_i}\, dx \tag{10.1}$$

where:

$M_i^P(x)$ is the bending moment in the *i*th member induced by the actual load

$\bar{M}_i(x)$ is the bending moment in the *i*th member caused by a unit load applied on the point and in the direction where the displacement is to be calculated

Now, if a structural member is added into the structure, the internal forces in the structure are consequently changed. The internal forces of the modified structure, induced by the load, can be expressed as the summation of the internal forces, $M_i^P(x)$, of the original structure, and the change of the internal force, $\Delta M_i^P(x)$, $(i=1, 2, \ldots, s)$. The displacement at the same location becomes [3]:

$$v = \sum_{i=1}^{s} \int_0^{L_i} \frac{M_i^P(x)\bar{M}_i(x)}{E_i I_i} dx + \sum_{i=1}^{s} \int_0^{L_i} \frac{\Delta M_i(x)\bar{M}_i(x)}{E_i I_i} dx = v_1 + v_2 \tag{10.2}$$

where

$$v_2 = \sum_{i=1}^{s} \int_0^{L_i} \frac{\Delta M_i(x)\bar{M}_i(x)}{E_i I_i} dx \tag{10.3}$$

If the member is positioned in the structure so that as many as possible of the terms $\Delta M_i(x)$ have the opposite signs to $M_i^P(x)$, v_2 will have the opposite sign to v_1. Therefore, the displacement in Equation 10.2 will be smaller than that in Equation 10.1. In other words, the function of the added member is to create additional internal forces that have the opposite directions to those induced by the actual load in the original structure. This reduces the magnitudes of the internal forces and creates a stiffer structure. The position of the member can be identified from the deflected shape of the structure, and the form can be determined by experience, intuitive understanding or by calculation. This is best illustrated through examples.

10.2.2 Ring and Tied Ring

A ring, with radius R, has a rigidity of EI and is subjected to a pair of vertical forces P at points B and D, as shown in Figure 10.1a [4]. Similar to studying the displacements and internal forces of a straight beam, it may be assumed that the ring experiences small deflections allowing the equilibrium equations to be established using the configurations before deformation. The deformation of the ring is dominated by bending. Due to symmetry, only the right top quarter of the ring needs to be considered for analysis (Figure 10.1b). Also due to the double symmetry of the loaded ring, the rotations at points A, B, C and D must be symmetric, and only zero rotations at these locations satisfy the condition. By cutting the ring at any section defined by θ (Figure 10.1c), the bending moment at the section can be written using the equilibrium condition as:

$$M_P(\theta) = M_P(0) + \frac{PR}{2}(1 - \cos\theta) \tag{10.4}$$

where $M_P(\theta)$ at $\theta=0$ is unknown, but can be determined using the condition that there is no relative rotation between points B and C. Applying a pair of unit moments at B and C to the unloaded quadrant (Figure 10.1b), the bending moment along the ring is a constant:

$$\bar{M}_{BC}(\theta) = 1 \tag{10.5}$$

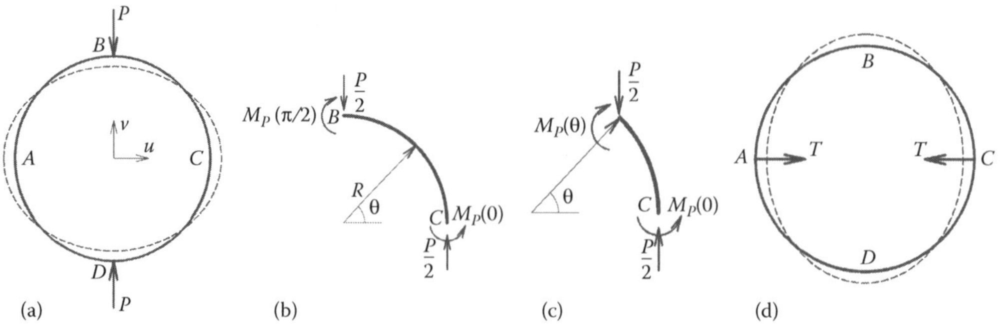

FIGURE 10.1
A ring. (a) A ring subjected to a pair of vertical loads. (b) Free-body diagram for a quadrant of the ring. (c) Equilibrium at any section of the ring. (d) The ring subjected to a pair of horizontal forces.

Thus, the relative rotation between B and C is:

$$\theta_B - \theta_C = \frac{1}{E1} \int_0^{\pi/2} M_P(\theta)\bar{M}_{BC}(\theta)R d\theta = \int_0^{\pi/2} \left[M_P(0) + \frac{PR}{2}(1 - \cos\theta) \right] R d\theta = 0 \qquad (10.6)$$

giving

$$M_P(0) = PR\left(\frac{1}{\pi} - \frac{1}{2} \right) \qquad (10.7)$$

Substituting Equation 10.7 into Equation 10.4 leads to

$$M_P(\theta) = PR\left(\frac{1}{\pi} - \frac{\cos\theta}{2} \right) \qquad (10.8a)$$

Similarly, when the force P is replaced by a unit load, Equation 10.8a becomes

$$\bar{M}_P(\theta) = R\left(\frac{1}{\pi} - \frac{\cos\theta}{2} \right) \qquad (10.8b)$$

Substituting Equation 10.8 into Equation 10.1, the relative vertical deflection between points B and D of the ring is

$$v_1 = \frac{2}{EI} \int_0^{\pi/2} M_P(\theta)\bar{M}_P(\theta)R d\theta = \frac{PR^3}{EI}\left(\frac{\pi}{8} - \frac{1}{\pi} \right) = 0.0744\frac{PR^3}{EI} \qquad (10.9)$$

The deformed shape of the ring subject to the concentrated loads P can thus be formed, and is as shown in Figure 10.1a, where points B and D deform inwards, while points A and C move outwards. To produce the deformations in the opposite directions to the deformations shown in Figure 10.1a, a pair of horizontal forces, T, are applied at points A and C of the ring in the inward directions, as shown in Figure 10.1d. It can be noted that the forcing condition shown in Figure 10.1d can be obtained by rotating the ring and forces shown in Figure 10.1a through 90° anticlockwise. Thus, the bending moments in any section θ of the top right quarter of the ring due to the pair of horizontal forces, T, and a pair of unit horizontal forces can be written using Equation 10.8:

$$M_T(\theta) = TR\left(\frac{1}{\pi} - \frac{\cos(\theta - \pi/2)}{2} \right) = TR\left(\frac{1}{\pi} - \frac{\sin\theta}{2} \right) \qquad (10.10a)$$

$$\bar{M}_T(\theta) = R\left(\frac{1}{\pi} - \frac{\sin\theta}{2} \right) \qquad (10.10b)$$

The vertical displacement at point B due to the pair of horizontal forces, T, can be evaluated using the second term in Equation 10.2, as follows:

$$v_2 = \frac{2}{EI} \int_0^{\pi/2} M_T\bar{M}_P R d\theta = \frac{TR^3}{EI}\left(\frac{1}{4} - \frac{1}{\pi} \right) = -0.0683\frac{TR^3}{EI} \qquad (10.11)$$

It can be seen that the pair of forces, T, produce outward displacements between points B and D, which offset some of the displacements induced by the vertical loads P. The negative value is because $M_P(\theta)$ and $M_T(\theta)$ have opposite signs. Figure 10.2 compares the normalised moments, $M_P(\theta)/PR$ (solid line) and $M_T(\theta)/TR$ (dashed line) along the top right quarter of the ring (between 0 and $\pi/2$).

To provide the pair of forces T, a tie may be added to connect points A and C across the diameter of the ring horizontally, as shown Figure 10.3. It may be assumed that the tied ring experiences small deformations, as was the case for the untied ring. The tied ring shown in Figure 10.3a will be stiffer than the original ring shown in Figure 10.1a. This is because the bending moment and the relative vertical displacement between points B and D of the tied ring become

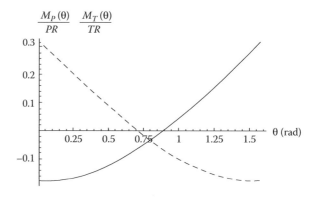

FIGURE 10.2
Comparison of the normalised bending moments induced by P and T, respectively (solid line: $M_P(\theta)/PR$; dashed line: $M_T(\theta)/TR$).

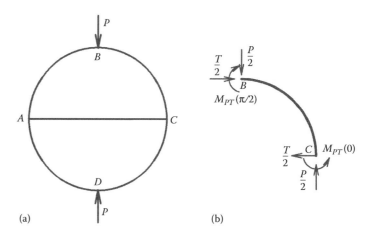

FIGURE 10.3
A tied ring. (a) A tied ring subjected to a pair of concentrated vertical loads. (b) Free-body diagram of the upper right part of the tied ring.

$$M_{PT}(\theta) = PR\left(\frac{1}{\pi} - \frac{\cos\theta}{2}\right) + TR\left(\frac{1}{\pi} - \frac{\sin\theta}{2}\right) \quad (10.12)$$

$$v = v_1 + v_2 = \left(\frac{\pi}{8} - \frac{1}{\pi}\right)\frac{PR^3}{EI} + \left(\frac{1}{4} - \frac{1}{\pi}\right)\frac{TR^3}{EI} = \left(0.0744P - 0.0683T\right)\frac{R^3}{EI} \quad (10.13)$$

where the force T can be determined using the compatibility condition for the horizontal displacement between points A and C, that is, the relative displacement between points A and C of the ring, which is equal to the extension of the wire:

$$u_1 + u_2 + u_T = 0 \quad (10.14)$$

where:

u_1 and u_2 are the relative horizontal displacements between points A and C induced by the concentrated load P and the horizontal forces T, respectively

μ_T is the extension of the wire

The horizontal displacements can be calculated in a similar manner to the vertical displacement:

$$u_1 = \frac{2}{EI} \int_0^{\pi/2} M_P(\theta)\bar{M}_T(\theta)ds$$

$$= \frac{2}{EI} \int_0^{\pi/2} PR(\cos\theta/2 - 1/\pi)R(\sin\theta/2 - 1/\pi)Rd\theta$$

$$= \frac{PR^3}{EI}\left(\frac{1}{4} - \frac{1}{\pi}\right) \tag{10.15}$$

$$u_2 = \frac{2}{EI} \int_0^{\pi/2} M_T(\theta)\bar{M}_T(\theta)ds$$

$$= \frac{2}{EI} \int_0^{\pi/2} TR^2(\sin\theta/2 - 1/\pi)^2 Rd\theta = \frac{TR^3}{EI}\left(\frac{\pi}{8} - \frac{1}{\pi}\right) \tag{10.16}$$

The extension of the tie is

$$u_T = \frac{(T/2)(2R)}{E_T A_T} = \frac{TR}{E_T A_T} \tag{10.17}$$

Substituting Equations 10.15 through 10.17 into Equation 10.14 leads to

$$\frac{PR^3}{EI}\left(\frac{1}{4} - \frac{1}{\pi}\right) + \frac{TR^3}{EI}\left(\frac{\pi}{8} - \frac{1}{\pi}\right) + \frac{TR}{E_T A_T} = 0 \tag{10.18}$$

By introducing the nondimensional rigidity ratio β:

$$\beta = \frac{E_T A_T R^2}{EI} \tag{10.19}$$

and substituting it into Equation 10.18, the internal force of the wire is

$$T = \frac{2(4 - \pi)\beta}{8\pi + (\pi^2 - 8)\beta} P \tag{10.20}$$

Equations 10.19 and 10.20 indicate that the tension in the wire is a function of the ratio of the rigidities of the wire to the ring and the radius of the ring. The relationship between T/P and β is plotted in Figure 10.4. It can be seen that T/P increases significantly up to approximately $\beta = 50$ and T/P increases slowly when $\beta > 200$.

Substituting Equation 10.20 into Equation 10.12 gives

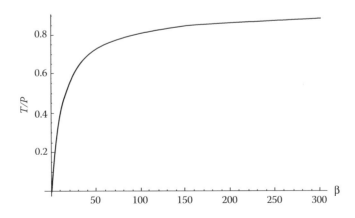

FIGURE 10.4
The relationship between T/P and β.

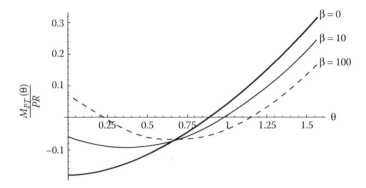

FIGURE 10.5
Bending moment of the tied ring.

$$M_{PT}(\theta) = PR(1/\pi - \cos\theta/2) + \frac{2(4-\pi)\beta}{8\pi + (\pi^2 - 8)\beta} PR(1/\pi - \sin\theta/2)$$

$$= \left[\frac{1}{\pi} - \frac{\cos\theta}{2} + \frac{2(4-\pi)\beta}{8\pi + (\pi^2 - 8)\beta}\left(\frac{1}{\pi} - \frac{\sin\theta}{2} \right) \right] PR$$

$$= \left[\frac{1}{\pi} - \frac{\cos\theta}{2} + \frac{(2-\pi\sin\theta)(4-\pi)\beta}{(8\pi + (\pi^2 - 8)\beta)\pi} \right] PR \qquad (10.21)$$

Equation 10.21 indicates that the moment of the tied ring is a function of the location θ and the rigidity ratio β. Figure 10.5 gives the relation between $M_{PT}(\theta)/PR$ and θ when $\beta = 0$, 10 and 100, respectively. It can be observed from Figure 10.5 that the magnitudes of the moments are reduced due to the addition of the tie and the increase of the rigidity β.

Substituting Equation 10.20 into Equation 10.13 leads to

$$v = \left(\frac{\pi}{8} - \frac{1}{\pi} \right)\frac{PR^3}{EI} + \left(\frac{1}{4} - \frac{1}{\pi} \right)\frac{2(4-\pi)\beta}{8\pi + (\pi^2 - 8)\beta}\frac{PR^3}{EI}$$

$$= \left[\left(\frac{\pi}{8} - \frac{1}{\pi} \right) - \frac{(4-\pi)^2\beta}{16\pi^2 + 2\pi(\pi^2 - 8)\beta} \right]\frac{PR^3}{EI} \qquad (10.22)$$

The ratio of the relative vertical displacements between points B and D of the tied ring (Equation 10.22) to the equivalent displacement of the untied ring (Equation 10.9) is

$$\frac{v}{v_1} = \frac{\left[\left(\frac{\pi}{4} - \frac{2}{\pi} \right) - \frac{(4-\pi)^2\beta}{16\pi^2 + 2\pi(\pi^2 - 8)\beta} \right]\frac{PR^3}{EI}}{\left(\frac{\pi}{4} - \frac{2}{\pi} \right)\frac{PR^3}{EI}}$$

$$= 1 - \frac{4(4-\pi)^2\beta}{[8\pi + (\pi^2 - 8)\beta](\pi^2 - 8)} \qquad (10.23)$$

Figure 10.6 shows that the displacement ratio changes with the rigidity ratio β. It can be observed from Figure 10.6 that the use of the wire effectively reduces the relative vertical displacement between points B and D.

It has been shown that the larger the tension force in the wire, the smaller the bending moment and the smaller the vertical displacement. On the other hand, the tension force T will not increase significantly when β is larger than 200. Therefore, Equation 10.19 can be used to design the rigidity of the tie.

When $E_T A_T \to \infty$, that is, $\beta \to \infty$, the maximum tension force and the minimum vertical deflection between points B and D from Equation 10.20 and 10.23, respectively, become

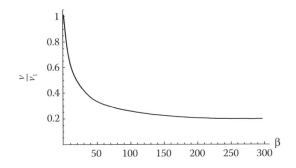

FIGURE 10.6
The ratio of the vertical displacements of the rings with and without the tie.

$$\lim_{\beta\to\infty} T = \lim_{\beta\to\infty} \frac{2(4-\pi)\beta}{8\pi+(\pi^2-8)\beta}P = \frac{2(4-\pi)}{(\pi^2-8)}P = 0.9183P \tag{10.24}$$

$$\frac{\lim_{\beta\to\infty} v}{v_1} = 1 - \frac{4(4-\pi)^2\beta}{\left[8\pi+(\pi^2-8)\beta\right](\pi^2-8)}$$

$$= 1 - \frac{4(4-\pi)^2}{(\pi^2-8)^2} = 1 - 0.8432 = 0.1568 \tag{10.25}$$

Equation 10.25 and Figure 10.6 show that the tied ring is much stiffer than the untied ring and in the extreme case, $\beta \to \infty$, the stiffness of the tied ring is over six times that of the untied ring.

EXAMPLE 10.1

A rubber ring has a radius $R=67.5$ mm and a circular cross section with diameter $d=22$ mm, as shown in Figure 10.7a. A similar ring has 15 bronze wires, twisted and tied horizontally through the centre of the ring, as shown in Figure 10.7b. Each of the wires has a diameter of 0.11 mm. Young's modulus values for the rubber and the bronze wires are 5 and 10,000 N mm^{-2}, respectively. Calculate the vertical displacements of the untied ring and the tied ring when a vertical load P of 22.3 N is applied to the top of the two rings.

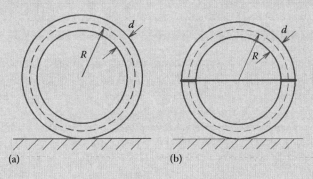

(a)   (b)

FIGURE 10.7
Example 10.1. (a) A rubber ring. (b) A tied rubber ring.

SOLUTION

For the rubber ring:

$$EI = \frac{E\pi d^4}{64} = 5 \times \frac{\pi \times 22^4}{64} = 57,495 \, \text{N mm}^{-1}$$

$$v_1 = 0.0744 \frac{PR^3}{EI} = 0.0744 \frac{22.3 \times 67.5^3}{57,495} = 8.87 \, \text{mm}$$

For the tied rubber ring:

$$E_T A_T = 100,000 \times 0.055^2 \pi \times 15 = 14,255 \, \text{N}$$

$$\beta = \frac{E_T A_T R^2}{EI} = \frac{14,255 \times 67.5^2}{57,495} = 1,130$$

$$v = \left[\left(\frac{\pi}{8} - \frac{1}{\pi} \right) - \frac{(4-\pi)^2 \beta}{16\pi^2 + 2\pi(\pi^2 - 8)\beta} \right] \frac{PR^3}{EI}$$

$$= \left[\left(\frac{\pi}{8} - \frac{1}{\pi} \right) - \frac{(4-\pi)^2 \times 1,130}{16\pi^2 + 2(\pi^2 - 8) \times 1,130} \right] \frac{22.3 \times 67.5^3}{57,495} = 1.48 \, \text{mm} = 0.167 v_1$$

For the two rings, the ratio of the displacements of the tied rubber ring to the untied rubber ring is 0.167. In other words, the vertical stiffness of the tied rubber ring is about six times that of the untied rubber ring. The ratio of 0.167 is slightly larger than 0.157 given in Equation 10.25, when the stiffness of the wire is infinite.

10.3 MODEL DEMONSTRATIONS

10.3.1 Pair of Rubber Rings

This pair of models demonstrates that *a tied ring is much stiffer than a similar ring without a tie, due to reduced internal forces.*

Figure 10.8 shows two rubber rings, one with and one without a wire tied across the diameter. The dimensions and material properties of the rings are described in Example 10.1 where the calculated vertical displacements of the rings are given. The same weight of 22.3 N is placed on the top of each of the two rings and the reduced deformation of the tied ring is apparent and its increased stiffness can be seen and felt. This may be explained since the force in the wire increases as the applied load increases, and produces a bending moment (Equation 10.10a) in the ring in the opposite direction to the bending moment (Equation 10.8a) caused by the external load (Figure 10.2). In this way, the force in the wire balances part of the bending moments in the ring due to the vertical load, reducing the internal forces in the ring, thus making it stiffer.

Examples of when this concept is used in practice are tied arches and tied-pitched roofs. The ties help to balance horizontal forces and reduce horizontal displacements, thus effectively increasing the structural stiffness.

10.3.2 Post-Tensioned Plastic Beams

This demonstration shows that *a beam with profiled post-tensioned wires is clearly stiffer than the same beam with straight post-tensioned wires.*

Three 500 mm long plastic tubes are stiffened by a pair of steel wires whose location differs in each tube, see Figure 10.9.

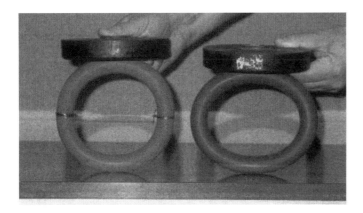

FIGURE 10.8
Comparison of the deformations of two rubber rings.

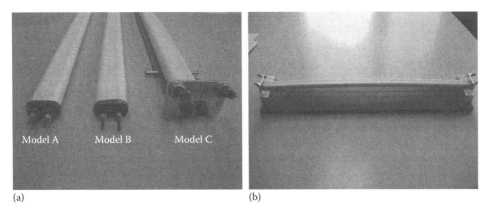

Model A   Model B   Model C

(a)   (b)

FIGURE 10.9
Three post-tensioned plastic beams with different positions of wires. (a) Locations of wires.
(b) A plastic beam stiffened by externally profiled wires.

- Model A: The pair of wires is positioned at the neutral plane of the tube.
- Model B: The pair of wires is positioned between the neutral plane and the bottom of the tube.
- Model C: The pair of wires is placed externally with a profiled shape, rather than being straight. Two small metal bars, longer than the width of the tube, are placed underneath the tube to create the desired profile for the wires.

Small metal plates are placed at the ends of the tubes to provide the supports to fix the wires. The wires are fixed into screws and the screws pass through the holes in the end plates and are fixed using nuts. By turning the nuts, the tension in the wires and hence the forces in the tubes are established. Figure 10.9a shows the details of the ends of the three tubes and the locations of the wires. Supporting the beams at their two ends produces three post-tensioned, simply supported beams. The structural behaviour of the three model beams can be described as follows:

- Model A: As the wires are placed in the neutral plane of the tube, the tube itself is in compression and the wires are in tension.
- Model B: As the wires are placed under the neutral plane of the tube, the tube is subjected to both compression and bending (bending upwards) and the wires are in tension.
- Model C (as shown in Figure 10.9b): The tensions in the wires provide upward forces through the two diverters to the beam which will partly balance any downward loads. In other words, the upward forces are equivalent to two spring supports to the beam. Thus, Model C is expected to be stiffer than Models A and B.

By positioning each model in turn on the supports shown in Figure 10.9b and then pressing the centre of each beam downwards, it can be felt that Model C is obviously stiffer than Models A and B.

The three models all form self-balanced systems, but only Model C has enhanced bending stiffness. A practical example of a stiffened floor using profiled post-tensioned cables will be given in Section 10.4.4.

10.4 PRACTICAL EXAMPLES

10.4.1 Raleigh Arena

The roof structure of the Raleigh Arena (Figure 10.10) consists of carrying (sagging) cables, and stabilising (hogging) cables which are supported by a pair of inclined arches. The structure forms, at least in part, a self-balanced system, which effectively reduces the internal forces in the arches. The carrying cables apply large forces to the arches and some of the vertical components of the forces are transmitted to external columns. Significant portions of the bending moments and the horizontal components of the shear and compressive forces in the arches are self-balancing at the points of contact between the two arches. Most of the horizontal components of the remaining shear and compressive forces in the lower parts of the arches are balanced by underground ties, which have a similar function to the wire tie in the ring used in the demonstration. The reduced internal forces not only allow the use of less material but also lead to a stiffer structure.

Another example is a stadium in Nanjing, China. The roof of the stadium is supported by a pair of arches, which are inclined outwards symmetrically and do not support each other. Figure 10.11 shows one of the arches. The pair of arches generates large horizontal forces of 13,000 kN at their supports. In order to avoid these horizontal forces at the ends of the arches being applied to the pile foundations, which are on soft soil, eight post-tension cables with a diameter of 25 mm and a length

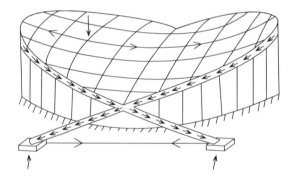

FIGURE 10.10
The force paths of the Raleigh Arena. (From Bobrowski, J. *Journal of Structural Engineer*, 64A, 5–12, 1986.)

FIGURE 10.11
A stadium in Nanjing, China.

of 400 m are placed underground to link the two ends of each arch to balance the large horizontal forces. The cables have the same function as demonstrated in the tied rubber ring.

10.4.2 Zhejiang Dragon Sports Centre

The Zhejiang Dragon Sports Centre in China (Figure 10.12) was built in 2000. The stadium has a diameter of 244 m and a cantilever roof that spans 50 m, providing unobstructed viewing for the spectators. The roof structure adopts double layer lattice shells that are supported by internal and external ring beams. Cables carry the internal ring beams back to the support towers located at two ends of the stadium. The cables are used as elastic supports to the roof, reducing the internal forces (bending moments) and increasing the stiffness of the shell roof.

The towers are subjected to large forces from the cables, which transmit the weight of the cantilever roof and the loads on the roof to the towers. These forces in turn cause large bending moments in the 85 m tall cantilever towers. To reduce the bending moments in the towers, post-tension forces were applied to the backs of the towers, providing the bending moments opposing the effects of the cables. In this way, the bending moments in the towers are reduced. This leads to a saving of material and a stiffer structure.

The bending moments in one of the towers are illustrated in Figure 10.13. Figure 10.13a qualitatively shows the elevation of the tower and the forces applied on the tower together with the bending moments in the tower. The maximum bending moment occurs at the base of the tower and is noted as M_C. Figure 10.13b shows the action of the post-tension forces applied to the tower and the corresponding bending moments. As the distance between the vertical line of action of the compression forces and the neutral axis of the tower varies linearly, the bending moment increases linearly from

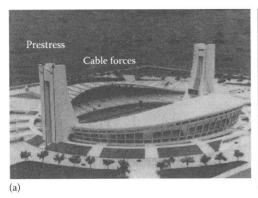
(a)

(b)

FIGURE 10.12
Zhejiang Dragon Sports Centre. (Courtesy of Professor Jida Zhao, China Academy of Building Research, China.)

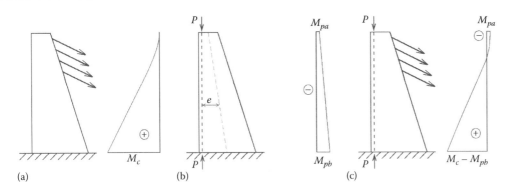
(a)   (b)   (c)

FIGURE 10.13
Illustration of the bending moments in a tower of the Zhejiang Dragon Sports Centre. (a) Bending moment due to cable forces. (b) Bending moment due to compression forces. (c) Bending moment due to cable forces and compression forces.

M_{pa} at the top to M_{pb} at the bottom of the tower, in the opposite direction to those induced by the cable forces. Finally, Figure 10.13c gives the combinations of the two sets of forces on the structure and the resulting bending moments. Thus, the bending moments due to the combined forces are $-M_{pa}$ at the top and $M_c - M_{pb}$ at the base. The reduced bending moments due to the action of the post-tensioning of the tower is obvious. According to Equation 10.2, the reduced forces will result in smaller displacements – that is, a stiffer structure.

10.4.3 Cable-Stayed Bridge

There are many long-span cable-suspended and cable-stayed bridges in the world. Cables are used for bridges not only because they are light and have high tensile strength, but also because they can create self-balanced systems in the structures and because they provide elastic supports to reduce the effective span of the decks. The latter reason is more significant.

Figure 10.14 shows a cable-stayed bridge in Lisbon. The stayed cables act as elastic supports to the bridge decks, which effectively reduce the internal bending moments in the decks, allowing the large clear spans. As the internal forces become smaller, the bridge deck becomes stiffer and can span a greater clear distance, producing a more economical design.

This behaviour of the bridge can also be considered using the concept of direct force paths. Due to the use of cables, the loads acting on the bridge decks are not transmitted to their supports primarily through bending actions. Rather, loads are transmitted mainly through the tensile forces in cables to the support tower (Figure 10.14), with the horizontal components of the cable forces induced by the self-weight of the bridge being self-balanced due to the symmetry of the structure. Vertical components of the cable forces pass directly through the tower to the foundation.

As a partially self-balanced system is created, forces are transmitted in a relatively straightforward manner from the deck to the supports, producing a stiffer and more economical design.

10.4.4 Floor Structure Experiencing Excessive Vibration

A floor in a factory on which machines were operated on a daily basis, experienced severe vibrations, causing significant discomfort for the workers. It was found that resonance occurred when the machines operated. The solution to the problem was to avoid the resonance by increasing the stiffness of the floor and hence its natural frequency.

It was not feasible to stiffen the floor by positioning additional column supports; however, the resonance problem was solved in a simple and economical manner by the Institute of Building Structures, China Academy of Building Research, Beijing, through stiffening the floor using external post-tensioned tendons, as shown in Figure 10.15.

FIGURE 10.14
A cable-stayed bridge in Lisbon.

FIGURE 10.15
A floor structure experiencing excessive vibration then stiffened using profiled post-tensioned cables. (Courtesy of Professor Jida Zhao, China Academy of Building Research, China.)

Due to the profile of the tendons and the post-tension forces applied, additional upward forces, or elastic supports, are provided at the points where the steel bars react against the concrete beams which support the floor. This reduces the internal forces (bending moments) in the beams, making the floor system stiffer with increased natural frequencies. With the natural frequencies avoiding the operating frequency of the machine, resonance did not occur and the dynamic response of the floor was significantly reduced. The static behaviour of the structure is demonstrated in Section 10.3.2.

10.4.5 Pitched Roof

Steel tendons are frequently used in pitched roofs of conservatories that have relatively large spans. Figure 10.16 shows the pitched roof of a conservatory in which a tendon (steel bar) is placed horizontally between the two sides of the roof. The tendon is supported at its centre by a vertical member from the roof ridge to avoid any sagging. The function of the tendon is to balance the horizontal force components transmitted from the pitched roof due to loading. This reduces lateral forces to the window frames that have relatively low lateral stiffness.

As discussed in Section 10.2, smaller internal forces will lead to a stiffer structure; the action of the tendon not only reduces the forces acting on the window frames but it also makes the roof structure stiffer.

FIGURE 10.16
A tendon placed in the roof of a conservatory. (Courtesy of Dr. L. Ding.)

PROBLEMS

1. Figure 10.17a shows a three-pin arch with a radius of R and Figure 10.17b shows the same arch but a tendon is added to replace the horizontal support at the right end of the arch. Calculate the internal forces of the two arches and the support reaction forces; compare the results and comment on the implication of the use of the tendon.

2. To examine the effect of the tendon in the pitched roof shown in Figure 10.16, two idealised plane models, without and with a tendon, of the conservatory are considered. Model 1, shown in Figure 10.18a, has two inclined bar members, AC and BC, representing the pitched roof, which are supported by two PVC columns AD and BE. The lower ends of the columns are fixed at their bases. Model 2, shown in Figure 10.18b, is the same as Model 1 except that a steel tendon is added to link A and B. Points A, B and C are considered as pin connections. The cross-sectional and material properties for the steel tendon and PVC columns are $A_{ST} = 100$ mm^2, $E_{ST} = 200$ kN mm^{-2}, $I_{PVC} = 1.0 \times 108$ mm^4 and $E_{PVC} = 20$ kN mm^{-2}. Answer the following questions.

   a. Sketch the likely deflected shapes of Models 1 and 2.
   b. Model 1: A 1 kN vertical load is applied at C; determine the forces acting on column AD. Calculate the maximum bending moment and maximum lateral displacement of column AD.
   c. Model 2: The same vertical forces act at C; determine the forces applied to the column AD. Calculate the maximum bending moment and maximum lateral displacement of column AD.
   d. Compare the displacements and internal forces for the two models and comment on the results in terms of structural efficiency.

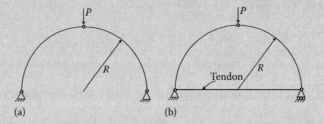

FIGURE 10.17
Two three-pin arches. (a) Without a tendon. (b) With a tendon.

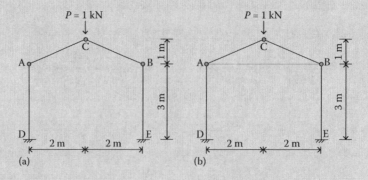

FIGURE 10.18
A pitched roof without and with a tendon. (a) Model 1. (b) Model 2.

REFERENCES

1. Hibbeler, R. C. *Mechanics of Materials*, 6th edn, Singapore: Prentice-Hall, 2005.
2. Gere, J. M. *Mechanics of Materials*, Belmont, CA: Thomson Books/Cole, 2004.
3. Ji, T. Concepts for designing stiffer structures, *The Structural Engineer*, 81, 36–42, 2003.
4. Seed, G. M. *Strength of Materials*, Edinburgh: Saxe-Coburg, 2000.
5. Bobrowski, J. Design philosophy for long spans in buildings and bridges, *Journal of Structural Engineer*, 64A, 5–12, 1986.

CHAPTER 11

CONTENTS

Buckling

11

11.1 DEFINITIONS AND CONCEPTS

Buckling of columns: When a slender structural member is loaded with an increasing axial compression force, the member deflects laterally and fails by combined bending and compression rather than by direct compression alone. This phenomenon is called *buckling*.

Critical load: The critical load of a structure is the load which creates the borderline between the stable and unstable equilibrium of the structure or it is the load that causes buckling of the structure.

Lateral torsional buckling of beams: Lateral torsional buckling is a phenomenon that occurs in beams which are subjected to vertical loading but suddenly deflect and fail in the lateral and rotational directions.

- The buckling load of a column is proportional to the flexural rigidity of the cross section EI and the inverse of the column length squared, L^2. Increasing the value of the second moment of area of the section I or reducing the length L or both, will increase the critical load.
- The buckling load of a column can be predicted through a bending test.

11.2 THEORETICAL BACKGROUND

11.2.1 Basics of Buckling

Section 2.3.2 illustrates the stable equilibrium and unstable equilibrium when using a ruler on two round pens and on one round pen, respectively. The stable and the unstable equilibrium can also be illustrated in a different manner. Figure 11.1a shows a rigid bar that has a pin support at its lower end and an elastic spring support with stiffness k at the other end. An axial compressive load P is applied at the top end of the bar. When the top of the bar is displaced horizontally by an amount Δ, there will be a disturbing moment, $P\Delta$, and a restoring moment, $kL\Delta$, about O. Thus,

$$P\Delta < kL\Delta \rightarrow \text{stable equilibrium}$$

$$P\Delta > kL\Delta \rightarrow \text{unstable equilibrium}$$

The critical condition occurs when

$$P_{cr}\Delta = kL\Delta \quad \text{or} \quad P_{cr} = kL$$

P_{cr} is termed the *critical load*, which is the borderline between stable and unstable equilibrium [1].

Consider two pin-ended columns that have the same cross section, but one is short (Figure 11.2a), that is, its length is not significantly larger than the dimensions of its cross section, and the other is long (Figure 11.2b), that is, its length is significantly larger than the dimensions of its cross section. Normally, they are termed *stocky* and *slender* struts, respectively. When the compressive loads P applied on the ends of the two columns increase, the following can be observed:

- The stocky strut will shorten *axially* by an amount Δ_a, which is proportional to the load applied. When the load reaches P_y, the product of the area of the strut and the compressive yield stress of the material, the material will deform plastically. The column 'squashes' and P_y is called the *squash load*. After removing the load, the column retains some permanent

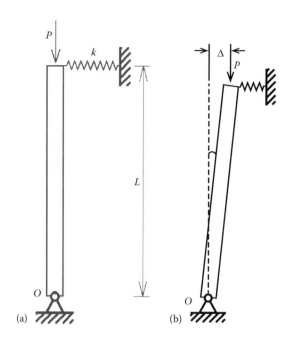

FIGURE 11.1
(a,b) Stable and unstable equilibrium.

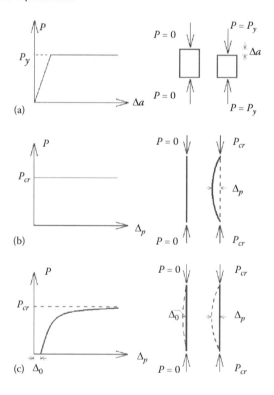

FIGURE 11.2
Comparison of the behaviour of columns. (a) Short stocky strut. (b) Long slender strut without imperfection. (c) Long slender strut with imperfection. (From Kirby, P. A. and Nethercot, D. A. *Design for Structural Stability*, London: Granada, 1979.)

deformation. The relationship between the load and the axial deformation is characterised in Figure 11.2a.

- The slender strut without imperfection will remain straight for a perfect strut when P is small. When the load reaches the critical load P_{cr}, the product of the area of the strut and the *critical* stress, the strut will suddenly move excessively in a lateral direction. This mode of failure is termed *buckling*, in which the critical stress is elastic and is often smaller or much smaller than the elastic limit stress. The relationship between the load and the lateral deformation at the mid-height of the strut is characterised in Figure 11.2b.

- In practice, a slender strut is unlikely to be perfectly straight. If an initial imperfection is considered to be defined by a lateral deformation at the mid-height of the strut with a value of Δ_0 under zero axial load, the lateral deformation will grow at an increasing rate as the load increases. At a certain load, which is normally smaller than or equal to the critical load P_{cr}, the member fails due to excessive lateral deformation. The relationship between the load and deformation is characterised in Figure 11.2c.

Clearly, the short stocky strut and the long slender strut have different modes of failure. The former depends on the yield stress of the material in compression and the area of the section but is independent of the elastic modulus, the length of the strut and the shape of the section, while the latter is independent of the yield stress but depends on the length of the strut, the elastic modulus and the shape and area of the section.

11.2.2 Buckling of a Column with Different Boundary Conditions

Figure 11.3a shows a column with pin joints at each end and it is assumed that the column is straight when unloaded. An axial compressive load is applied through the longitudinal axis of the column with increasing magnitude until the column takes up the deformed shape shown in Figure 11.3b. For this deformation state, the equilibrium equation is to be established to capture the nature of the buckling phenomenon. If Figure 11.3b rotates anticlockwise through 90°, it becomes a beam in bending subject to a pair of end forces. Thus, the theory of beam bending in Chapter 8 can be used to establish basic equations for the behaviour of the member. In order to obtain the differential equation of equilibrium, consider a free-body diagram from the beam, as shown in Figure 11.3c. At the distance y from the top joint, the displacement is u, the bending moment is $M = Pu$ and the compressive force is P. These forces and the load P at the top maintain the free-body in equilibrium. Using Equation 8.2 gives

$$EI\frac{d^2u}{dx^2} = -Pu \quad \text{or} \quad \frac{d^2u}{dx^2} + \frac{P}{EI}u = 0 \tag{11.1}$$

Solving Equation 11.1 [1,2] gives the expression of the buckling load of the simply supported column as follows:

$$P_{cr} = \frac{\pi EI}{L^2} \tag{11.2}$$

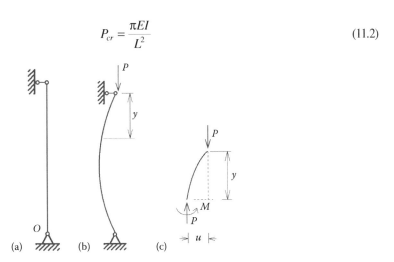

FIGURE 11.3
(a–c) Buckling of a pin-ended column.

Equation 11.2 indicates that the buckling load is proportional to the rigidity of the cross section EI and the inverse of the column length squared, L^2. Increasing the value of I or reducing the length L or both, will increase the critical load.

The buckling load of the column with different boundary conditions can be derived in a similar manner. The expressions for the buckling loads will have the same form as Equation 11.2 and can be shown in a unified formula:

$$P_{cr} = \frac{\pi^2 EI}{(\mu L)^2}$$

(11.3)

where μL is the effective length of the column considering different boundary conditions. Table 11.1 lists and compares the buckling loads of a column with four different boundary conditions.

The results in Table 11.1 show that the critical load of a slender strut with two fixed ends is four times that of a simply supported column and the critical load of a simply supported column is four times that of a cantilever. When the compressive load on a slender strut approaches its critical load, the effective stiffness of the column tends to zero.

More detailed information for the buckling of a column can be found in [1–3].

11.2.3 Lateral Torsional Buckling of Beams

The primary function of a beam carrying vertical loading is to transfer loads by means of bending action; and the beam deforms within its plane until it reaches the full plastic yield stress over the whole of its section at failure. This is only true if the beam is restrained from out-of-plane movement. If no lateral restraint is provided, the beam under load bends in the vertical plane passing through its neutral axis. Gradually increasing the load results in a situation where the beam may suddenly deflect sideways and twist before the full yield stress is reached. This mode of failure is known as *lateral torsional buckling*, in which collapse is initiated as a result of lateral deflection and twisting [3–5].

Consider a uniform, straight and elastic beam without any initial bow or twist (Figure 11.4), simply supported in both the y–z plane and the x–z plane. A pair of moments is applied at the two ends of the beam and in the y–z plane of the beam with increasing magnitude until the beam takes up the deformed shape shown in Figure 11.4. For this deformation state, the equilibrium equation is to be established to capture the nature of the lateral torsional buckling phenomenon. Figure 11.4 also shows the geometrical relationship of section A–A before and after lateral deformation, illustrating

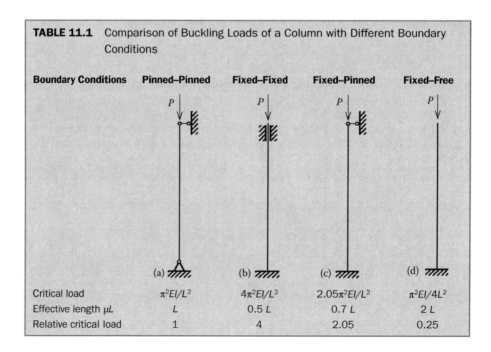

TABLE 11.1 Comparison of Buckling Loads of a Column with Different Boundary Conditions

Boundary Conditions	Pinned–Pinned	Fixed–Fixed	Fixed–Pinned	Fixed–Free
	(a)	(b)	(c)	(d)
Critical load	$\pi^2 EI/L^2$	$4\pi^2 EI/L^2$	$2.05\pi^2 EI/L^2$	$\pi^2 EI/4L^2$
Effective length μL	L	$0.5\,L$	$0.7\,L$	$2\,L$
Relative critical load	1	4	2.05	0.25

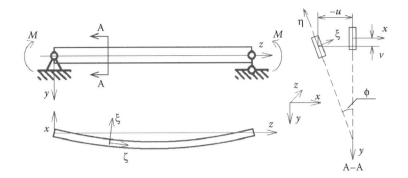

FIGURE 11.4
Lateral buckling of a rectangular beam.

that lateral buckling involves both a lateral deflection u and a twist φ about an axis parallel to the z-axis. Both deformations need to be considered to determine the critical load.

The differential equation of equilibrium is [3]

$$\frac{d^2\varphi}{dx^2} + \frac{M^2}{EI_yGJ}\varphi = 0 \tag{11.4}$$

where:
  I_y is the second moment of area of the beam in respect of the y-axis
  G and J are the shear modulus and the polar second moment of area or torsional constant of the section, which have been defined in Chapter 7

For a rectangular section, $J = hb^3/3$ in which b and h are the width and height of the beam.

Equation 11.4 has the same format as Equation 11.1. Using the same solution process or simply the analogy, the critical moment, similar to the critical load for buckling of a column, can be expressed as

$$M_{cr} = \frac{\pi\sqrt{EI_yGJ}}{L} \tag{11.5}$$

Equation 11.5 shows the coupled nature of the deformation involved in the buckled shape (Figure 11.4) due to the presence of EI_y, the lateral bending rigidity, and GJ, the torsional rigidity.

For rectangular beams with depth h and width b, $I_y = hb^3/12$ and $J = hb^3/3$, and substituting these expressions into Equation 11.5 gives

$$M_{cr} = \frac{\pi hb^3\sqrt{EG}}{6L} \tag{11.6}$$

Equation 11.6 indicates that *the buckling moment is proportional to the height of the beam, the third power of the width of the beam and the inverse of the span of the beam*. Therefore, increasing the width of the beam will be more effective than increasing the height for increasing the critical moment.

Equation 11.5 or Equation 11.6 is obtained when the beam is subjected to a constant bending moment. When a concentrated load is applied at the centroid of the free end of a cantilever, the bending moment in the vertical plane at any cross section with distance z from the fixed origin is

$$M = P(L - z) \tag{11.7}$$

Substituting Equation 11.7 into Equation 11.4 and solving the equation leads to [4]

$$M_{cr} = 1.28\frac{\pi\sqrt{EI_yGJ}}{L} = 0.67\frac{hb^3\sqrt{EG}}{L} \tag{11.8}$$

Equation 11.8 will be verified in Section 11.3.3.

EXAMPLE 11.1

Consider three brass cantilever beams, A, B and C, which have the same length of 270 mm but have the following different section sizes: 12.7 × 0.397 mm (1/2 × 1/64 in.), 12.7 × 0.794 mm (1/2 × 1/32 in.) and 25.4 × 0.397 mm (1 × 1/64 in.). The elastic modulus and the shear modulus of the material are 100 and 40 GPa, respectively [2]. Calculate the buckling moments of the three beams and the maximum concentrated loads which may be applied at the free ends of the beams.

SOLUTION

Equation 11.8 can be used directly to calculate the critical moment of cantilever beam A as follows:

$$M_{cr} = 0.67 \frac{hb^3 \sqrt{EG}}{L} = \frac{0.67(12.7)(0.397)^3 \sqrt{100,000 \times 40,000}}{270} = 125 \text{ mm}$$

The critical load at the free end of the cantilever beam is

$$P_{cr} = \frac{M_{cr}}{L} = \frac{125}{270} = 0.462 \text{ N}$$

Beam B has twice the thickness of Beam A and Beam C has twice the height of Beam A; the critical loads for Beams B and C can be determined from Equation 11.8 and are 3.70 and 0.924 N, respectively. The three beams used in this example will be tested in Section 11.3.3 to examine the predictions from Equation 11.8.

When an I-section beam is subjected to torsion, axial deformation of the flange will occur. This type of deformation is called *warping* and the applied torque will be resisted by the shear stresses associated with pure torsion and the axial stresses associated with warping. In other words, considering the effect of warping, the member will have a larger capacity to resist the external torque or will have a larger buckling moment.

When the beam with the narrow rectangular section studied above is replaced by an I-section, the effect of warping needs to be considered and Equation 11.5 is extended to [3]

$$M_{cr} = \frac{\pi \sqrt{EI_y GJ}}{L} \sqrt{1 + \frac{\pi^2 EI_w}{L^2 GJ}} \tag{11.9}$$

where $I_w = I_y h^2/4$ is the warping constant.

If $I_w = 0$, Equation 11.9 reduces to Equation 11.5 and considering the effect of warping increases the buckling moment.

11.2.4 Relationship between Maximum Displacement and Buckling Load of a Straight Member

Consider a simply supported uniform beam, with a span of L and cross-sectional rigidity of EI, subjected to a concentrated load F at the centre of the beam shown in Figure 11.5a. The maximum deflection of the beam is:

$$\Delta_{\max} = \frac{FL^3}{48EI} \tag{11.10}$$

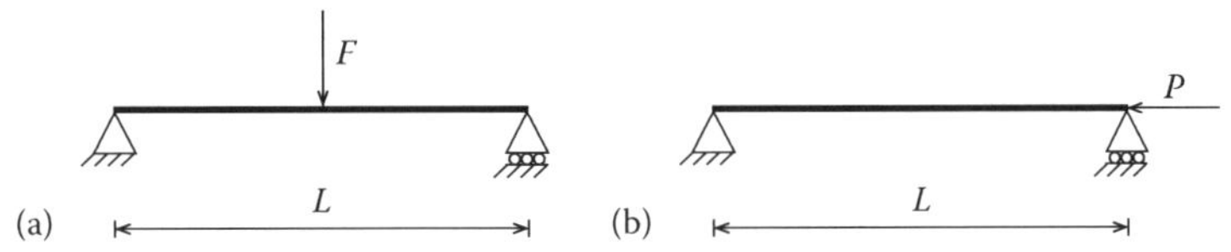

FIGURE 11.5
A simply supported beam subjected to different forces. (a) A simply supported beam subjected to a concentrated load at its centre. (b) A simply supported column subjected to a compressive force.

Now remove the vertical load of F from the beam and apply a compressive force P to the right-hand end of the beam along the longitudinal axis of the beam, as shown in Figure 11.5b. The critical load of the member is

$$P_{cr} = \frac{\pi^2 EI}{L^2} \tag{11.11}$$

It can be noted that Equations 11.10 and 11.11 contain the same term EI, through which a relationship between the critical load P_{cr} and the maximum displacement Δ_{max} can be established:

$$P_{cr} = \frac{FL\pi^2}{48\Delta_{max}} \tag{11.12}$$

Equation 11.12 indicates that *the buckling load of a strut can be predicted by a bending test of the same member through measuring the maximum deflection induced by a concentrated transverse load at the centre of the beam.* Equation 11.12 forms a basis for nondestructive testing to determine the buckling load of a strut. An experimental verification of Equation 11.12 will be given in Section 11.3.5.

EXAMPLE 11.2

A 700 mm long, steel member has a cross section of 25.3 × 3.11 mm. The member has pinned supports at its two ends and is subjected to a concentrated transverse force 2.32 N at its centre. Calculate the maximum displacement of the member using Equation 11.10 and the buckling load of the member using Equation 11.12. Take $E = 210 \times 10^3$ N mm^{-2}.

SOLUTION
The second moment of area of the cross section of the member:

$$I = \frac{bh^3}{12} = \frac{25.3 \times 3.11^3}{12} = 63.42 \text{ mm}^3$$

The maximum displacement, using Equation 11.10, is

$$\Delta_{max} = \frac{FL^3}{48EI} = \frac{2.32 \times 700^3}{48 \times 210 \times 10^3 \times 63.42} = 1.243 \text{ mm}$$

The buckling load, using Equation 11.12, is

$$P_{cr} = \frac{FL\pi^2}{48\Delta_{max}} = \frac{0.2362 \times 9.81 \times 700\pi^2}{48 \times 1.243} = 268.3 \text{ N}$$

This member will be tested to examine a possible nondestructive test method for predicting the buckling load in Section 11.3.5.

11.3 MODEL DEMONSTRATIONS

11.3.1 Buckling Shapes of Plastic Columns

These models demonstrate *the phenomenon of buckling and the buckling shapes of a column with different boundary conditions. Users can feel the magnitudes of the buckling forces of the plastic columns.* Column buckling behaviour can be demonstrated using slender members such as a thin plastic ruler:

1. Put one end of the ruler on the surface of a table and the other end in the palm of one's hand. Press axially on the top end of the ruler and gradually increase the compression force. The straight ruler will suddenly deflect laterally, as shown in Figure 11.6a. The deformation becomes larger with further application of the compressive force. This simulates the buckling of a column with two pinned ends.
2. Now hold the two ends of the ruler tightly to prevent any rotational and lateral movements of the two ends of the ruler. Then, gradually press axially on the ruler using the fingers until

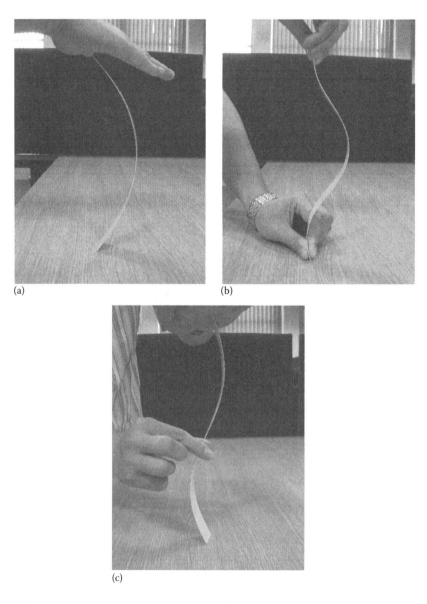

(a)   (b)

(c)

FIGURE 11.6
Buckling shapes and boundary conditions of a plastic ruler. (a) Two pinned ends. (b) Two fixed ends. (c) Two pinned ends with a lateral support at the middle of the rule.

the ruler deforms sideways, as shown in Figure 11.6b. This demonstrates a different buckling shape for a column with two ends fixed. One can clearly feel that a larger force is needed in this demonstration than in the previous demonstration.

3. If one intermediate lateral support is provided, so that the ruler cannot move sideways at this point, a larger compressive force will be required to make the ruler buckle in the shape shown in Figure 11.6c.

11.3.2 Buckling Loads and Boundary Conditions

This model demonstrates qualitatively and quantitatively *the relationships between the critical loads and boundary conditions of a column (Equation 11.3) and the corresponding buckling shapes.*

The model shown in Figure 11.6 contains four columns with the same cross section but each with different boundary conditions: (from left to right and from bottom to top) pinned–pinned, fixed–fixed, pinned–fixed and free–fixed. The model has load platens on the tops of the columns which allow weights to be added and applied to the columns. This model can demonstrate the different buckling or critical loads of the columns and the associated buckling shapes or modes.

1. The different boundary conditions should be observed by the user before applying loads to the columns (Figure 11.7a).
2. Weights are added incrementally to the platens until the columns buckle (Figure 11.7b), at which time the loads, including the weights and the platens, are the critical loads of the columns and the shapes of the columns are the buckling modes.
3. With all columns loaded (Figure 11.7b), the relative buckling loads and buckling modes can be observed and compared for the different boundary conditions. Table 11.2 gives the measured critical loads of the four columns.

It is to be expected that there will be some differences between the predicted and the test critical loads, but significant error occurs in the fixed–free model (Table 11.2). Several trials showed that the free end at the bottom of the strut was not really free to move laterally due to a small amount of

(a)   (b)

FIGURE 11.7
Critical loads, buckling shapes and boundary conditions of columns. (a) Four columns with different boundary conditions. (b) Critical loads and buckling shapes.

TABLE 11.2 Comparison of Buckling Loads of a Column with Different Boundary Conditions

Boundary Conditions	Pinned–Pinned	Fixed–Fixed	Fixed–Pinned	Fixed–Free
Predicted critical load	$\pi^2 EI/L^2$	$4\pi^2 EI/L^2$	$2.05\pi^2 EI/L^2$	$\pi^2 EI/4L^2$
Relative critical load	1	4	2.05	0.25
Test critical load (N)	5.61	20.8	13.0	4.05
Relative test critical load	1	3.71	2.31	0.72

friction. More importantly, the column is shorter than the others, which can be seen in Figure 11.6b. Real structures rarely behave in exactly the same manner as theoretical models.

11.3.3 Lateral Buckling of Beams

This set of models demonstrates *the behaviour of lateral buckling of a narrow rectangular beam with different sizes of section and thus the validity of Equation 11.8.*

The buckling behaviour of the three brass cantilevers described in Example 11.1 is demonstrated through simple tests. One end of each of the brass strips is fixed to a wooden block through screws, creating cantilevers, as shown in Figure 11.8.

Model A: 270 mm long cantilever with a 12.7 × 0.397 mm rectangular cross section: Hold one end of the cantilever (Figure 11.8a) firmly and apply a vertical concentrated load at the free end. Increase the load gradually until the beam moves sideways and twists (when the loading is about 0.49 N). This type of deformation typifies the form of instability called *lateral torsional buckling.*

Model B: 270 mm long cantilever with a 12.7 × 0.397 mm rectangular cross section: The effect of increasing the height of the section: Repeat the type of test carried out for Model A. When the concentrated load reaches approximately 0.98 N, the cantilever starts to move sideways and twist (Figure 11.8b). This value is twice that of the critical load for Model A. As predicted by Equation 11.8, doubling the height of the narrow beam doubles the buckling moment or the critical load.

Model C: 270 mm long cantilever with a 12.7 × 0.397 mm rectangular cross section: The effect of increasing the width of the section: Repeat the type of test carried out for Models A and B. When the concentrated load reaches approximately 3.82 N, the cantilever starts to move sideways and twists (Figure 11.8c). The load is now nearly eight times that which caused lateral torsional buckling of Model A. It is about four times that which caused lateral torsional buckling of Model B, although the cantilever uses the same amount of material as Model B. Following Equation 11.8, the buckling load of Model C should be exactly eight times that of Model A and four times that of Model B.

Model D: Model A with a lateral support: The effect of lateral restraint: Repeat the type of test carried out for Model A, but this time hold the cantilever at midspan using a finger and thumb, to prevent lateral movement. The loading can increase significantly without lateral buckling. This indicates that the lateral restraint effectively increases the lateral torsional buckling capacity of the cantilever.

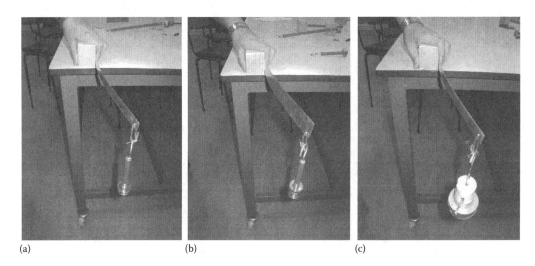

(a)   (b)   (c)

FIGURE 11.8
Lateral torsional buckling behaviour. (a) Model A. (b) Model B. (c) Model C.

TABLE 11.3	Comparison between the Calculated and Measured Lateral Buckling Loads		
Boundary Conditions	**Model A**	**Model B**	**Model C**
Theoretical critical load (N)	0.462	0.924	3.7
Relative theoretical critical load	1	2	8
Test critical load (N)	0.49	0.98	3.83
Relative test critical load	1	2	7.8

Table 11.3 summarises the theoretical values obtained from Equation 11.8 and the experimental values of the lateral buckling loads. There is close agreement between the two sets of values.

11.3.4 Buckling of an Empty Aluminium Can

This demonstration shows that *local buckling that occurred on a drink can causes the global collapse of the can*.

The phenomenon of buckling is not limited to columns. Buckling can occur in many kinds of structures and can take many forms. Figure 11.9a shows that two empty drink cans can carry a standing person of 65 kg. However, this is close to the critical loads of the two cans. When the standing person slightly moved his body, which caused a small shift of some of the body weight from one can to the other, the thin cylindrical wall of one can buckled and the cans then collapsed completely, as shown in Figure 11.9b.

11.3.5 Buckling Load of a Straight Member Predicted through a Bending Test

This experiment demonstrates that *the buckling load of a strut can be predicted through a nondestructive bending test of the strut*.

A 700 mm long, 25.3 mm wide and 3.11 mm thick steel member is used for bending tests. It is expected that the buckling load of the member can be predicted using Equation 11.12. A bending

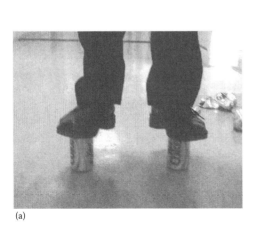
(a)

(b)

FIGURE 11.9
(a,b) Buckling failure of empty drink cans.

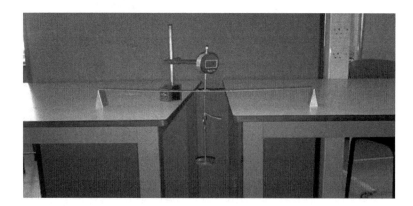

FIGURE 11.10
Bending test.

TABLE 11.4   Deflection and the Predicted Buckling Load for Experiment 1

Weight (kg)	Measured Deflection (mm)	Predicted Buckling Load Using Equation 11.12 (N)
0	0	—
0.1	0.56	252.1
0.1454	0.80	256.6
0.1908	1.05	256.6
0.2362	1.27	262.5

test of the member with pinned supports at its two ends was conducted as shown in Figure 11.10. A concentrated load was applied at the centre of the beam and a digital gauge was used to measure the displacement at the centre of the beam due to loading. Four measurements were taken with gradually increased load. The loads, measured displacements and predicted buckling loads using Equation 11.12 are listed in Table 11.4.

A buckling test was also conducted. Figure 11.11 shows the member placed in a buckling test rig with pinned supports at the two ends of the member. Weights were gradually added on

FIGURE 11.11
Buckling test.

the top platform until the member buckled. The buckling load was 263.2 N which can be compared with the theoretical prediction of 268.3 N (Example 11.2) and the bending test prediction of 262.5 N.

11.4 PRACTICAL EXAMPLES

11.4.1 Buckling of Bracing Members

Steel bracing members are normally slender and are not ideal for use in compression. Therefore, they are usually designed to transmit tension forces only. For cross-braced panels, which are often used to resist lateral loads, one bracing member will be in tension and the other in compression. When the loading is applied in the opposite direction, the member previously in tension will be subjected to compression and the other member becomes a tension member.

Figure 11.12a shows one of two bracing members in a panel of a storage rack which has buckled. Comparing the value of the material stored and the cost of the bracing members, the use of more substantial bracing members would have easily been justified. Figure 11.12b shows two buckled bracing members in a building.

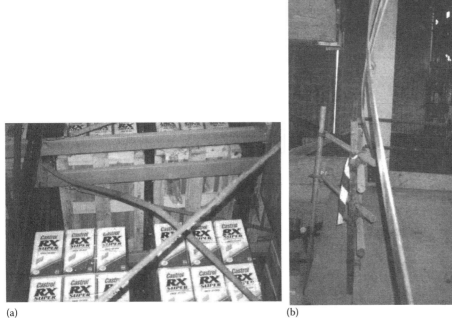

(a)　　　　　　　　　　　　　　　　　　(b)

FIGURE 11.12
Buckling of bracing members. (a) Buckling of a bracing member in a storage rack. (Courtesy of Dr. A. Mann, Jacobs.) (b) Buckling of two bracing members in a building.

FIGURE 11.13
Buckling of a box girder. (Courtesy of Dr. A. Mann, Jacobs.)

11.4.2 Buckling of a Box Girder

Figure 11.13 shows a box girder that buckled during an earthquake. The girder was subjected to large compressive forces and the material was also squashed.

11.4.3 Prevention of Lateral Buckling of Beams

Figure 11.14 shows a system where additional members are provided to prevent the lateral and torsional buckling of cantilever beams through reducing the beam length. Cellular beams are used as supporting structures for the cantilever roof of a grandstand. The lower parts of the beams are subjected to compressive forces, but the lateral supports from the roof cladding on the top of the beams may not be effective at preventing lateral and torsional buckling at the bottom parts of the beams; thus, additional members are placed perpendicular to the beams.

Figure 11.15 shows the lateral supports provided to two arch bridges. The loads on the bridges are transmitted primarily through compression forces in arches to the supports of the arches. Thus, lateral supports are provided to prevent any possible lateral buckling.

FIGURE 11.14
Additional supports provided to prevent lateral torsional buckling. (Courtesy of Westok Ltd.)

(a)   (b)

FIGURE 11.15
(a,b) Members are provided to prevent the lateral buckling of arches.

11.4.4 Bi-Stability of a Slap Band

Bi-stability is a fundamental phenomenon in nature. A slap band that is *bi-stable* can rest load in two equilibrium states. By default, the band will be in the state with the minimum energy, as shown on the right side in Figure 11.16. A transition from the state of minimal energy to the other state of equilibrium, that has the maximum energy, requires some form of activation energy. If one stretches the band into a straight position, it can remain as shown on the left side in Figure 11.16. If a small disturbance is introduced to this state of maximum energy, the band will relax into the state of the lowest energy again, that is, it will roll up.

Reflective slap bands, as shown in Figure 11.16, have high visibility and have become a safety product, which can be wrapped round wrists, arms and legs. Due to the interesting nature of bi-stability and their low cost, slap bands are also used as promotion products.

FIGURE 11.16
A slap band in its two equilibrium states with the largest and smallest stored energies.

PROBLEMS

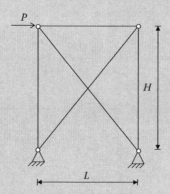

FIGURE 11.17
A braced frame.

1. Use a slender plastic ruler to demonstrate the buckling phenomena shown in Figure 11.5.
2. A simply supported beam is subjected to a uniformly distributed load. Qualitatively, what is the relationship between the lateral buckling load q and the span L of the beam?
3. Figure 11.17 shows a braced frame with a height of H and a width of L in which all joints are considered as pin connected. If the two bracing members have an area A, a second moment area I and a modulus of elasticity E, estimate the buckling load of the frame, P_{cr}.

REFERENCES

1. Benham, P. P., Crawford, R. J. and Armstrong, C. G. *Mechanics of Engineering Materials*, Harlow: Addison Wesley Longman, 1998.
2. Gere, J. M. *Mechanics of Materials*, 6th edn, Belmont, CA: Thomson Books/Cole, 2004.
3. Kirby, P. A. and Nethercot, D. A. *Design for Structural Stability*, London: Granada, 1979.
4. Allen, H. G. and Bulson, P. S. *Background of Buckling*, New York: McGraw-Hill, 1980.
5. Timoshenko, S. P. and Gere, J. M. *Theory of Elastic Stability*, New York: McGraw-Hill, 1961.

CHAPTER 12

CONTENTS

Prestress

12

12.1 DEFINITIONS AND CONCEPTS

Prestressing is a technique that generates stresses in structural elements before they are loaded. This can be used to reduce particular unwanted stresses or displacements or both, which would develop due to external loads, or to generate particular shapes of tension structures.

Prestressing techniques can be used to achieve:

- A redistribution of internal forces: A reduction of the maximum internal forces will permit the design of lighter structures.
- Avoidance of cracks: Keeping a member in constant compression means that there will be no cracks.
- Stiffening a structure or a structural element.

12.2 THEORETICAL BACKGROUND

Barrels are made from separated wooden staves, which are kept in place by metal bands/hoops, as shown in Figure 12.1a. The metal bands are slightly smaller in diameter than the diameter of the barrel, and are forced into place over the staves. This forces the staves together forming a watertight barrel, that is, a structure. From free-body diagrams of a typical stave and half the metal band, as shown in Figure 12.1b and c, it can be seen that circular prestressing (compression) forces are applied along the vertical sides of the stave and the tensile force F on half of the metal band due to internal pressure, and this is balanced by the circular hoop pressure. When liquid is added to the barrel, the liquid applies tension forces along the sides of each stave, but this force is, and must be, smaller than the pre-added compressive forces, otherwise the barrel will leak.

Pretensioning or post-tensioning or both, is widely used in concrete elements, creating prestressed concrete. Prestressed concrete may be considered as essentially a concrete structure with the tendons supplying the prestress to the concrete, or as steel and concrete acting together, with the steel taking the tension and the concrete taking the compression, so that the two materials form a resisting couple against the external actions [2,3]. Here, the focus is on the first aspect.

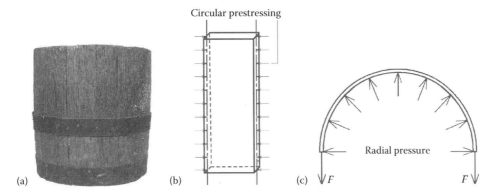

FIGURE 12.1
A barrel and its internal forces. (a) A barrel (From Hemera Technologies. Photo-Objects 50,000 Volume 2, Canada, 2001.) (b) One wooden stave. (c) A half of a circular hoop.

12.2.1 Concentrically Prestressed Beams

Consider a simply supported beam prestressed by a tendon through its neutral axis and loaded by external loads, as shown in Figure 12.2a. Due to the pretension force F, a uniform compressive stress, σ_c, occurs across the section of an area A and this is

$$\sigma_c = \frac{F}{A} \tag{12.1}$$

The stress distribution is shown in Figure 12.2c. If M is the maximum moment at the centre of the beam induced by the external load, the normal stress at any point y across the section is

$$\sigma_b = \frac{My}{I} \tag{12.2}$$

where:

    y is the distance from the neutral axis
    I is the second moment of area of the section about its neutral axis

The stress distribution defined in Equation 12.2 is illustrated in Figure 12.2d. Thus, the resulting normal stress distribution on the section is:

$$\sigma = \frac{F}{A} + \frac{My}{I} \tag{12.3}$$

which is shown in Figure 12.2e. If there is no tensile stress in any section for the given prestress and load conditions, a beam comprising separate blocks and a tendon shown in Figure 12.2b is similar to the beam in Figure 12.2a. The prestress provides compressive stress on the sections of the beam which removes or reduces the tensile stress induced by external loads.

12.2.2 Eccentrically Prestressed Beams

If the location of the tendon in Figure 12.2a is placed eccentrically, with respect to the neutral axis of the beam, by a distance of e, as shown in Figure 12.3a, an additional moment will be induced due to the eccentricity e. The resulting normal stress is given as follows:

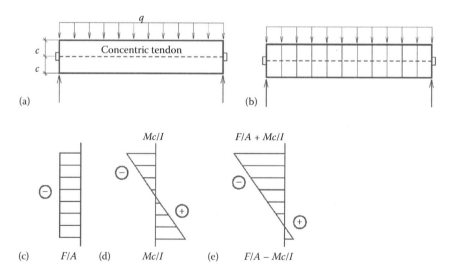

FIGURE 12.2
(a–e) A centrally prestressed beam.

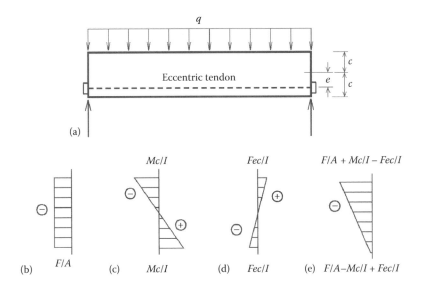

FIGURE 12.3
(a–e) An eccentrically prestressed beam.

$$\sigma_e = \frac{Fey}{I} \tag{12.4}$$

The normal stress distribution across a typical section is illustrated in Figure 12.3d. From Figure 12.3c and d, it can be seen that σ_e is in the opposite direction to that induced by the load (Equation 12.2), making the stress distribution across the section more even.

12.2.3 Externally Prestressed Beams

Prestressing tendons can also be used externally, and the tendons can be bent or curved. Figure 12.4a and b shows a similar beam with two tendons with two bends placed externally and symmetrically. If a free-body diagram is drawn for the beam and the tendons, it can be seen that the tendons provide not only the direct compressive forces but also a pair of upward forces (Figure 12.4c), which can effectively balance part of the external load. To simplify the discussion, it is assumed that there is no friction loss along the tendon due to the sharp bend and that the deviation produced by the bends is small in comparison with the length of the beam [2,3]. The upward force P is equal to the

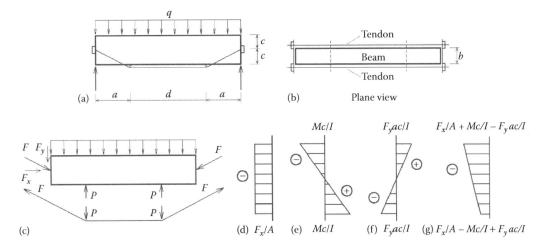

FIGURE 12.4
(a–g) An externally prestressed beam.

component of the pretension force in the vertical direction, F_y. The resulting normal stress, σ_p, due to the pair of vertical upward forces on the beam, is

$$\sigma_p = \frac{F_y a y}{I} \qquad (12.5)$$

The normal stresses induced by the compressive force F_x, the moment M due to the external load, the upward forces F_y and their combination are shown qualitatively in Figure 12.4d through g.

EXAMPLE 12.1

A simply supported prestressed concrete rectangular beam with a span of 8 m, a width of $b=0.3$ m and a height of $2c=0.6$ m, is subjected to a uniform vertical loading of $q=20$ kN m^{-1}. Three ways of using prestressing are considered as follows:

- Case A: A prestressing tendon is placed at the neutral axis of the beam with a force of 200 kN, as shown in Figure 12.2a.
- Case B: A prestressing tendon is placed at $e=0.2$ m below the neutral axis with a force of 200 kN, as shown in Figure 12.3a.
- Case C: A pair of prestressing tendons are placed with bends at $a=1.5$ m and $c=0.3$ m, as shown in Figure 12.4a. The total force in the two tendons is 200 kN.

Calculate the maximum and minimum stresses in the central section of the beam for the three designs.

SOLUTION

The horizontal and vertical components of the tendon force applied at the ends of the beam for case C are, respectively (Figure 12.4b):

$$F_x = 200 \times \frac{1.5}{\sqrt{1.5^2 + 0.3^2}} = 196 \text{ kN}$$

$$F_y = P = 200 \times \frac{0.3}{\sqrt{1.5^2 + 0.3^2}} = 39.2 \text{ kN}$$

The moments induced by the uniform load, the eccentricity and the upward forces are thus:

$$M = \frac{qL^2}{8} = \frac{20 \times 8^2}{8} = 160 \text{ kNm}$$

$$M_e = F_e = 200 \times 0.2 = 40 \text{ kNm}$$

$$M_p = F_y a = 39.2 \times 1.5 = 58.8 \text{ kNm}$$

Therefore, the maximum normal stresses induced by F (and F_x), M, M_e and M_p at the centre of the beam are:

$$\sigma_c = \frac{F}{A} = \frac{200,000}{300 \times 600} = 1.11 \text{ N mm}^{-2}$$

$$\sigma_a = \frac{F_x}{A} = \frac{196,000}{300 \times 600} = 1.09 \text{ N mm}^{-2}$$

$$\sigma_q = \frac{Mc}{bh^3/12} = \frac{160 \times 0.3 \times 10^9 \times 12}{0.3 \times 0.6^3 \times 10^{12}} = 8.89\,\text{N mm}^{-2}$$

$$\sigma_e = \frac{M_e c}{bh^3/12} = \frac{40 \times 0.3 \times 10^9 \times 12}{0.3 \times 0.6^3 \times 10^{12}} = 2.22\,\text{N mm}^{-2}$$

$$\sigma_p = \frac{M_p c}{bh^3/12} = \frac{58.8 \times 0.3 \times 10^9 \times 12}{0.3 \times 0.6^3 \times 10^{12}} = 3.27\,\text{N mm}^{-2}$$

The resulting stresses at the bottom and top fibres in the section at the centre of the beam are:

Case A:
   Bottom: $-\sigma_{c1} + \sigma_q = -1.11 + 8.89 = 7.78$ N mm^{-2}
   Top: $-\sigma_{c1} - \sigma_q = -1.11 - 8.89 = -10.0$ N mm^{-2}

Case B:
   Bottom: $-\sigma_c + \sigma_q - \sigma_e = -1.11 + 8.89 - 2.22 = 5.56$ N mm^{-2}
   Top: $-\sigma_c - \sigma_q + \sigma_e = -1.11 - 8.89 + 2.22 = -7.78$ N mm^{-2}

Case C:
   Bottom: $-\sigma_{ct} + \sigma_q - \sigma_p = -1.09 + 8.89 - 3.27 = 4.53$ N mm^{-2}
   Top: $-\sigma_{ct} - \sigma_q + \sigma_p = -1.09 - 8.89 + 3.27 = -6.71$ N mm^{-2}

The results show that the eccentrically placed tendon is more effective than the centrally placed tendon in reducing the stress levels; and the externally placed profiled tendons may be even more effective. The other benefit of using the externally placed profiled tendons is to increase the stiffness of the beam. A practical example is given in Section 10.4.4.

Prestressing can also be used to create tension structures in which the internal forces of the structures are tensile; for example, fabric membranes, prestressed cable nets and cable beams in the form of trusses or girders.

High-tensile steel cables can transmit large axial forces in tension. The ever-increasing spans and elegance of modern suspension and cable-stayed bridges and cable roofing structures are the most obvious examples in which large loads are supported and transmitted by members and cables in tension. The Raleigh Arena in North Carolina, mentioned in Section 10.4.1, is well known for its cable roof structure supported by a pair of inward-inclined arches [4,5].

12.3 MODEL DEMONSTRATIONS

12.3.1 Prestressed Wooden Blocks Forming a Beam and a Column

This model demonstrates *the effect of prestressing which makes separate wooden blocks act as a beam or column.*

Figure 12.5a shows a number of separated wooden cubes which are linked using a metal wire through small holes at their centres. One end of the wire passes over a support post while the other end is constrained horizontally by a metal post preventing the cubes from falling down. The structure formed cannot support even its own weight and the cubes simply hang on the wire. If weights are attached to the right-hand end of the wire, the wire is tightened. The metal post and the

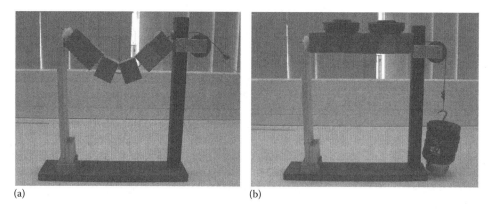

(a)   (b)

FIGURE 12.5
(a,b) Effect of prestressing (1).

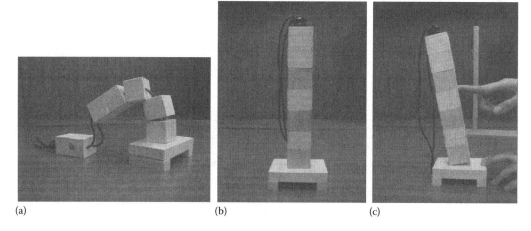

(a)   (b)   (c)

FIGURE 12.6
(a–c) Effect of prestressing (2).

increased compressive prestress between the cubes enable them to form a structure that can now carry an external load, as shown in Figure 12.5b.

A loose elastic string, with one end fixed to a base, passes through the central holes of a pile of wooden blocks as shown in Figure 12.6. The effect of prestress can be demonstrated as follows:

1. Push the column from one side; it will topple, as shown in Figure 12.6a.
2. Reform the column and tighten the elastic string anchoring it to the top block, as shown in Figure 12.6b.
3. Hold the base of the model and again push one side of the column. This time the blocks act as a single member which cannot be easily toppled, as shown in Figure 12.6c.

12.3.2 Toy Using Prestressing

This model demonstrates *the effect of tension in strings which makes a toy stable and upright.*

A popular toy (Figure 12.7a) makes use of prestressing. The mechanism for the prestressing is illustrated in Figure 12.7b and c. Strings are threaded through the legs of the toy and attached to the base. A spring is used to push the base down and create tension in the strings to hold the toy upright and stable (Figure 12.7b).

When the base is pushed upwards, the tension in the string is released. The sections of the legs are no longer held in place and become unstable and the toy collapses (Figure 12.7c). When the force is released, the spring makes the base move down, the strings become taught and the toy straightens and becomes stable.

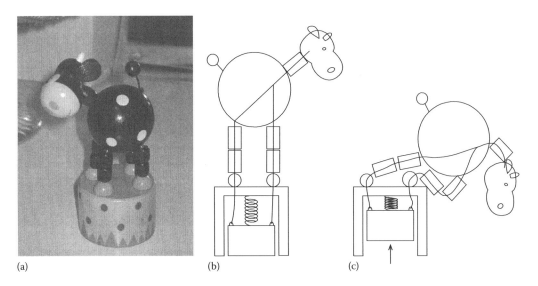

FIGURE 12.7
(a–c) Prestressing used in a toy. (Courtesy of Miss G. Christian.)

12.4 PRACTICAL EXAMPLES

12.4.1 Centrally Post-Tensioned Column

Figure 12.8 shows three piles of stones which form stone columns in a park. This landmark is similar to the wooden block column model shown in Figure 12.6. A steel bar, with one end anchored into a foundation below the stones, is threaded through premade central holes in the stones. The bar is then tensioned to make the stone blocks act together similar to a single member.

FIGURE 12.8
Centrally post-tensioned stone columns.

FIGURE 12.9
An eccentrically post-tensioned beam.

12.4.2 Eccentrically Post-Tensioned Beam

In order to reduce the maximum sagging bending moment of a beam and its maximum deflection when loaded, prestressing can be used to produce initial hogging bending moments and upward deflections, which will offset parts of the bending moments and deflections induced by the subsequent downward vertical loading.

To produce a hogging bending moment and upward deflection of a beam, the post-tensioned steel bar needs to be placed below the neutral axis of the cross section of the beam. The position of the tendon should be as low as possible to produce larger hogging moments and deflections or to allow smaller post-tension forces.

Figure 12.9 shows a composite beam that has a span of over 10 m [7]. The cross section of the beam is T-shaped and the post-tensioning tendon is placed at a position much lower than the neutral axis of the section.

12.4.3 Spider's Web

Spiders have been able to produce silk for the last 400 million years and have been building orb webs for the last 180 million years. The webs have evolved to arrest the flight and capture fast-moving and relatively large insects [8,9]. Figure 12.10 shows two spiders' webs that consist of radial threads, capture spirals and anchor threads.

The anchor threads span from the supports and suspend the radial threads. The radial threads are then overlaid by the sticky capture spirals. The prestressing is applied during construction by simply pulling the individual strands tighter around their supports.

12.4.4 Cable-Net Roof

Cable-net roofs are ideally suited for use with long-span structures such as gymnasia. Figure 12.11a shows half of the cable-net roof used for the Sichuan Provincial Gymnasium, which was built in 1988 and is able to seat over 10,000 spectators [10].

FIGURE 12.10
Spiders' webs. (Courtesy of Dr. A. S. K. Kwan, Cardiff University.)

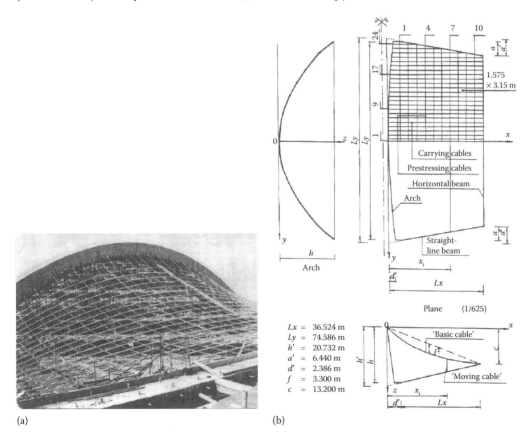

Lx = 36.524 m
Ly = 74.586 m
h' = 20.732 m
a' = 6.440 m
d' = 2.386 m
f = 3.300 m
c = 13.200 m

FIGURE 12.11
(a,b) Cable-net roof of the Sichuan Provincial Gymnasium.

The stiffness of the cable-net roof was established by prestressing. By applying forces at the ends of prestressing (hogging) cables, tension forces are induced in both carrying (sagging) cables and prestressing cables. The prestressing and carrying cables are normally placed perpendicularly to each other (Figure 12.11b). The level of prestress applied should ensure that no prestressing cables are slack for any load configuration.

PROBLEMS

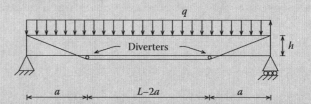

FIGURE 12.12
A beam stiffened by a profiled cable tendon.

1. To simulate the behaviour of concrete and reinforced concrete beams, make two ice beams using your home freezer, one without and one with string or similar, frozen into the lower part of the beam. Test the load capacity of the two beams.
2. The investigation of the problem presented in Section 10.4.4 (Figure 10.15) can be simplified as a study of a beam stiffened by a profiled cable tendon shown in Figure 12.12. The reinforced concrete beam has a span of L, a height of h and a rigidity of $E_c I_c$ and the tendon has an area of A and a modulus of elasticity of E_s. The beam is subjected to a uniformly distributed load q. For a qualitative analysis, it is assumed that the cable has a constant force of $0.1\,qL$ and the deformation compatibility condition between the beam and the cable is ignored.
   a. Draw the bending moment diagrams of the beam with and without the cable tendon using the free-body diagram shown in Figure 12.4c.
   b. Calculate the maximum displacements of the beam with and without the cable tendon when $L=10$ m, $h=0.5$ m and $a=1.2$ m.
   c. Determine the optimum location of the diverters, a, that minimises the deflection of the beam with the cable tendon when $L=10$ m and $h=0.5$ m.
   d. Discuss the findings from the above analysis (b and c).

REFERENCES

1. Hemera Technologies. Photo-Objects 50,000 Volume 2, Canada, 2001.
2. Lin, T. Y. *Design of Prestressed Concrete Structures*, New York: John Wiley, 1955.
3. Nawy, E. G. *Prestressed Concrete, A Fundamental Approach*, 2nd edn, Upper Saddle River, NJ: Prentice-Hall, 1995.
4. Buchholdt, H. A. *Introduction to Cable Roof Structures*, Cambridge: Cambridge University Press, 1985.
5. Lewis, W. J. *Tension Structures: Form and Behaviour*, London: Thomas Telford, 2003.
6. Ji, T. and Bell, A. J. *Enhancing the Understanding of Structural Concepts—A Collection of Students' Coursework*, Manchester: The University of Manchester, 2007.
7. Bailey, C. G., Currie, P. M. and Miller, F. R. Development of a new long span composite floor system, *The Structural Engineer*, 84, 32–38, 2006.
8. Lin, L. H., Edmonds, D. T. and Vollrath, F. Structural engineering of an orbspider's web, *Nature*, 373, 146–168, 1995.
9. Ji, T. *Cable Structures—A Collection of Students' Coursework*, Manchester: UMIST, 2001.
10. Ji, T. Design and analysis of orthogonal cable-net roofs with complex shapes, *Proceedings of International Symposium on Membrane Structures and Space Frames*, Vol. 2, Osaka, Japan, 1986.

CHAPTER 13

CONTENTS

Horizontal Movements of Structures Induced by Vertical Loads

13

13.1 CONCEPTS

- Vertical loads acting on structures can induce horizontal movements. Exceptions are symmetric structures subject to symmetric vertical loads and (rarely) antisymmetric structures subject to antisymmetric loads.
- The magnitudes of the horizontal movements of structures in response to vertical loads depend on the load distribution and the structural geometry.
- When the frequency of a dynamic vertical load is close to one of the lateral natural frequencies of a structure, resonance in the lateral direction of the structure can occur.

13.2 THEORETICAL BACKGROUND

When a structure moves horizontally, it is usually considered that this is in response to horizontal loads. However, vertical loads can also induce horizontal movements. This is because structures are three-dimensional and movements in the orthogonal directions are often coupled. For some structures, such horizontal movements can be a significant design consideration, especially when dynamic response is important.

Horizontal movements may result from the following:

- Horizontal loading, for example wind loading which will generate translational movement of tall buildings.
- Loading that, although primarily vertical, has a horizontal component, for example walking. The Millennium Footbridge in London is a structure where significant horizontal movements were induced by people walking [1].
- Vertical loading acting on asymmetric structures. Due to the structural geometry, vertical loads can induce both vertical and horizontal movements, that is, vertical motion is coupled with the horizontal response. An example is a simple frame that has two columns with different lengths subject to vertical loading; this will be studied in Section 13.2.1.3.
- Vertical loading acting asymmetrically on structures. Due to their location, vertical loads can induce both vertical and horizontal movements. An example is that of a train crossing a bridge and producing horizontal movements orthogonal to the rails; this will be discussed in Section 13.4.3.

This chapter considers the last two situations where vertical loading can generate horizontal movements of frame structures [2].

13.2.1 Static Response

13.2.1.1 Symmetric system

Consider a simple symmetric frame with no horizontal forces, which is subjected to a concentrated vertical load as shown in Figure 13.1a. The beam has a length L and a rigidity EI_b and the two columns have the same length h and a rigidity EI_c.

If the axial deformations of the columns and the beam can be considered to be negligible, the structure has three degrees of freedom: the horizontal displacement, u, and the rotations θ_A and θ_B at the connections of the beam and columns. Thus, the equations of static equilibrium of the frame are given by

$$\frac{EI_c}{h^3}\begin{bmatrix} 24 & 6h & 6h \\ 6h & 4h^2(\alpha\beta+1) & 2h^2\alpha\beta \\ 6h & 2h^2\alpha\beta & 4h^2(\alpha\beta+1) \end{bmatrix}\begin{Bmatrix} u \\ \theta_A \\ \theta_B \end{Bmatrix} = \begin{Bmatrix} 0 \\ M_A \\ M_B \end{Bmatrix} \tag{13.1}$$

where

$$\alpha = \frac{h}{L}; \quad \beta = \frac{EI_b}{EI_c} \tag{13.2}$$

The fixed-end moments of the beam produced by the vertical loading are M_A and M_B. By convention, the positive sign occurs when the end moment induces a clockwise rotation. As the coefficient matrix in Equation 13.1 is fully populated, the horizontal displacement is coupled with the rotations.

Expanding the first row of Equation 13.1 gives

$$u = -\frac{h(\theta_A + \theta_B)}{4} \tag{13.3}$$

Therefore, u is zero when $\theta_A = -\theta_B$. This occurs when symmetric loads are applied to the beam. Solving Equation 13.1 gives the horizontal movement of the frame due to the vertical load:

$$u = \frac{-(M_A + M_B)}{4(6\alpha\beta+1)\alpha L}\frac{h^3}{EI_c} \tag{13.4}$$

The negative sign indicates that the movement of the frame is to the left if the sum of the moments is positive.

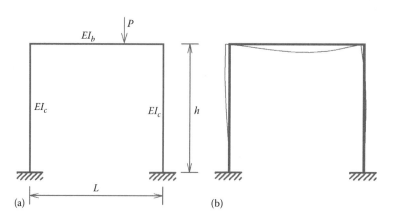

FIGURE 13.1
(a,b) A symmetric frame subjected to an asymmetrical vertical load.

Now consider a horizontal load that will induce the same horizontal movement as the vertical load. This will be termed an *equivalent horizontal load*. If a horizontal force of F to the left is applied at one of the beam/column connections instead of the vertical load, the solution of Equation 13.1 gives the horizontal displacement as

$$u = \frac{-(3\alpha\beta + 2)F}{12(6\alpha\beta + 1)} \frac{h^3}{EI_c} \qquad (13.5)$$

For the same horizontal displacement at the beam–column connections of the frame, equating Equations 13.4 and 13.5 gives the equivalent horizontal load:

$$F = \frac{12(M_A + M_B)(6\alpha\beta + 1)}{4(6\alpha\beta + 1)(3\alpha\beta + 2)\alpha L}$$

$$= \frac{(M_A + M_B)}{LP_{TV}} \frac{3}{(3\alpha\beta + 2)\alpha} P_{TV} = C_L C_S P_{TV} = C_{LS} P_{TV} \qquad (13.6)$$

in which

$$C_L = \frac{M_A + M_B}{LP_{TV}} \qquad (13.7)$$

$$C_S = \frac{3}{(3\alpha\beta + 2)\alpha} \qquad (13.8)$$

$$C_{LS} = C_L C_S \qquad (13.9)$$

where:

P_{TV} is the total vertical load, for the present case $P_{TV} = P$
C_L is a load factor that relates to the type and distribution of vertical loads
C_S is a structure factor that is a function of structural form (α) and the distribution of member rigidities (β)
C_{LS} is an equivalent horizontal load factor

It can be seen that the smaller the values of α and β, the larger the structure factor C_S. It should also be noted that, for this case, the load factor and the structure factor are independent.

Equation 13.6 indicates that the equivalent horizontal load can be expressed as a product of the load factor, the structure factor and the vertical load. Table 13.1 provides the values of the load factor C_L for several vertical load distributions on a beam with two fixed ends. Table 13.2 shows the structure factors C_S for a range of geometry and rigidity ratios.

Consider a particular case where $\alpha = 1$, $\beta = 1$ and a concentrated load P acts at a quarter of the length of the horizontal member of the frame as shown in Figure 13.1a. The moments in Equation 13.1 are

$$M_A = -\frac{3Ph}{64} \quad M_B = \frac{9Ph}{64} \qquad (13.10)$$

Substituting these into Equation 13.6 gives

$$F = \frac{63P}{1130} = 0.05625P \qquad (13.11)$$

In this case, the horizontal movement of the frame at the beam–column connections induced by the vertical load P is equal to that of a horizontal load of 5.625% of P applied on one of the corner connections in the direction of the movement of the frame.

TABLE 13.1   Load Factor (C_L) for Different Load Distributions on a Uniform Beam with Two Fixed Ends

Load Distribution	M_A	M_B	C_L
Uniformly distributed load over full length	$-qL^2/12$	$qL^2/12$	0
Concentrated load acting at a quarter of the span from the right	$-3PL/64$	$9PL/64$	3/32
Uniformly distributed load over half of the span from the right	$-5qL^2/192$	$11qL^2/192$	1/16
Uniformly distributed load over three-quarters of the span from the right	$-63qL^2/1024$	$81qL^2/1024$	3/128

TABLE 13.2   Structure Factor (C_S) for Different Ratios of Length and Rigidity for a Symmetric Frame

	$\beta=0.5$	$\beta=1$	$\beta=2$
$\alpha=0.5$	2.182	1.714	1.200
$\alpha=1$	0.857	0.600	0.375
$\alpha=2$	0.300	0.188	0.107

TABLE 13.3   Equivalent Horizontal Load Factor (C_{LS}) for a Symmetric Frame

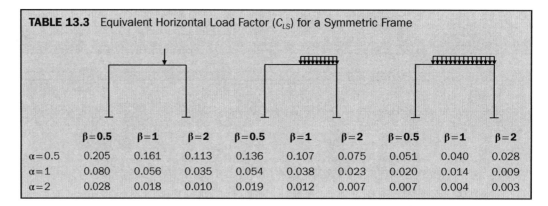

	$\beta=0.5$	$\beta=1$	$\beta=2$	$\beta=0.5$	$\beta=1$	$\beta=2$	$\beta=0.5$	$\beta=1$	$\beta=2$
$\alpha=0.5$	0.205	0.161	0.113	0.136	0.107	0.075	0.051	0.040	0.028
$\alpha=1$	0.080	0.056	0.035	0.054	0.038	0.023	0.020	0.014	0.009
$\alpha=2$	0.028	0.018	0.010	0.019	0.012	0.007	0.007	0.004	0.003

EXAMPLE 13.1

Consider the frame shown in Figure 13.1a with $h=6$ m, $L=6$ m, $E = 30\times10^9$ N m^{-2}, $I_b =I_c=0.25^4/12=3.255\times10^{-4}$ m^4 and for two loading cases:

1. A vertical concentrated load, $P=100$ kN, acts at one-quarter of the length of the beam from the right end of the frame
2. A horizontal concentrated load, $F=0.05625\,P=5.625$ kN, acts on the left beam–column connection towards the left

Calculate the horizontal displacements induced by the two loading cases.

SOLUTION

Figure 13.1b shows the deformed shape of the frame subject to the concentrated vertical load. The horizontal displacements determined by solving Equation 13.5 and by using the finite element method are –7.406 mm and –7.405 mm, respectively.

The lateral displacement induced by the equivalent lateral load is also –7.406 mm from the computer analysis.

The combined effect of the load and structure factors are shown in Table 13.3, which provides the equivalent horizontal load factors C_{LS} for $\alpha=0.5$, 1 and 2, and $\beta=0.5$, 1 and 2.

From Tables 13.1 through 13.3 the following can be observed:

- The magnitude of the horizontal displacement induced by vertical loads (Equation 13.4) or the equivalent horizontal load (Equation 13.6) depends on the load distribution and the structural form.
- The structure factors C_S are significantly larger than the load factors C_L.
- The smaller the values of α and β, the larger the equivalent horizontal load factor – that is, if the frame is relatively short and has a relatively large span or the rigidity of the beam is smaller than that of the column or both, the frame will have a relatively large horizontal load factor. Hence, the equivalent horizontal load for a taller frame is smaller than that for a shorter frame if both are subjected to the same vertical loading.
- The height/length ratio α is more significant than the rigidity ratio β when determining the magnitude of the horizontal movement.
- The horizontal movement of a frame due to vertical loads is zero when $M_A=-M_B$, that is, when concentrated loads act on the beam/column joints or when a symmetric load is applied to the beam.

It is useful to explain why the frame moves to its left when the vertical load is applied on the right half of the frame.

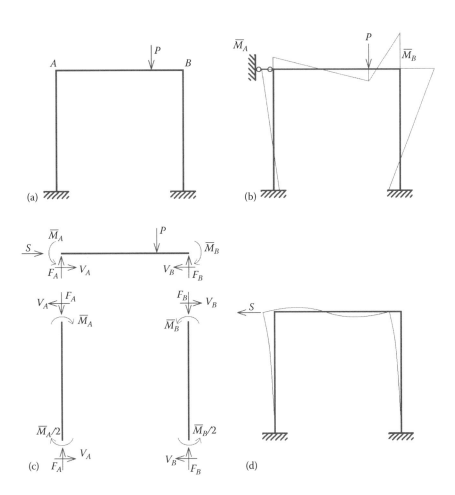

FIGURE 13.2
(a–d) Conceptual analysis of the lateral movements of a simple frame subjected to an asymmetric vertical load.

Applying a lateral constraint at the left-hand beam–column joint, the bending moment diagram of the frame can be drawn for no lateral movements of the frame as shown in Figure 13.2b, where $\overline{M}_B > \overline{M}_A$ due to the location of the load. The free-body diagram of the beam and two columns of the frame can be plotted (Figure 13.2c) based on the bending moment diagram, where V_A, F_A and V_B, F_B are the internal shear forces and axial forces on the connections A and B, respectively, and S is the force in the horizontal support at A assumed to be towards the right.

Considering the equilibrium of the beam in the horizontal direction gives:

$$S + V_A - V_B = 0$$

As $M_B > M_A$, then $V_B > V_A$. Thus, S is positive, acting in the assumed direction. As there is no horizontal restraint on the beam–column connection in the actual frame, the action of S must be removed. Following the principle of superposition, this can be achieved by adding a force $-S$ on the left beam/column connection, as shown in Figure 13.2d. Obviously, $-S$ causes the frame to move to its left. Adding the two loading conditions shown in Figure 13.2b and d together forms the actual frame with the vertical load (Figure 13.2a) and adding together the displacements of the two loading cases shows horizontal deflections of the frame.

13.2.1.2 Antisymmetric system

If the left column of the frame, shown in Figure 13.1a, is rotated through 180° about its connection to the beam, the frame becomes antisymmetric, as shown in Figure 13.3a. The equivalent horizontal load can be found, in a manner similar to that in Section 13.2.2.1, as

$$F = \frac{(M_B - M_A)}{4(2\alpha\beta+1)\alpha L} \frac{12(2\alpha\beta+1)}{(\alpha\beta+2)}$$

$$= \frac{(M_B - M_A)}{LP_{TV}} \frac{3}{(\alpha\beta+2)\alpha} P_{TV} = C_L C_S P_{TV} = C_{LS} P_{TV} \tag{13.12}$$

where

$$C_L = \frac{M_B - M_A}{LP_{TV}} \tag{13.13}$$

$$C_S = \frac{3}{(\alpha\beta+2)\alpha} \tag{13.14}$$

where C_{LS} is defined by Equation 13.9. Equations 13.12 through 13.14 have the same form as Equations 13.6 through 13.8. For comparison, similar tables for the load factor, the structure factor and the equivalent horizontal load factor of the antisymmetric frame are given in Tables 13.4 through 13.6.

In addition to the conclusions drawn from Section 13.2.1.1 which are also valid for the antisymmetric system, the following can be deduced:

- The load factor C_L and structure factor C_S for the antisymmetric systems are significantly larger than those for the symmetric system. Hence, the magnitude of the horizontal movement due to vertical loads depends primarily on the structural form.
- Equation 13.12 indicates that the antisymmetric frame has no horizontal movement when $M_A = M_B$, which requires a particular distribution of antisymmetric vertical loading. For any other vertical loading there will be a resulting horizontal movement.

EXAMPLE 13.2

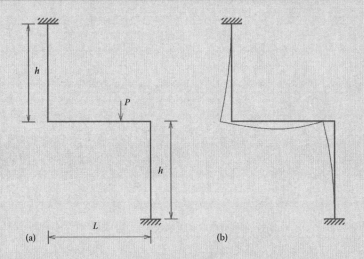

FIGURE 13.3
(a,b) An antisymmetric frame with an asymmetrical load.

Consider the frame shown in Figure 13.3a with similar properties to those used for Example 13.1, that is, $h=6$ m, $L=6$ m, $E=30\times10^9$ Nm^{-2}, $I_b=I_c=0.25^4/12=3.255\times 10^{-4}$ m^4 and $P=100$ kN (acting at one-quarter the length of the beam from the right end). Calculate the horizontal movement of the antisymmetric frame at the beam–column connection.

SOLUTION

The equivalent horizontal load can be evaluated using Equation 13.12 and is 18.75 kN. Computer analysis is used to determine the horizontal displacements induced first by the vertical load of 100 kN, and then by a horizontal load of 18.75 kN, applied at the beam–column connection, towards the left. The displacements are found to be identical and have a value of −34.56 mm. Figure 13.3b shows the deformed shape of the frame subject to the concentrated vertical load.

TABLE 13.4 Load Factor (C_L) for Different Load Distributions for an Antisymmetric System

Load Distribution	M_A	M_B	C_L
Uniformly distributed load over full length	$-qL^2/12$	$qL^2/12$	1/6
Concentrated load acting at a quarter of the span from the right	$-3PL/64$	$9PL/64$	3/16
Uniformly distributed load over half of the span from the right	$-5qL^2/192$	$11qL^2/192$	1/6
Uniformly distributed load over three-quarters of the span from the right	$-63qL^2/1024$	$81qL^2/1024$	3/16

TABLE 13.5 Structure Factor (C_S) for Different Ratios of Length and Rigidity for an Antisymmetric Frame

	$\beta=0.5$	$\beta=1$	$\beta=2$
$\alpha=0.5$	2.667	2.400	2.000
$\alpha=1$	1.200	1.000	0.750
$\alpha=2$	0.500	0.375	0.250

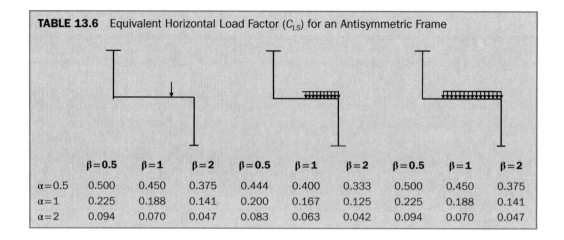

TABLE 13.6 Equivalent Horizontal Load Factor (C_{LS}) for an Antisymmetric Frame

	β=0.5	β=1	β=2	β=0.5	β=1	β=2	β=0.5	β=1	β=2
α=0.5	0.500	0.450	0.375	0.444	0.400	0.333	0.500	0.450	0.375
α=1	0.225	0.188	0.141	0.200	0.167	0.125	0.225	0.188	0.141
α=2	0.094	0.070	0.047	0.083	0.063	0.042	0.094	0.070	0.047

13.2.1.3 Asymmetric system

If the lengths of the columns of the frame shown in Figure 13.1a are different, the frame becomes asymmetric, as shown in Figure 13.4a. The ratios given in Equation 13.9 are redefined as follows:

$$\alpha = \frac{h_1}{L} \quad \beta = \frac{EI_b}{EI_c} \quad \gamma = \frac{h_1}{h_2} \tag{13.15}$$

and the equivalent horizontal load becomes:

$$F = \frac{3\left[\alpha\beta\left(2-\gamma^2\right)+2\gamma\right]M_A + 3\left[\alpha\beta\left(2\gamma^2-1\right)+2\gamma^2\right]M_B}{\alpha L\left[4\left(\alpha\beta+1\right)\gamma+\alpha\beta\left(3\alpha\beta+4\right)\right]P_{TV}}P_{TV} = C_{LS}P_{TV} \tag{13.16}$$

where

$$C_{LS} = \frac{3\left[\alpha\beta\left(2-\gamma^2\right)+2\gamma\right]M_A + 3\left[\alpha\beta\left(2\gamma^2-1\right)+2\gamma^2\right]M_B}{\alpha L\left[4\left(\alpha\beta+1\right)\gamma+\alpha\beta\left(3\alpha\beta+4\right)\right]P_{TV}} \tag{13.17}$$

C_{LS} is an equivalent horizontal load factor, which is a function of the load distribution, location and structural form. In contrast to the symmetric and antisymmetric frames considered in the previous two sections, the load factor and the structure factor are coupled for the asymmetric frame.

Consider $\gamma = 3/2$. The equivalent horizontal load factors for the same loading cases, length ratios α and rigidity ratios β, as for the symmetric and antisymmetric frames, are given in Table 13.7.

Comparing the results in Tables 13.3 and 13.7, it can be seen that the equivalent horizontal load factors for the asymmetric frame are significantly larger than those for the symmetric frame. This again shows that structural form affects the magnitudes of the horizontal movements of frame structures subject to vertical loads.

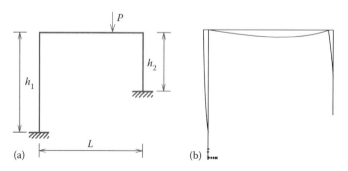

FIGURE 13.4
(a,b) An asymmetrical frame with an asymmetrical load.

EXAMPLE 13.3

Consider the frame shown in Figure 13.4a with $h_1 = L = 6.0$ m, $h_2 = 4$ m, $E = 30 \times 10^9$ Nm^{-2}, $I_b = I_c = 3.255 \times 10^4$ m^4 and $P = 100$ kN (acting at one-quarter the length of the beam from the right end). Calculate the horizontal movement of the asymmetric frame at the beam/column connection.

SOLUTION

The equivalent horizontal load for this case is evaluated using Equation 13.16 and is 15.73 kN. The horizontal displacements of the frame, obtained from computer analysis, for the vertical loading and a horizontal load of 15.73 kN applied to the left at the left-hand beam/column connection have the same value of −10.59 mm. Figure 13.4b shows the deformed shape of the frame.

13.2.1.4 Further comparison

Table 13.8 summarises the ranges of the equivalent horizontal load factors for the three types of frame subject to three types of vertical loading for variations of α and β between 0.5 and 2. From Table 13.8 it can be seen that

- The equivalent horizontal load factors for the antisymmetric frame have the largest values, but this type of structure may not be common.
- The equivalent horizontal load factors of the asymmetric frame are at least double those of the symmetric frame for the same loading conditions.

TABLE 13.7   Equivalent Horizontal Load Factor (C_{LS}) for an Asymmetric System

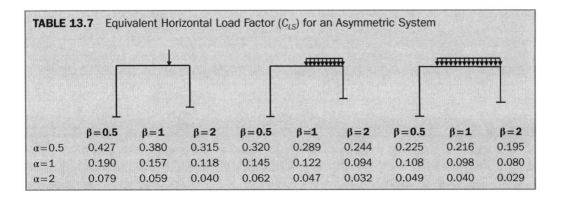

	$\beta=0.5$	$\beta=1$	$\beta=2$	$\beta=0.5$	$\beta=1$	$\beta=2$	$\beta=0.5$	$\beta=1$	$\beta=2$
$\alpha=0.5$	0.427	0.380	0.315	0.320	0.289	0.244	0.225	0.216	0.195
$\alpha=1$	0.190	0.157	0.118	0.145	0.122	0.094	0.108	0.098	0.080
$\alpha=2$	0.079	0.059	0.040	0.062	0.047	0.032	0.049	0.040	0.029

TABLE 13.8   Comparison of the Ranges of the Equivalent Horizontal Load Factor (C_{LS})

	Symmetric Frame	Antisymmetric Frame	Asymmetric Frame
Concentrated load acting at one-quarter of the span from the right	0.205–0.010	0.500–0.047	0.427–0.040
Uniformly distributed load over half of the span from the right	0.136–0.007	0.444–0.042	0.320–0.032
Uniformly distributed load over three-quarters of the span from the right	0.051–0.003	0.500–0.047	0.225–0.029

13.2.2 Dynamic Response

When a structure is subjected to dynamic loading, resonance may occur with a consequent, and potentially significant, increase in response (see Chapter 17). The possibility of vertical dynamic loading resulting in a resonant horizontal response, therefore, must be considered.

Consider the frame discussed in Section 13.2.1.1, and shown in Figure 13.1a, subjected to a simple sinusoidal vertical load, $P(t)$, with maximum amplitude P_0:

$$P(t) = P_0 \sin 2\pi f_p t \tag{13.18}$$

where:

f_p is the frequency of the load
t is time

The mass densities for the columns and the beam are assumed to be $\bar{m}$ and $10\,\bar{m}$, respectively, with the high density of the beam representing added loads that may arise from floors supported by the beam. The equation of undamped forced vibrations of the frame can be shown [2]:

$$\frac{\bar{m}h}{420}\begin{bmatrix} 4512 & 22h & 22h \\ 22h & 44h^2 & -30h^2 \\ 22h & -30h^2 & 44h^2 \end{bmatrix}\begin{Bmatrix} \ddot{u} \\ \ddot{\theta}_A \\ \ddot{\theta}_B \end{Bmatrix} + \frac{EI_c}{h^3}\begin{bmatrix} 24 & 6h & 6h \\ 6h & 8h^2 & 2h^2 \\ 6h & 2h^2 & 8h^2 \end{bmatrix}\begin{Bmatrix} u \\ \theta_A \\ \theta_B \end{Bmatrix} = \begin{Bmatrix} 0 \\ M_A \\ M_B \end{Bmatrix}\sin(2\pi f_p t) \tag{13.19}$$

The elements in the coefficient (mass) matrix for accelerations are obtained in the same manner as those in the coefficient (stiffness) matrix for displacements. The mode shapes and natural frequencies of the structure can be found by solving the eigenvalue problem associated with Equation 13.19. Taking the mass density, $\bar{m}$, equal to 150 kg m^{-1} and other data as used in Example 13.1, the three natural frequencies of the frame are 1.39 Hz, 5.00 Hz and 14.5 Hz and the corresponding mode shapes are shown in Figure 13.5. The first mode shows the dominating horizontal movements of the frame while the two other modes give symmetric and antisymmetric rotations of the two connections of the frame, respectively. The response in the first mode of the frame is

$$A_1(t) = \frac{\phi_{21}M_A + \phi_{31}M_B}{K_1}\frac{1}{1-(f_p/f_1)^2}\sin 2\pi f_p t \tag{13.20}$$

where:

$A_1(t)$ is the amplitude of the horizontal motion of the frame
$\phi_{21}M_A + \phi_{31}M_B$ is the load for the first mode, which acts in the horizontal direction

Equation 13.20 indicates that if the load for the first mode is not equal to zero and the load frequency (f_p) is close to the fundamental natural frequency (f_1), the vertical load will induce near resonant vibration of the frame in the horizontal direction. This conclusion can be verified numerically.

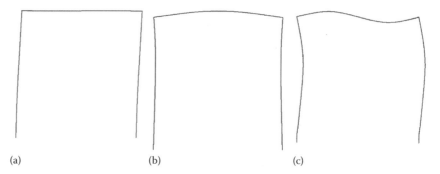

(a) (b) (c)

FIGURE 13.5
Mode shapes of the symmetric frame. (a) Horizontal movement. (b) Symmetric movement. (c) Antisymmetric movement.

EXAMPLE 13.4

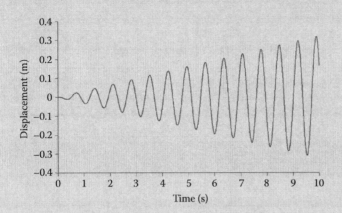

FIGURE 13.6
Resonant response of the frame.

Consider the frame defined in Example 13.1 with the additional data as follows:

$$\bar{m} = 2400 \text{ kg m}^3 \times (0.25 \text{ m})^2 = 150 \text{ kg m}^{-1}, f_p = 1.39 \text{ Hz},$$

$$P_0 = 100 \text{ kN and } P(t) = P_0 \sin 2\pi f_p t$$

Calculate the dynamic displacements of the frame in the lateral direction using Equation 13.19.

SOLUTION

Dynamic analysis was carried out by computer with the damping set to zero. Figure 13.6 shows the time history of the horizontal motion of the frame, up to 10 s, due to the vertical harmonic load. A typical resonance situation is encountered.

Although the example is simple, it illustrates the important phenomenon that:

If the frequency of a vertical load is close to one of the horizontal natural frequencies of a structure, resonance in the horizontal direction can occur as a result of vertical excitation.

This situation should be recognised in the design of structures.

The necessary condition for no horizontal movement of the frame occurs when the vertical loads are applied either symmetrically on the beam or at the beam–column joints. For any other distributions of vertical dynamic loads, resonance can occur with motion in the horizontal direction.

13.3 MODEL DEMONSTRATIONS

13.3.1 Symmetric Frame

This model demonstrates that *a symmetric frame deforms laterally when it is subjected to a vertical load applied asymmetrically.*

Figure 13.7 shows a simple symmetric plastic frame unloaded (a) and carrying an asymmetrically concentrated load positioned close to the right-hand column (b). It can be observed that the horizontal member deflects vertically and the loaded frame moves to the left. Note that the movement is to the left for the load placed to the right of the centre line of the frame. If the vertical load

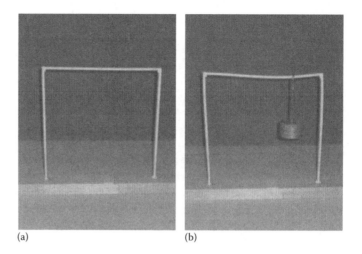

(a)　　　　　　　　　　　(b)

FIGURE 13.7
(a,b) A symmetric frame subjected to an asymmetric load.

was a harmonic dynamic load and its frequency matched the natural frequency of the frame in its horizontal direction, resonance with significant horizontal movements of the frame would occur.

13.3.2 Antisymmetric Frame

This model demonstrates that *an antisymmetric frame deforms laterally when it is subjected to a vertical load.*

Figure 13.8 shows a simple antisymmetric plastic frame unloaded (a) and carrying an asymmetrically concentrated load positioned close to the right-hand column (b). It can be observed that the horizontal member deflects vertically and the loaded frame moves to its left. Again, the movement is to the left for the load placed to the right of the centre line of the frame.

13.3.3 Asymmetric Frame

This model demonstrates that *an asymmetric frame deforms laterally when it is subjected to a vertical load.*

Figure 13.9 shows a simple asymmetric plastic frame unloaded (a) and carrying a concentrated load positioned close to the right-hand column (b). The horizontal member deflects vertically and it

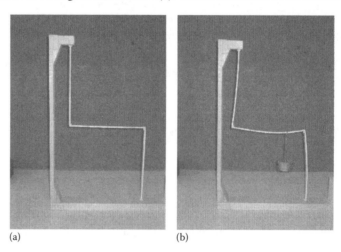

(a)　　　　　　　　　　　(b)

FIGURE 13.8
(a,b) An antisymmetric frame subjected to an asymmetric load.

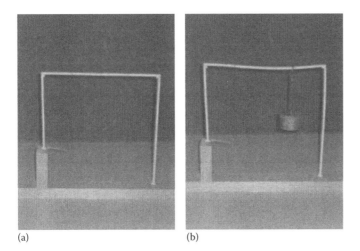

(a)   (b)

FIGURE 13.9
(a,b) An asymmetric frame subjected to an asymmetric load.

can be observed that the loaded frame moves to the left. Again, the movement is to the left for the load placed to the right of the centre line of the frame.

13.4 PRACTICAL EXAMPLES

13.4.1 Grandstand

Figure 13.10 shows coupled vertical and front-to-back movement of the cross section of a grandstand in one typical mode of vibration. It can be seen that the front-to-back movements of the grandstand are larger than the vertical movements of the two tiers for this particular mode. This mode shape indicates that resonance in the front-to-back direction would occur if one of the frequencies of vertical loading on a tier was close to the natural frequency associated with the mode, even though the vertical movement will be small. It was observed at a pop concert that a stand moved much more significantly in the front-to-back direction than in the vertical direction although the human loading was primarily in the vertical direction.

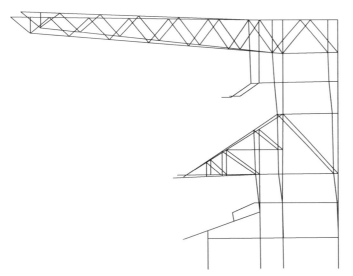

FIGURE 13.10
Typical mode of vibration of a frame model of a cantilever grandstand showing coupled vertical and front-to-back movements. (From Mandal, P. and Ji, T. *Struct Building: Proceed Inst Civil Eng.*, 157, 173–184, 2004.)

13.4.2 Building Floor

A 9×6 m test panel of a large composite floor is shown in Figure 13.11. The structural response of the panel was measured for group of 64 students jumping in time to a musical beat (Figure 13.12). At the centre of the test floor panel, the vertical acceleration was recorded for just over 16 s, as was the horizontal acceleration in the direction orthogonal to the direction in which the students

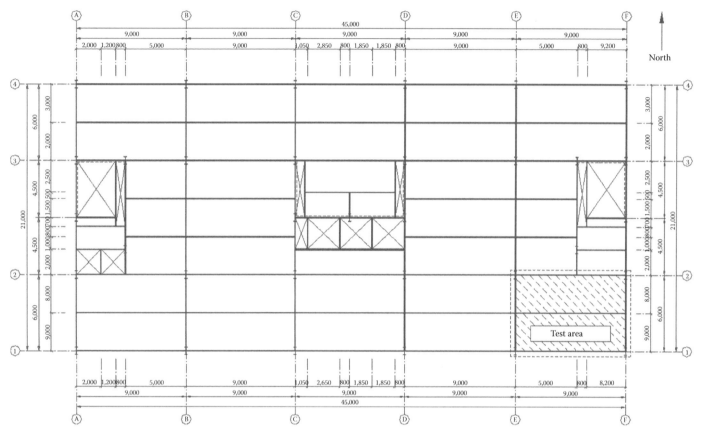

FIGURE 13.11
A plan of a floor used for crowd jumping tests at its corner panel.

FIGURE 13.12
Sixty-four students jumping on a floor in response to music.

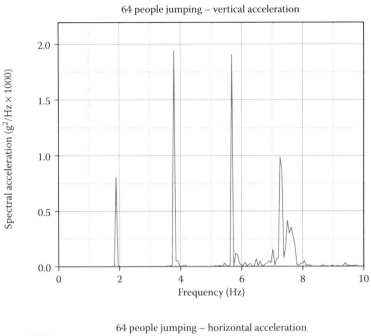

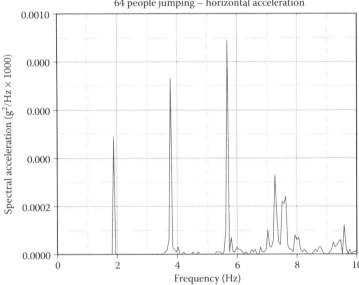

FIGURE 13.13
The acceleration spectrum in the vertical and horizontal directions for 64 students jumping on a floor.

were facing. The peak vertical acceleration was 0.48 g and the corresponding horizontal acceleration was 0.03 g. The autospectra for these records are shown in Figure 13.13 and the characteristic response can be seen in both directions. The test area was part of the much larger flooring system (Figure 13.11) and the vertical human loading was thus applied asymmetrically on the structure as a whole, which induced the horizontal motion of the whole building system.

13.4.3 Railway Bridges

Horizontal movements of some railway bridges have been observed as trains passed over them. Because of the increasing speed of trains, a number of bridges have had to be reassessed for safety. As there are often two or more rail tracks on a bridge, the loading from any one train is effectively applied in an asymmetrical manner on the structure generating lateral horizontal movements of the bridge. There will also be some horizontal forces generated by lateral movement of the railway vehicles, even along straight tracks. With the increasing speed of trains, the vertical loading frequency will increase and this may be a problem if resonance occurs, that is, if one of the train load frequencies in the vertical direction is close to one of the lateral natural frequencies of the bridge. Therefore, it is necessary to check horizontal as well as vertical natural frequencies of bridges to ensure that both are above the likely loading frequencies associated with trains running at higher speeds.

PROBLEMS

1. A bracing member is added, from the left support to the top right joint, to the frame shown in Figure 13.1. The bracing member has an area A and the same modulus of elasticity E as the other members. A vertical concentrated load, $P = 100$ kN, acts at one-quarter of the length of the beam from the right end of the frame. Answer the following questions.
   a. Derive the equation of equilibrium of the frame.
   b. Derive the expressions of the load factor C_L and the structure factor C_S.
   c. Compare the load factors and structure factors between this frame and the frame shown in Figure 13.1, and draw qualitative conclusions.
   d. Assuming that $A = 0.005$ m^2 and $E = 200 \times 10^9$ N m^{-2}, calculate the lateral displacement and compare this with the result in Example 13.1.

REFERENCES

1. Dallard, P., Fitzpatrick, T., Flint, A., Low, A., Smith, R. R. and Willford, M. Pedestrian induced vibration of footbridges, *The Structural Engineer*, 78, 13–15, 2000.
2. Ji, T., Ellis, B. R. and Bell, A. J. Horizontal movements of frame structures induced by vertical loads, *Structures and Buildings: The Proceedings of the Institution of Civil Engineers*, 156, 141–150, 2003.
3. Mandal, P. and Ji, T. Modelling the dynamic behaviour of a cantilever grandstand, *Structures and Buildings: The Proceedings of the Institution of Civil Engineers*, 157, 173–184, 2004.

PART II

DYNAMICS

CHAPTER 14

CONTENTS

Energy Exchange

14

14.1 DEFINITIONS AND CONCEPTS

Conservative systems: A system is said to be conservative if no energy is added or lost from the system during movement. This is an idealised system, but one that can be used to aid the solution of many problems. In real structures, internal friction forces will do work and damping will dissipate energy.

    Conservation of energy means that the total energy at two different positions or at two different times is the same in a conservative system.

    Conservation of momentum indicates that the momentum of a system is the same at two different times when the resulting external force is zero between those two times.

- Energy can be transformed from one form to another; for instance, mechanical energy can be changed into electrical energy.
- For a conservative system, the total energy is constant and a body, once moved, will continue to move or to vibrate. During motion there is a constant exchange between potential energy and kinetic energy.
- For a nonconservative system, energy has to be added to maintain motion.

14.2 THEORETICAL BACKGROUND

Gravitational potential energy of a mass is defined as the work done against gravity to elevate the mass a distance above an arbitrary reference position where the gravitational potential energy is defined to be zero. Thus, the potential energy is

$$U_g = mgh \tag{14.1}$$

where:
- m is the mass of the body
- g is the gravitational acceleration
- h is the height of the mass above the reference position

    Elastic potential energy is the potential energy found in the deformation of an elastic body, such as a spring or a deformable beam. The elastic potential energy in a spring with stiffness k when it is displaced with a distance of u is defined as

$$U_e = \tfrac{1}{2} k u^2 \tag{14.2}$$

    This is the total work required to take the mass from its original position to a displacement of u. For a deformable beam with rigidity EI, the elastic potential energy due to bending of the beam is expressed as

$$U_e = \frac{1}{2} \int_0^L EI \left(\frac{d^2 v(x)}{dx^2} \right)^2 dx \tag{14.3}$$

where $v(x)$ is the deformation of the beam along its length.

The elastic potential energy defined by Equation 14.3 is also called the *strain energy* of the beam. The strain energy due to shear forces can be disregarded if the length of a beam is much greater than the depth (say the ratio of the span to the depth of the beam is larger than 8) [1]. Strain energy or elastic potential energy is always positive, regardless of the direction of the displacement.

Kinetic energy of a mass of m with a velocity of $\dot{u}$ is defined as

$$T = \tfrac{1}{2} m \dot{u}^2 \tag{14.4}$$

This is equal to the work required to move the mass from a state of rest to a velocity of $\dot{u}$. As is the case for the strain energy, the kinetic energy is always positive regardless of the direction of motion. For a vibrating beam with a distributed mass of $m(x)$, the kinetic energy is

$$T = \frac{1}{2} \int \bar{m}(x) \dot{v}^2 dx \tag{14.5}$$

All the forms of energy are scalar quantities, with SI units of Newton metre (Nm) or joules (J).

Conservative systems: For a conservative system, the total energy is constant. In other words, if the energy of the system is calculated at two different locations or two different times, the values of energy at the two locations/times are the same, that is,

$$U_1 = U_2 \quad \text{or} \quad U_1 - U_2 = 0 \tag{14.6}$$

Equation 14.6 shows the *principle of conservation of energy* and indicates that *energy can be transformed from one form to another whilst keeping the total energy constant*. The principle of conservation of energy (Equation 14.6) leads to a useful method of analysis.

The basic equation of motion of a mass m moving in a particular direction is

$$\sum F = m\ddot{v} = \frac{d}{dt}(m\dot{v}) \quad \text{or} \quad \sum F = \frac{d}{dt}(G) = \dot{G} \tag{14.7}$$

where:
F is the force acting on the mass
$\ddot{v}$ is the acceleration of the mass

The product of the mass and velocity is defined as the **momentum** $G = m\dot{v}$.

Equation 14.7 indicates that *the resultant of all forces acting on the mass equals its rate of change of momentum with respect to time*. Momentum has SI units of kg m s^{-1} or Ns.

Equation 14.7 can be extended to other types of displacement, including rotation. If the resultant force on a mass or a system is zero during an interval of time, the momentum in that interval is constant, that is,

$$G_1 = G_2 \quad \text{or} \quad G_1 - G_2 = 0 \tag{14.8}$$

Equation 14.8 shows the *principle of conservation of momentum* and states that *the momentum at a time interval is constant if the resultant force is zero during that time interval*. This principle of the conservation of momentum (Equation 14.8) also leads to a method for solving some problems. In addition, Equation 14.8 can be applied to both conservative and nonconservative systems.

EXAMPLE 14.1

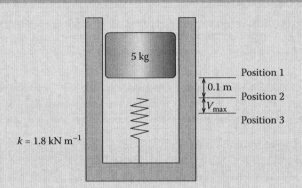

FIGURE 14.1
An energy conservation system. (From Meriam, J. L. and Kraige, L. G. *Engineering Mechanics: Dynamics*, 4th edn, John Wiley, New York, 1998.)

A 5 kg cylinder is released from rest in the position shown in Figure 14.1, and compresses a spring of stiffness $k = 1.8$ kN m^{-1}. Determine the maximum compression v_{max} of the spring and the maximum velocity $\dot{v}_{max}$ of the cylinder [1].

SOLUTION

Consider the system to be conservative in which the effect of air resistance and any friction is negligible. The cylinder moves down v_{max} before it rebounds due to the spring action.

Before releasing the cylinder from rest (Position 1), the total gravitational potential and elastic potential energies are:

$$U_{g1} = mgh = 5 \times 9.81 \times (0.1 + v_{max}) \quad \text{and} \quad U_{e1} = 0$$

The reference position is selected to be where the spring experiences its maximum deflection. The total energies at the reference position (Position 3) are

$$U_{g3} = 0 \quad \text{and} \quad U_{e3} = \tfrac{1}{2}kv_{max}^2 = 900$$

Using the condition of energy conservation (Equation 14.6) leads to

$$900v_{max}^2 = 5 \times 9.81(0.1 + v_{max}) \quad \text{or} \quad 900v_{max}^2 - 49.5v_{max}^2 - 49.5 = 0$$

Solving this equation gives $v_{max} = 0.264$ m.

The total gravitational potential and kinetic energies at the position immediately before the cylinder contacts the spring (Position 2) are

$$U_{g2} = mgh = 5 \times 9.81 \times 0.1 = 4.95 \text{ Nm} \quad \text{and} \quad T_2 = \tfrac{1}{2}m\dot{v}_{max}^2 = 2.5\dot{v}_{max}^2$$

Equating the total energy at Positions 1 and 2 gives

$$4.95 + 2.5\dot{v}_{max}^2 = 5 \times 9.81(0.1 + 0.264) = 17.9$$

Thus, the maximum velocity of the cylinder is

$$\dot{v}_{max} = \sqrt{17.9 - 4.95} = 3.59 \text{ m/s}$$

EXAMPLE 14.2

Consider a simply supported beam of length L, with a flexural rigidity of EI and a uniformly distributed mass of $\bar{m}$. The vibration in the fundamental mode of the beam is defined as

$$v(x,t) = A_0 \sin\frac{\pi x}{L}\sin\omega t \qquad (14.9)$$

where:
  A_0 is the vibration amplitude
  ω is the natural frequency of the vibration

Determine the expression for the natural frequency of the beam.

SOLUTION

The total energy of the system at anytime can be calculated, but the simplest case to consider is when the strain energy reaches its maximum while the kinetic energy is zero, or alternatively when the kinetic energy reaches its maximum while the strain energy is zero.
  At $t = \pi/(2\omega)$, the displacement and velocity of the beam are

$$v\left(x,\frac{\pi}{2\omega}\right) = A_0 \sin\omega\left(\frac{\pi}{2\omega}\right)\sin\frac{\pi x}{L} = A_0 \sin\frac{\pi x}{L}$$

$$\dot{v}\left(x,\frac{\pi}{2\omega}\right) = A_0\omega\cos\omega\left(\frac{\pi}{2\omega}\right)\sin\frac{\pi x}{L} = 0$$

Thus, the total energy is equal to the maximum strain energy. Using Equation 14.3 gives

$$U_{e,max} = \frac{1}{2}\int_0^L EI\left(\frac{d^2v}{dx^2}\right)^2 dx = \frac{1}{2}\int_0^L EI\left(-A_0\frac{\pi^2}{L^2}\sin\frac{\pi x}{L}\right)^2 = \frac{A_0^2 EI\pi^4}{2L^4}\frac{L}{2} = \frac{A_0^2 EI\pi^4}{4L^3}$$

At $t = \pi/\omega$, the displacement and velocity of the beam are

$$v(x,\pi/\omega) = A_0 \sin\omega\left(\frac{\pi}{\omega}\right)\sin\frac{\pi x}{L} = 0$$

$$\dot{v}(x,\pi/\omega) = A_0\omega\cos\omega\left(\frac{\pi}{\omega}\right)\sin\frac{\pi x}{L} = -A_0\omega\sin\frac{\pi x}{L}$$

The total energy of the system is equal to the maximum kinetic energy. Using Equation 14.5 gives

$$T_{max} = \frac{1}{2}\int_0^L \bar{m}\dot{v}^2 dx = \frac{1}{2}\int_0^L \bar{m}\left(-A_0\omega\sin\frac{\pi x}{L}\right)^2 = \frac{A_0^2\omega^2\bar{m}}{2}\frac{L}{2} = \frac{A_0^2\omega^2\bar{m}L}{4}$$

Using Equation 14.6, that is, equating the maximum strain energy to the maximum kinetic energy, gives

$$\omega^2 = \frac{EI\pi^4}{\bar{m}L^4}$$

A more powerful way of using energy concepts for solving problems is to use the Lagrange equation. For free vibration, this may be written as [2–4]

$$\frac{d}{dt}\left(\frac{\partial T}{\partial \dot{v}}\right) + \frac{\partial U}{\partial v} = 0 \tag{14.10}$$

where T and U are the kinetic and potential energies of the system, respectively. Example 14.2 is reanalysed to show the usefulness of the Lagrange method.

The motion of the beam (Equation 14.9) is written as

$$v(x,t) = A(t)\phi(x) = A(t)\sin\frac{\pi x}{L} \tag{14.11}$$

The kinetic energy of the system is

$$T = \frac{1}{2}\int_0^L \bar{m}\dot{v}^2 dx = \frac{1}{2}\int_0^L \bar{m}\left(\dot{A}(t)\sin\frac{\pi x}{L}\right)^2 dx = \frac{\dot{A}^2(t)\bar{m}}{2}\frac{L}{2} = \frac{\dot{A}^2(t)\bar{m}L}{4} \tag{14.12}$$

The elastic potential energy is

$$U = \frac{1}{2}\int_0^L EI\left(\frac{d^2v}{dx^2}\right)^2 dx = \frac{1}{2}\int_0^L EI\left(-A(t)\frac{\pi^2}{L^2}\sin\frac{\pi x}{L}\right)^2 = \frac{A^2 EI\pi^4}{2L^4}\frac{L}{2} = \frac{A^2(t)EI\pi^4}{4L^3} \tag{14.13}$$

Substituting the kinetic and elastic energies into Equation 14.10 leads to

$$\ddot{A}(t) + \frac{EI\pi^4}{\bar{m}L^4}A(t) = 0 \tag{14.14}$$

Substituting $A(t) = A_0\sin(\omega t)$ into Equation 14.14 gives

$$\omega^2 = \frac{EI\pi^4}{\bar{m}L^4}$$

This is the same as that obtained in Example 14.2.

Comparing the solution procedures using the principle of the conservation of energy and the Lagrange equation, the following can be noted:

- The Lagrange equation leads to an equation of motion and the natural frequency can be obtained directly from the definition, while the principle of conservation of energy leads to the solution of the natural frequency without producing the equation of motion.
- For the Lagrange equation, the kinetic and elastic potential energies only need to be represented at an arbitrary position. Using the principle of conservation of energy requires evaluating the kinetic and elastic potential energies at two different positions or times.

14.3 MODEL DEMONSTRATIONS

14.3.1 Moving Wheel

This demonstration shows *the energy exchange between several common forms of energy, including gravitational potential energy, kinetic energy, energy loss due to friction and electromagnetic energy.*

FIGURE 14.2
A moving wheel showing energy transformation.

Two parallel, parabolic plastic tracks supporting an aluminium wheel are shown in Figure 14.2. Three small magnets are placed on the perimeter of the aluminium wheel. Four batteries are located in the track support and an electromagnetic field is also provided in the base support.

Place the wheel at one end of the tracks and release it. The wheel rolls towards the middle of the tracks with increasing speed as the potential energy of the wheel changes to kinetic energy. When passing the middle part of the tracks, the wheel speeds up due to a charge of electromagnetic energy which is sufficient to compensate for the energy loss due to the friction between the wheel and the tracks. The wheel then moves upwards and the kinetic energy is converted to potential energy. When all the kinetic energy is converted, the wheel stops and then starts to roll back down the track for a new cycle of movement. The wheel will continue to roll backwards and forwards along the track as long as sufficient electromagnetic energy is provided to compensate for the energy loss due to friction.

14.3.2 Collision Balls

This demonstration shows *the use of the principles of conservation of energy and conservation of momentum.*

Figure 14.3 shows a Newton's cradle which consists of five identical stainless steel balls suspended from above and arranged in a row with each ball just touching its neighbour(s). When one pulls a ball back and releases it, it collides with the row of the remaining balls, ejecting one at the far end. When two balls are pulled back and released, they eject two balls at the far end as shown in Figure 14.3. Lifting and releasing three balls ejects three balls at the far end, and pulling four ejects four.

(a)

(b)

FIGURE 14.3
Collision balls: Energy conservation.

First, consider the collision between two identical balls of mass m; one moves with initial velocity $\dot{u}_{1i}$ and the other is at rest with $\dot{u}_{2i} = 0$. After the collision, the two balls move with velocities $\dot{u}_{1f}$ and $\dot{u}_{2f}$, respectively. Conservation of momentum requires

$$m\dot{u}_{1i} = m\dot{u}_{1f} + m\dot{u}_{2f} \quad \text{or} \quad \dot{u}_{1i} = \dot{u}_{1f} + \dot{u}_{2f}$$

Conservation of energy requires

$$\tfrac{1}{2}m\dot{u}_{1i}^2 = \tfrac{1}{2}m\dot{u}_{1f}^2 + \tfrac{1}{2}m\dot{u}_{2f}^2 \quad \text{or} \quad \dot{u}_{1i}^2 = \dot{u}_{1f}^2 + \dot{u}_{2f}^2$$

Squaring the first equation and subtracting the second equation leads to

$$\left(\dot{u}_{1f} + \dot{u}_{2f}\right)^2 - \left(\dot{u}_{1f}^2 + \dot{u}_{2f}^2\right) = 0 \quad \text{or} \quad \dot{u}_{1f}\dot{u}_{2f} = 0$$

As

$$\dot{u}_{2f} \neq 0, \dot{u}_{1f} = 0$$

This shows that after collision the first ball will be at rest [5,6].

Now consider the case of several balls. Let m_1 be the total mass of the balls launched with velocity $\dot{u}_{1i}$ and m_2 be the total mass of the balls ejected with velocity $\dot{u}_{2f}$. Noting that after collision, m_1 becomes stationary, that is, $\dot{u}_{1f} = 0$, and using conservation of momentum and conservation of energy gives

$$m_1\dot{u}_{1i} = m_2\dot{u}_{2f}$$

$$\tfrac{1}{2}m_1\dot{u}_{1i}^2 = \tfrac{1}{2}m_2\dot{u}_{2f}^2$$

Squaring the first equation and subtracting $2m_1$ times the second equation gives

$$\left(m_2 - m_1\right)m_2\dot{u}_{2f}^2 = 0$$

This equation shows that the number of balls ejected is equal to the number of balls launched if all the steel balls are the same size, as shown in Figure 14.3.

14.3.3 Dropping a Series of Balls

This model gives *an entertaining demonstration of conservation of energy.*

Figure 14.4 shows four rubber balls of different sizes placed together using a plastic bar passing through their centres with the smaller balls resting on the larger balls.

When the balls are lifted vertically through a height of about 0.15 m from a desk and then released, the smallest ball rebounds to hit the ceiling which is over 2 m above the desk. This observation can be explained using either the principle of conservation of energy or the principle of conservation of momentum.

The total mass of the four balls is 120 g and the mass of the smallest ball is 5 g. The gravitational potential energy of the balls before dropping them is estimated to be

$$mgh = 0.12 \times 9.81 \times 0.15 = 0.177 \text{ Nm}$$

FIGURE 14.4
Dropping a series of balls.

After the impact of the largest ball on the desk, the three larger balls bounce up almost together through about 0.03 m. According to the conservation of energy, the total gravitational potential energy before and after impact should be the same if no energy loss takes place, that is,

$$0.177 = 0.115 \times 9.81 \times 0.03 + 0.005 \times 9.81 \times h_1$$

From this equation, the predicted bounce height of the smallest ball is 2.92 m. Actually, there is some loss of energy due to the impact between the largest ball and the desk and between the adjacent balls, so the smallest ball would not bounce quite as high as 2.92 m.

This example can also be analysed using the principle of conservation of momentum.

14.4 PRACTICAL EXAMPLES

14.4.1 Rollercoasters

Figure 14.5a shows a rollercoaster which is raised from ground level to the top of the first tower using mechanical energy. This builds up a reservoir of potential energy in the rollercoaster.

When the rollercoaster is released at the top of the tower, it moves forward and down, the potential energy being converted rapidly to kinetic energy with the speed of the rollercoaster increasing accordingly. The speed reaches its maximum when the rollercoaster reaches the lowest point between two adjacent towers. The rollercoaster then moves up to the next tower, the kinetic energy changing back to potential energy and the speed of the rollercoaster reduces as its potential energy increases. The rest of the towers and the dips, twists and turns of the ride, serve to change the energy of the rollercoaster back and forth between potential energy and kinetic energy (Figure 14.5b).

(a)

(b)

FIGURE 14.5
Rollercoasters. (a) The first tower. (b) Twists and turns.

During the motion, some energy will be lost due to friction and, for this reason, the first tower must be higher than all the other towers so that sufficient energy is provided to overcome the energy loss.

14.4.2 Torch without a Battery

Torches are normally powered by batteries. However, Figure 14.6 shows an environmentally friendly torch which works on the principle of energy exchange.

When one shakes the torch, a strong magnet at the far right in the body of the torch (Figure 14.6) moves, passes through electrical wires backwards and forwards and produces an electric current. Part of the kinetic energy generated by the magnet is converted to electric energy. The electric energy then changes into chemical energy through an electronic circuit and is stored in an internal storage cell. When a user switches on the torch, the chemical energy in the storage cell converts to electric energy and the electric energy changes to light energy in the bulb.

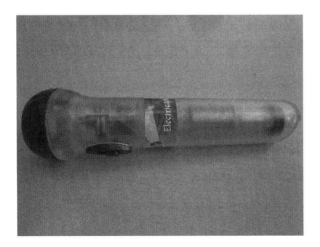

FIGURE 14.6
An environmentally friendly torch.

PROBLEMS

1. There is a height difference, *H*, between the highest point of a tower and the lowest point in the next valley of a rollercoaster track (Figure 14.5). A rollercoaster with a mass of *m* starts with a speed of *v* at the highest point of the tower. Assuming a conservative system, what is the speed of the rollercoaster when it reaches the valley?

2. Consider a uniform cantilever, with a length *L*, a flexural rigidity of *EI* and a uniformly distributed mass of $\bar{m}$. The vibration in the fundamental mode of the beam is estimated as

$$v(x,t) = \frac{A_0 x^2}{3L^4}(6L^2 - 4Lx + x^2)\sin\omega t$$

where A_0 is the vibration amplitude and ω is the natural frequency of the vibration. Determine the natural frequency of the cantilever using the Lagrange equation.

REFERENCES

1. Meriam, J. L. and Kraige, L. G. *Engineering Mechanics: Dynamics*, 4th edn, New York: John Wiley, 1998.
2. Beards, C. F. *Structural Vibration: Analysis and Damping*, London: Arnold, 1996.
3. Thomson, W. T. *Theory of Vibration and Applications*, London: Allan and Unwin, 1966.
4. Wang, G. *Applied Analytical Dynamics*, Beijing: High Education Press (in Chinese), 1981.
5. Sprott, J. C. *Physics Demonstrations: A Sourcebook for Teachers of Physics*, Madison, WI: The University of Wisconsin Press, 2006.
6. Ehrlich, R. *Turning the World Inside Out and 174 Other Simple Physics Demonstrations*, Princeton, NJ: Princeton University Press, 1990.

CHAPTER 15

CONTENTS

Pendulums

15

15.1 DEFINITIONS AND CONCEPTS

A simple gravity pendulum: A massless string with one end attached to a weight and the other end fixed. When an initial push is given, the pendulum will swing back and forth under the influence of gravity.

A rotational suspended system: A rigid body is suspended by two massless strings/hangers of equal length, fixed at the same point and the body rotates about the fixed point in the plane of the strings.

A translational suspended system: A rigid body is suspended by two parallel and vertical massless strings/hangers with equal length and the strings rotate about their own fixed points. Thus, the rigid body moves parallel to its static position in the plane of the strings.

- The natural frequency of a simple pendulum is independent of its mass and only relates to the length of the massless string.
- The natural frequency of a translational suspended system is independent of its mass and the location of its centre of mass, and is the same as that of an equivalent simple pendulum.
- The natural frequency of a rotational suspended system is dependent on the location of its centre of mass but is independent of the magnitude of its mass.
- An outward-inclined suspended system is a mechanism. When it is loaded vertically and asymmetrically, it will move sideways and rotate.
- The lateral natural frequency of a suspended bridge estimated using the simple beam theory will be smaller than the true lateral natural frequency.

15.2 THEORETICAL BACKGROUND

15.2.1 Simple Pendulum

Well known for its use as a timing device, the simple pendulum is a superb tool for teaching science and engineering. It can serve as a model for the study of a linear oscillator [1].

Consider a pendulum consisting of a concentrated mass, m, suspended from a pivotal point by an inextensible string of length, l, as shown in Figure 15.1. Several assumptions are used to investigate the linear oscillation of a pendulum, the principal ones being:

- Negligible friction, that is, the resulting torque on the system about the pivot is due solely to the weight of the mass.
- Small amplitude oscillations, which means that the sine of the rotation, θ, can be replaced by the rotation when specified in radians.
- The mass of the string is negligible and the mass of the body is concentrated at a point.
- The string is inextensible.

The equation of motion of the pendulum can be established using Newton's second law directly or using the Lagrange equation. The moment induced by the weight of the mass about the pivot is $mgl \sin \theta$, where g is the acceleration due to gravity. The moment of inertia of the mass about the pivot is ml^2 and the angular acceleration is $d^2\theta/dt^2$. Using Newton's second law gives

$$ml^2 \frac{d^2\theta}{dt^2} = -mgl\sin\theta \quad \text{or} \quad \frac{d^2\theta}{dt^2} + \frac{g}{l}\sin\theta = 0 \qquad (15.1)$$

Using the second assumption, $\sin \theta = \theta$ when θ is small and measured in radians. This leads to the equation of motion for small amplitude oscillations:

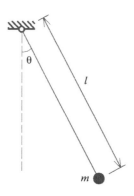

FIGURE 15.1
A simple pendulum.

$$\frac{d^2\theta}{dt^2} + \frac{g}{l}\theta = 0 \tag{15.2}$$

Substituting $\theta(t) = \theta_0 \sin(\omega t)$ into Equation 15.2 (or according to the definition given in Chapter 16) gives the expression for the natural frequency of the pendulum system as

$$f = \frac{\omega}{2\pi} = \frac{1}{2\pi}\sqrt{\frac{g}{l}} \tag{15.3}$$

Equation 15.3 indicates that *the natural frequency of the pendulum system is a function of the length l and is independent of the mass of the weight.*

15.2.2 Generalised Suspended System

Consider the generalised suspended system shown in Figure 15.2. This consists of a uniform rigid body symmetrically suspended by two massless and inextensible inclined strings/hangers. The top ends of the hangers are restrained by vertical and horizontal springs with stiffnesses k_x and k_y, respectively. The centroid of the rigid body is constrained by horizontal, vertical and rotational springs with stiffnesses of k_{ox}, k_{oy} and k_{or}, respectively.

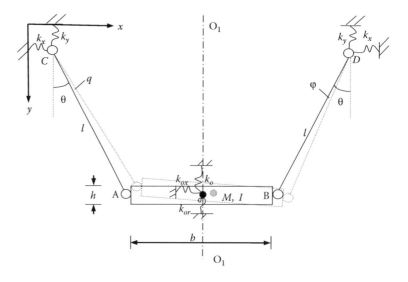

FIGURE 15.2
A generalised suspended system.

The rigid body has a length b, thickness h, mass M and moment of inertia I about its centroid. The two symmetrically inclined strings have the same length l. The lower end of each string has a pinned connection to one end of the rigid body and the upper end has a pinned connection to an elastic support. When in equilibrium under gravity loading, the angles of inclination of the strings are the same and are θ. Changing the angles allows the strings to be inclined outwards ($\theta > 0$, as shown in Figure 15.2), inwards ($\theta < 0$) or vertical ($\theta = 0$). These properties form the basis for this study.

Small amplitude vibrations of the system are considered so that linearisation of the movement of the system is reasonable. It is significant to note that:

- The small second-order quantities of the displacement should be considered when gravitational potential is concerned.
- Small amplitude vibrations of the system take place about its position of static equilibrium.
- The sway angle of the left string can be different to that of the right string during vibration.

This model can be used, for example to represent a cross section of a suspension bridge and then used to study its lateral and torsional movements. Here, the rigid body represents the bridge deck; k_x and k_y represent the constraints from the suspension cables; while k_{ox}, k_{oy} and k_{or} represent the actions on the deck section from its neighbouring elements.

Due to symmetry, the model shown in Figure 15.2 has one symmetric and two antisymmetric vibration modes, which can be studied separately. A detailed investigation of the dynamic characteristics of the system is given in [2]. The findings of some simple special cases are provided in the following subsections.

15.2.2.1 Symmetric (vertical) vibration

When $\theta = 0$ or $k_x = \infty$, as shown in Figure 15.3a and b, respectively,

$$\omega_v^2 = \frac{k_{oy} + 2k_y}{M} = \omega_{oy}^2 + \omega_y^2 \tag{15.4}$$

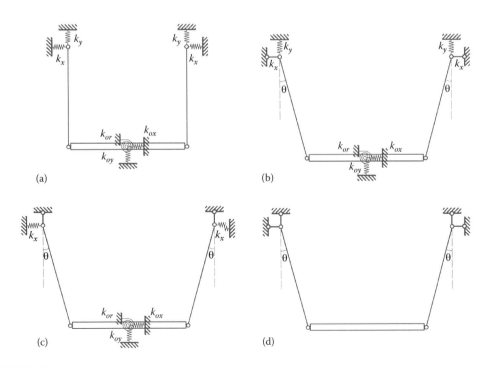

FIGURE 15.3
(a–d) Typical suspended systems.

where

$$\omega_{oy}^2 = \frac{k_{oy}}{M}$$

$$\omega_y^2 = \frac{2k_y}{M}$$

This is the natural frequency of a single-degree-of-freedom system consisting of a mass M and springs with stiffnesses k_{oy} and 2_{ky}.

When $k_y = \infty$, as shown in Figure 15.3c,

$$\omega_v^2 = \frac{k_{oy} + 2k_x/\tan^2\theta}{M} + \frac{\cos\theta}{\sin^2\theta}\frac{g}{l} = \omega_{oy}^2 + \frac{1}{\tan^2\theta}\omega_x^2 + \frac{\cos\theta}{\sin^2\theta}\omega_g^2 \qquad (15.5)$$

in which $\omega_x^2 = 2k_x/M$ and $\omega_g^2 = (g/l)\cdot\left(2k_x/\tan^2\theta\right)$ is the projected vertical stiffness of the two lateral springs due to the inclined strings/hangers and the third term in Equation 15.5 shows the suspension effect.

It can be observed from Equations 15.4 and 15.5 that

- When $\theta = 0$ (Figure 15.3a), that is, the strings/hangers are vertical, the horizontal stiffness at the hanging points has no effect on the symmetric vertical vibration of the rigid body.
- When $k_x = \infty$ (Figure 15.3b), the inclined angle has no effect on the symmetric vertical vibration.
- When $k_y = \infty$ (Figure 15.3c), both the horizontal stiffness k_x at the hanging points and gravity affect the natural frequency of the symmetric vertical vibration.
- When $k_y = \infty$, the natural frequency monotonically increases with decreases of the inclined angle and tends to infinity when θ is close to zero.
- When $k_y = \infty$, the natural frequency is the same for $-\theta$ and θ.

15.2.2.2 Antisymmetric (lateral and rotational) vibration

When $\theta = 0$ as shown in Figure 15.3a, the natural frequencies of the rotational and swaying vibrations of the suspended system are, respectively,

$$\omega_r^2 = \frac{b^2 k_y}{2I} + \omega_{or}^2 \qquad (15.6)$$

$$\omega_s^2 = \omega_{ox}^2 + \frac{\omega_g^2 \omega_x^2}{\omega_x^2 + \omega_g^2} \qquad (15.7)$$

where $\omega_{or} = \sqrt{k_{or}/I}$ and $\omega_{ox} = \sqrt{k_{ox}/M}$ are the natural frequencies of the rigid body in the rotational and horizontal directions without the actions of the hangers.

It can be seen from Equations 15.6 and 15.7 that

- The rotational vibration of the rigid body is independent of the horizontal stiffnesses k_x and k_{ox} and the vertical stiffness k_{oy}.
- The sway vibration is independent of the vertical stiffnesses k_y and k_{oy} and the rotational stiffness k_{or}.

Equation 15.7 can be rewritten as

$$\omega_s = \beta\omega_{ox} \qquad (15.8)$$

where:

$$\beta = \sqrt{1 + \frac{1}{\left(\omega_{ox}/\omega_g\right)^2 + k_{ox}/(2k_x)}} \tag{15.9}$$

β is the magnification factor for the suspension effect. It can be seen from Equations 15.8 and 15.9 that the lateral natural frequency of the system with suspension is always larger than that of the system without suspension. This indicates that *the lateral natural frequency of a suspended bridge estimated using simple beam theory will be smaller than the true lateral natural frequency.* Table 15.1 gives the magnification factors when ω_{ox}/ω_g varies between 0.01 and 100 and k_{ox}/k_x varies between 1 and 100. The results indicate that the larger the value of ω_{ox}/ω_g or the larger the value of k_{ox}/k_x or both, the smaller the suspension effect.

When $k_y = \infty$, $k_x = \infty$ and $k_{ox} = k_{oy} = k_{or} = 0$, as shown in Figure 15.3d, only sway vibration occurs. For a uniform rectangular rigid body whose moment of inertia about the centroid of the body is $I = M(b^2 + h^2)/12$. The solution is

$$\omega_s = \mu\omega_g \tag{15.10}$$

where

$$\mu = \sqrt{\frac{1 + \tan^2\theta\left(1 + 2l\sin\theta/b\right)}{\cos\theta\left[1 + \tan^2\theta\left(1 + h/b\right)/3\right]}} \tag{15.11}$$

where μ is the magnification factor related to the inclined angle and the other design parameters.

As the denominator in the square root is always positive, a real solution of Equation 15.11 requires that

$$1 + \tan^2\theta\left(1 + \frac{2\,l\sin\theta}{b}\right) > 0 \tag{15.12}$$

From Equations 15.11 and 15.12 and parametric analysis [2], the following can be noted.

- When $\theta > 0$, the natural frequency monotonically increases with increasing l/b. However, when $\theta < 0$, the natural frequency monotonically decreases with increasing l/b.
- The natural frequency is independent of the mass of the rigid body. Increasing the thickness–length ratio h/b results in a decrease of the natural frequency. However, for relatively small inclined angles such as $|\theta| < 40°$, the effect of h/l on the natural frequency is very small.
- The critical angle is $\theta = -\arcsin(\sqrt[3]{b/2l})$ which can be obtained from Equation 15.12. Only when the inclined angle is larger than the critical angle will the system oscillate. Otherwise, the system is in a state of unstable equilibrium, as will be demonstrated in Section 15.3.1.

TABLE 15.1   Values of β for Varying Values of ω_{ox}/ω_g and k_{ox}/k_x

ω_{ox}/ω_g	$k_{ox}/k_x = 1$	$k_{ox}/k_x = 5$	$k_{ox}/k_x = 25$	$k_{ox}/k_x = 100$
0.01	1.732	1.183	1.039	1.010
0.1	1.721	1.183	1.039	1.010
1	1.291	1.134	1.036	1.010
10	1.005	1.005	1.004	1.003
100	1	1	1	1

- The critical angle can only be negative. When $b < l$, there is always a critical angle at which the natural frequency of the system becomes zero. However, when $b \geq 2l$, Equation 15.12 always holds. Therefore, the necessary condition for an unstable equilibrium state is that the two strings cross each other.

15.2.3 Translational and Rotational Suspended Systems

When $\theta = 0$, the system shown in Figure 15.3d becomes a translational suspended system in which the motion of the rigid body is around the two points where the two strings are fixed. Any position of the rigid body is therefore parallel to its original position.

In this case, the magnification factor μ is equal to unity and Equation 15.8 reduces to Equation 15.1, that is, the natural frequency of the translational suspended system can be calculated using the equation for a simple pendulum.

One interesting feature of the system is that the system has a constant natural frequency defined by Equation 15.1 even if a mass is added to the rigid body. The size of the mass and the location of the centre of mass do not affect the natural frequency of the system.

When the two fixed points of the two strings meet, it becomes a rotational suspended system in which the motion of the rigid body is around the single fixed point. Its natural frequency can be calculated using Equation 15.10. The rotational suspended system has some different dynamic characteristics from the translational suspended system.

The natural frequencies of the translational and rotational suspended systems and the effect of the added mass on the two systems will be examined through demonstration models in Section 15.3.2.

15.3 MODEL DEMONSTRATIONS

15.3.1 Natural Frequency of Suspended Systems

This model demonstration *verifies Equation 15.10 and shows the unstable equilibrium state at which the suspended system will not oscillate.*

A suspended system consists of a uniform hollow aluminium bar with a square section (a rigid body) and two symmetric strings to suspend the bar. The bar has a length of $b = 0.45$ m, a total mass of $M = 0.094$ kg and a moment of inertia about its centroid of $I = 0.0016$ kg m^2. The lengths and the angles of the strings can be varied.

Three typical suspension forms include vertical (Figure 15.4a, $\theta = 0$), outward-inclined (Figure 15.4b, $\theta > 0$) and inward-inclined (Figure 15.4c, $\theta < 0$) suspension systems. In the tests, an initial lateral displacement is applied to the bar and the bar is suddenly released to generate free vibrations. The number of oscillations is counted and a stopwatch is used to record the duration of the vibrations. The swaying natural frequencies of seven different forms of suspension were measured and are listed in Table 15.2, together with the theoretical predictions obtained using Equation 15.10. It can be seen that there is good agreement between the measured and calculated lateral natural frequencies.

Figure 15.4d shows an unstable equilibrium state of the system where the two strings cross each other. An additional horizontal string is applied to prevent out-plane movement of the system. When a small movement is applied to the system, the aluminium bar moves to, and balances at, a new position without oscillation.

15.3.2 Effect of Added Masses

The models demonstrate that *the natural frequency of a translational suspended system is independent of its mass and the location of the centre of the mass, and is the same as that of an equivalent simple pendulum.*

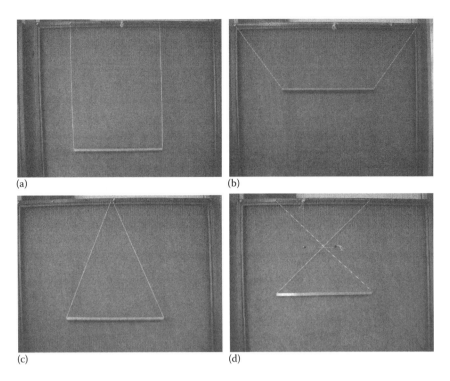

(a) (b) (c) (d)

FIGUFRE 15.4
Typical forms of a suspension system used for demonstration and tests. (a) Vertical suspension system. (b) Outward-inclined suspension system. (c) Inward-inclined suspension system. (d) Unstable equilibrium state.

TABLE 15.2 Comparison of the Measured and Predicted Lateral Natural Frequencies

Parameters		Experiment (Hz)	Theory (Hz)	Error (%)
$\theta = 0°$	$l = 0.415$ m	0.778	0.774	−0.5
$\theta = -10.15°$	$l = 0.431$ m	0.773	0.769	−0.5
$\theta = -22.28°$	$l = 0.825$ m	0.543	0.536	−1.3
$\theta = -30.69°$	$l = 0.485$ m	0.718	0.717	−0.1
$\theta = -52.70°$	$l = 0.584$ m	UES[a]	UES[a]	–
$\theta = 24.66°$	$l = 0.405$ m	0.930	0.929	−0.1
$\theta = 48.40°$	$l = 0.324$ m	1.67	1.71	2.4

[a] UES: unstable equilibrium state.

Figure 15.5 shows another simple pendulum system. A plate is suspended by four strings through four holes at the corners of the plate. The other ends of the four strings are fixed to two cantilever frames as shown in Figure 15.5. The strings are vertical when viewed from an angle perpendicular to the plane of the frame and are inclined when viewed in the plane of the frame.

When the plate moves in the plane of the steel frames, it forms a translational pendulum system in which the plate remains horizontal during its motion. Figure 15.6 shows two similar translational suspended systems in which the masses sway in the plane of the supporting frames. Eight magnetic bars and a steel block are placed on the plate of one suspended system to raise the centre of mass of the system. On applying the same displacement to the two plates in the planes of the frames and releasing them simultaneously, it can be observed that the two suspended systems with different masses and different centres of mass sway at the same frequency. This demonstrates that *the natural frequency of a translational suspended system is independent of its mass and the location of the centre of mass.*

FIGURE 15.5
A model of a translational suspended system and a rotational suspended system.

FIGURE 15.6
Effects of mass and the centre of mass.

The suspended systems in Figure 15.6 can also be used as two identical rotational suspended systems when the plates sway perpendicular to the planes of the frames. Applying the same displacements to the two plates in the direction perpendicular to the frames and then releasing them simultaneously, it can be observed that the plate with the added weights oscillates faster than the other plate.

Figure 15.7 shows two arrangements of eight identical magnetic bars. In one case, the eight bars stand on the plate (Figure 15.7a), and in the other case, four magnetic bars are placed on the top and

(a)   (b)

FIGURE 15.7
(a,b) Effect of the centre of mass.

TABLE 15.3 Comparison of the Times for 30 Oscillations of the Suspended Systems

	Translational Suspended Systems (s)	Rotational Suspended Systems (s)
Empty (Figure 15.5)	34.3, 34.4	34.4, 34.5
With full weights (Figure 15.6)	34.3, 34.5	32.3, 32.4
With magnetic bars placed vertically (Figure 15.7a)	34.2, 34.2	33.4, 33.3
With magnetic bars placed horizontally (Figure 15.7b)	34.2, 34.3	34.3, 34.3

bottom surfaces of the plate, respectively (Figure 15.7b). The two cases have the same amount of mass but different locations of the centres of mass. Conducting the same experiment as before for the rotational suspended systems, it can be seen that the system with standing bars oscillates faster than that with horizontal bars.

The systems shown in Figures 15.5 and 15.7b have different masses but the same location of the centre of mass. If the oscillations of the two systems shown are generated by giving the same initial displacements in the plane perpendicular to the frames, it can be observed that the two systems oscillate at the same frequency.

The two sets of experiments demonstrate that *the natural frequency of a rotational suspended system is dependent on the location of its centre of mass but is independent of the magnitude of its mass.*

Table 15.3 compares the times recorded for 30 oscillations of the translational and rotational suspended systems with added masses. Each case was tested twice.

15.3.3 Static Behaviour of an Outward-Inclined Suspended System

This demonstration shows that *an outward-inclined suspended system is a mechanism that moves sideways if a vertical load is applied asymmetrically on the plate.*

Consider two suspended systems, where a steel plate is suspended by two vertical strings and the same plate is suspended by two symmetrically outward-inclined strings, as shown in Figure 15.8.

Place similar weights asymmetrically on the two plates, as shown in Figure 15.8b. It can be observed that the vertically suspended system does not experience any lateral movement while the outward-inclined suspended system undergoes both lateral and rotational movements.

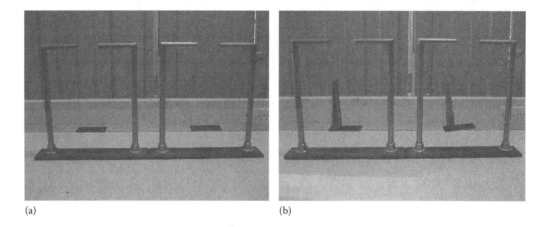

(a)   (b)

FIGURE 15.8
(a,b) Static behaviour of an outward-inclined suspended system.

For the vertically suspended system, the vertical and lateral movements are independent and the vertical load only induces vertical deformations of the strings with little rotation due to the difference of the elastic elongation of the strings. For the plate suspended by the two outward-inclined strings, the horizontal and rotational movements of the plate are coupled. The movements relate to the geometry of the system rather than the elastic elongation of the strings.

15.4 PRACTICAL EXAMPLES

15.4.1 Inclined Suspended Wooden Bridge in a Playground

As demonstrated in Section 15.3.3, an inclined suspended system moves sideways if a vertical load is applied asymmetrically. The phenomenon should be avoided in engineering structures, as it may lead to unsafe structures. However, Figure 15.9 shows an outward-inclined suspended wooden footbridge in a playground. The bridge is purposely built in such a way that the bridge

(a)   (b)

FIGURE 15.9
(a,b) An inclined suspended wooden bridge in a playground.

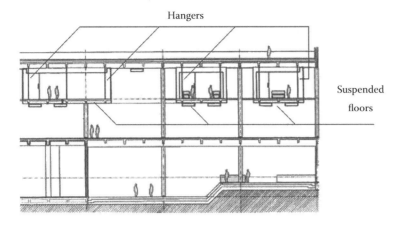

Hangers

Suspended

floors

FIGURE 15.10
A translational suspended floor. (Courtesy of Professor M. Kawaguchi.)

wobbles when a child walks on the bridge, creating excitement and a challenge for crossing the bridge.

15.4.2 Seismic Isolation of a Floor

As the natural frequency of a translational suspended system is only governed by its length and is independent of its mass, the concept of a pendulum seismic isolator was developed and used in Japan.

As shown in Figure 15.10, a floor suspended from the girders of a building frame in the form of a translational suspended system was adopted for the exhibition rooms of an actual museum for pottery and porcelain in Japan. The area of the suspended floor is about 1000 m², and its mass is about 1000 t. Hinges having universal joints were used for the upper and lower ends of the hangers. The hangers were 4.5 m long, producing a natural period of more than 4 s, which was considered sufficiently long for seismic isolation.

15.4.3 Foucault Pendulum

Figure 15.11 shows a large brass pendulum that swings over the lower foyer in the Manchester Conference Centre at the University of Manchester. The pendulum was set up as a tribute to Jean Bernard Leon Foucault.

Jean Bernard Leon Foucault (1819–1868), a French physicist, demonstrated the earth's rotation using his famous pendulum. He suspended a 28 kg bob with a 67 m wire from the dome of the Pantheon in Paris. Foucault was acclaimed by witnesses for proving that the earth does indeed spin on its axis. As the plane of the pendulum oscillation remained unchanged with the stars, observers could understand that the Pantheon moved around the pendulum and not vice versa! Foucault's pendulum was the first dynamical proof of the earth's rotation in an easy-to-see experiment.

FIGURE 15.11
The Foucault pendulum.

PROBLEMS

1. Figures 15.8b and 15.9 show that outward-inclined suspended systems would experience both sway and rotational movements when vertical loads are not placed symmetrically.
   a. What is the implication of the observations for bridge design?
   b. How can it be ensured that a footbridge using an outward-inclined suspended system will not experience swaying movements (wobbling)?
2. Different behaviour of suspended systems can be observed through conducting the following simple experiments.
   a. Make a simple suspended system, as demonstrated in Section 15.3.1.
   b. Measure the natural frequency of the vertical, outward-inclined and inward-inclined suspended systems, as illustrated in Figure 15.4, using a stopwatch.
   c. Examine the behaviour of the suspended system in the unstable equilibrium state shown in Figure 15.4d.
   d. Calculate the natural frequencies of the test systems using Equations 15.10 and 15.11, and compare the calculated and measured results.

REFERENCES

1. Matthews, M. R., Gauld, C. F. and Stinner, A. *The Pendulum: Scientific, Historical, Philosophical and Educational Perspectives*, Dordrecht: Springer, 37–47, 2005.
2. Zhou, D. and Ji, T. Dynamic characteristics of a generalised suspension system, *International Journal of Mechanical Sciences*, 50, 30–42, 2008.

CHAPTER 16

CONTENTS

Free Vibration

16

16.1 DEFINITIONS AND CONCEPTS

Free vibration: A structure is said to undergo free vibration when it is disturbed from its static stable equilibrium position by an initial displacement and/or initial velocity and then allowed to vibrate without any external excitation.

Period of vibration: The time required for an undamped system to complete one cycle of free vibration is the natural period of vibration of the system.

Natural frequency: The number of cycles of free vibration of an undamped system in one second is termed **the natural frequency of the system**, and is the inverse of the period of vibration.

A single-degree-of-freedom system: If the displacement of a system can be uniquely determined by a single variable, this system is called a *single-degree-of-freedom* (SDOF) system. Normally it consists of a mass, a spring and a damper. The square of the natural frequency of an SDOF system is proportional to the stiffness of the system and the inverse of its mass.

A generalised SDOF system: Consider a discrete system that consists of several masses, springs and dampers, or a continuous system that has distributed mass, stiffness and damping. If the shape or pattern of its displacements is known or assumed, the displacements of the system can then be uniquely determined by its magnitude (a single variable). This system is termed a *generalised SDOF system*. The analysis developed for an SDOF system is applicable to a generalised SDOF system.

- For a structure with a given mass, the stiffer the structure, the higher the natural frequency.
- The larger the damping ratio of a structure, the quicker the decay of its free vibration.
- The higher the natural frequency of a structure, the quicker the decay of its free vibration.
- The fundamental natural frequency reflects the stiffness of a structure. Thus, it can be used to predict the displacement of a simple structure. Also, the displacement of a simple structure can be used to estimate its fundamental natural frequency.

16.2 THEORETICAL BACKGROUND

16.2.1 Single-Degree-of-Freedom System

Consider an SDOF system, as shown in Figure 16.1a, that consists of the following:

- A mass, m (kg or $Ns^2 m^{-1}$), whose motion is to be examined.
- A spring with stiffness k (Nm^{-1}). The action of the spring tends to return the mass to its original position of equilibrium. Thus, the direction of the elastic force applied on the mass is opposite to that of the motion. Hence, the force on the mass is $-ku$ (N).
- A damper (dashpot) that exerts a force whose magnitude is proportional to the velocity of the mass. The constant of proportionality c is known as the *viscous damping coefficient* and has units of Nsm^{-1}. The action of the damping force tends to reduce the velocity of motion. Thus, the direction of the damping force applied on the mass is opposite to that of the *velocity* of the motion. The force is expressed as $-c\dot{u}$ (N).

Considering the free-body diagram in Figure 16.1b, the equation of motion of the system can be obtained using Newton's second law as

$$m\ddot{u} = -c\dot{u} - ku \qquad (16.1)$$

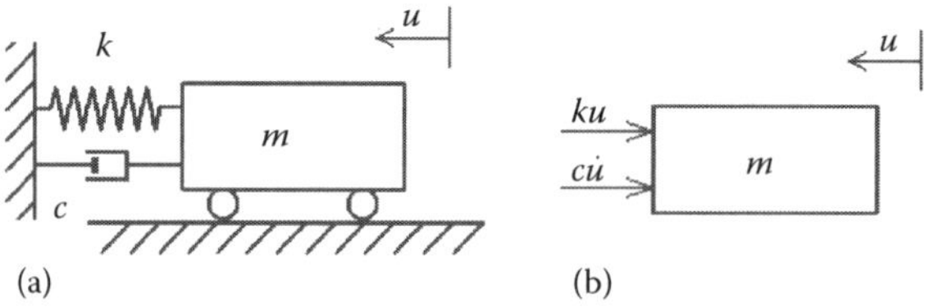

FIGURE 16.1
Free vibration of an SDOF system. (a) An SDOF system. (b) Free-body diagram.

or

$$m\ddot{u} + c\dot{u} + ku = 0 \qquad (16.2)$$

The frequency of oscillation of the system, called its *natural frequency*, in hertz, is

$$f = \frac{\omega}{2\pi} = \frac{1}{2\pi}\sqrt{\frac{k}{m}} \qquad (16.3a)$$

For viscous damping, it can be shown that

$$c = 2\xi m\omega \qquad (16.3b)$$

where ξ is the damping ratio. Substituting Equation 16.3 into Equation 16.2 gives

$$\ddot{u} + 2\xi\omega\dot{u} + \omega^2 u = 0 \qquad (16.4)$$

The solution of Equation 16.4 is in the form of [1,2]

$$u = Ae^{s_1 t} + Be^{s_2 t} \qquad (16.5a)$$

$$s_{1,2} = \left[-\xi \pm \sqrt{\xi^2 - 1} \right]\omega \qquad (16.5b)$$

where A and B are constants that can be determined from the initial conditions. It can be seen from Equation 16.5 that the response of the system depends on whether the damping ratio ξ is greater than, equal to or smaller than unity.

Case 1: $\xi = 1$, that is, critically damped systems.
In this special case, there are two identical roots from Equation 16.5b:

$$s_1 = s_2 = -\omega \qquad (16.6)$$

Substituting Equation 16.6 into Equation 16.5a gives

$$u = (A + Bt)e^{-\omega t} \qquad (16.7)$$

If the mass is displaced from its position of static equilibrium at time $t = 0$, the initial conditions $u(0)$ and $\dot{u}(0)$ can be used to determine the two constants in Equation 16.7. Thus, Equation 16.7 can be written as

$$u = \left[u(0)(1 + \omega t) + \dot{u}(0)t \right]e^{-\omega t} \qquad (16.8)$$

Equation 16.8 can be illustrated graphically in Figure 16.2a for the values of $u(0) = 1$ cm, $\dot{u}(0) = 1$ cm s^{-1} and $\omega = 1$ rad s^{-1}. This shows that the response of the critically damped system does not oscillate about its equilibrium position, but returns to the position of equilibrium asymptotically, as dictated by the exponential term in Equation 16.8.

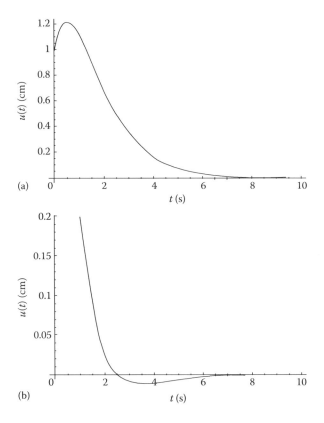

FIGURE 16.2
(a,b) Free vibration of a critically damped system.

It is of interest to know the condition if the response crosses the zero-deflection position. Solving Equation 16.8 at $u=0$ gives the condition that the response crosses the horizontal axis once at time:

$$t = -\frac{u(0)}{\dot{u}(0)+u(0)\omega} \tag{16.9}$$

Consider a case where $u(0)=1$ cm, $\dot{u}(0)=-1.4$ cm s^{-1} and $\omega=1$ rad s^{-1} and where t must be larger than zero. Substituting the values into Equation 16.9 gives the solution $t=2.5$ s. Figure 16.2b shows the curve defined by Equation 16.8, for this particular case. It can be seen that the response crosses the zero-deflection position at 2.5 s then monotonically approaches the zero-deflection position.

Case 2: $\xi>1$, that is, overcritically damped systems.
In this special case, the solution given in Equation 16.5 can be rewritten as [2–4]

$$u(t) = \left[A\sinh \bar{\omega}t + B\cosh \bar{\omega}t\right]e^{-\xi\omega t} \tag{16.10}$$

where $\bar{\omega} = \omega\sqrt{\xi^2-1}$
The constants A and B can be determined using the initial conditions $u(0)$ and $\dot{u}(0)$ and Equation 16.10 becomes

$$u(t) = \left[\frac{\dot{u}(0)+u(0)\xi\omega}{\bar{\omega}}\sinh \bar{\omega}t + u(0)\cos h\bar{\omega}t\right]e^{-\xi\omega t} \tag{16.11}$$

Equation 16.11 shows that the response reduces exponentially, which is similar to the motion of the critically damped system. For example, using the same data $u(0)=1$ cm, $\dot{u}(0)=1$ cm s^{-1} and $\omega=1$ rad s^{-1} as were used for producing Figure 16.2a, but with $\xi=2$. The free vibration of the over-critically damped system is shown in Figure 16.3.

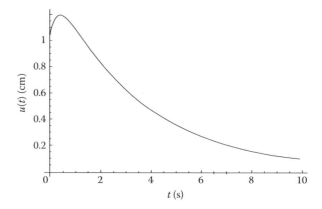

FIGURE 16.3
Free vibration of an overcritically damped system.

Comparing the displacement time histories in Figures 16.2a and 16.3, it can be noted that the overcritically damped system takes a longer time to return to the original equilibrium position than the critically damped system. The movement of an overcritically damped system will be demonstrated in Section 16.3.3.

Case 3: $\xi < 1$, that is, undercritically damped systems.
In this special case, the two roots in Equation 16.5b become

$$s_{1,2} = -\xi\omega \pm i\omega_D \tag{16.12}$$

where

$$\omega_D = \omega\sqrt{1-\xi^2} \tag{16.13}$$

ω_D is the damped natural frequency of the system in free vibrations. The solution given in Equation 16.5 becomes

$$u(t) = \left[A\cos\omega_D t + B\sin\omega_D t\right]e^{-\xi\omega t} \tag{16.14}$$

Using the initial conditions $u(0)$ and $\dot{u}(0)$, the constants can be determined leading to

$$u(t) = \left[\frac{\dot{u}(0) + u(0)\xi\omega}{\omega_D}\sin\omega_D t + u(0)\cos\omega_D t\right]e^{-\xi\omega t} \tag{16.15}$$

or

$$u(t) = d\cos(\omega_D t - \theta)e^{-\xi\omega t} \tag{16.16}$$

where

$$d = \sqrt{u(0)^2 + \left(\frac{\dot{u}(0) + u(0)\xi\omega}{\omega_D}\right)^2} \tag{16.17a}$$

$$\sin\theta = \frac{\dot{u}(0) + u(0)\xi\omega}{\omega_D d} \tag{16.17b}$$

$$\cos\theta = \frac{u(0)}{d} \tag{16.17c}$$

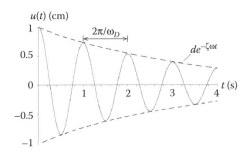

FIGURE 16.4
Free vibration of an undercritically damped system.

$$\theta = \tan^{-1}\left\{\frac{\dot{u}(0)+u(0)\xi\omega}{\omega_D u(0)}\right\} \tag{16.18}$$

Figure 16.4 shows a typical free vibration of an undercritically damped SDOF system defined by Equation 16.16.

From Figure 16.4 and Equation 16.15 it can be seen that

- The damped system oscillates about the position of equilibrium with a damped natural frequency of ω_D.
- The oscillation decays exponentially.
- The rate of the exponential decay depends on the product of the damping ratio ξ and the natural frequency of ω.

This last point indicates that *a system with a higher natural frequency will decay more quickly than a similar system with a lower natural frequency in free vibration.*

EXAMPLE 16.1

Two lightly damped SDOF systems have the same damping ratio $\xi = 0.05$ (or 5% critical) but different natural frequencies of $\omega_1 = 2\pi$ rad s^{-1} and $\omega_2 = 4\pi$ rad s^{-1}, respectively. Applying the same initial displacement $u(0) = 1.0$ cm to the two systems and releasing them simultaneously will generate free vibrations. Calculate the exponential decay at $t = 3$ s and plot the vibration time histories.

SOLUTION

For the first system: $-\xi\omega_1 t = -0.05 \times 2\pi \times 3 = -0.3\pi = -0.942$

The exponential decay is $e^{-\xi\omega_1 t} = e^{-0.942} = 0.390$

For the second system $-\xi\omega_2 t = -0.05 \times 4\pi \times 3 = -0.6\pi = -1.885$

The exponential decay is $e^{-\xi\omega_2 t} = e^{-1.885} = 0.152$

Figure 16.5 shows the curves of free vibrations of the two systems using Equation 16.15. The results show that *the higher the natural frequency, the quicker the decay of its free vibration.*

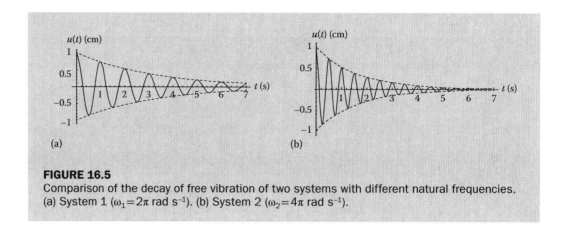

FIGURE 16.5
Comparison of the decay of free vibration of two systems with different natural frequencies.
(a) System 1 ($\omega_1 = 2\pi$ rad s^{-1}). (b) System 2 ($\omega_2 = 4\pi$ rad s^{-1}).

16.2.2 Generalised Single-Degree-of-Freedom System

Consider the vibration of a particular mode of a structure, the vibration can be described by an SDOF system. Thus, it is necessary to calculate the generalised properties of the system, the generalised mass, damping and stiffness.

Consider a continuous system, such as the simply supported beam shown in Figure 16.6a. For a particular mode of vibration $\varphi(x)$, which may be known or, if not, be assumed with a maximum value of unity, the vibration of the mode can be described by the variable $z(t)$ at the centre of the beam and the movement of the beam at coordinate x as $z(t)\varphi(x)$. The vibration of the beam $v(x, t)$ in the particular mode can then be represented by

$$v\left(x,t\right) = z\left(t\right)\varphi\left(x\right) \tag{16.19}$$

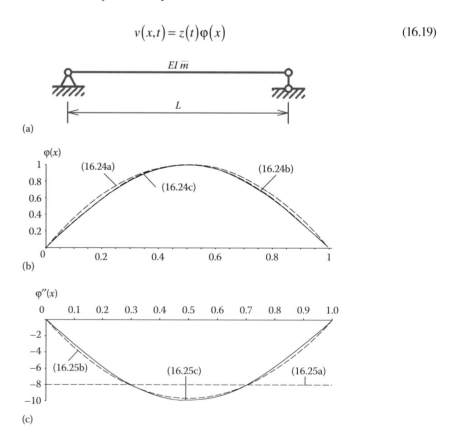

FIGURE 16.6
A simply supported beam with three fundamental mode shapes considered. (a) A simply supported beam. (b) Comparison of three mode shapes. (c) Comparison of the second derivatives of the three mode shapes.

The generalised properties for the vibration of the system in a particular mode are [2]

$$\text{Modal mass: } M_m = \int_0^L \bar{m}(x)\varphi(x)^2\,dx \tag{16.20a}$$

$$\text{Modal stiffness: } K_m = \int_0^L EI(x)\varphi''(x)^2\,dx \tag{16.20b}$$

$$\text{Damping coefficient: } C_m = 2\xi M_m \omega \tag{16.21}$$

The equation of motion of the system (Equation 16.2) then becomes

$$M_m\ddot{z} + C_m\dot{z} + K_m z = 0 \tag{16.22}$$

The natural frequency of the vibration mode of the structure can be obtained from Equation 16.22 as

$$\omega^2 = \frac{K_m}{M_m} = \frac{\int_0^L EI(x)\varphi''(x)^2\,dx}{\int_0^L \bar{m}(x)\varphi(x)^2\,dx} \tag{16.23}$$

The accuracy of the natural frequency of the vibration mode depends on the quality of the known or assumed mode shape $\varphi(x)$. In order to obtain a good estimation of the natural frequency, the assumed shape $\varphi(x)$ should satisfy as many boundary conditions of the structure as possible. In general:

1. For a pinned support at $x=0$, the displacement and bending moment at the support should be zero, that is,

$$\varphi(0) = 0 \quad \text{and} \quad \varphi''(0) = 0$$

2. For a fixed support at $x=0$, the displacement and rotation at the support should be zero, that is,

$$\varphi(0) = 0 \quad \text{and} \quad \varphi'(0) = 0$$

3. For a free end at $x=0$, the bending moment and shear force at the free end should be equal to the applied load P and bending moment M, that is,

$$EI\varphi''(0) = -M \quad \text{and} \quad EI\varphi'''(0) = -P$$

If there is no applied loading at the free end, $M=P=0$.

For a simple structure, a good approximation of the fundamental mode shape is the shape of the static deflection of the structure when it is subjected to distributed loads proportional to the mass distribution of the structure. However, this is not a good approximation for more complicated structures, for example a bridge with more than one span.

EXAMPLE 16.2

Figure 16.6a shows a simply supported beam with a length of L, a uniformly distributed mass of $\bar{m}$ and a flexural rigidity of EI. Three different mode shapes will be considered in this analysis. They are

$$\varphi_1(x) = 4\left(\frac{x}{L}\right)\left(\frac{1-x}{L}\right) \tag{16.24a}$$

$$\varphi_2(x) = \frac{16x}{5L^4}\left(L^3 - 2Lx^2 + x^3\right) \tag{16.24b}$$

$$\varphi_3(x) = \sin\frac{\pi x}{L} \tag{16.24c}$$

Calculate and compare the modal masses, modal stiffnesses and natural frequencies determined using the three mode shapes.

SOLUTION

Figure 16.6b compares the three mode shapes used in Equation 16.24 where the horizontal axis is x/L, varying between 0 and 1. It can be seen that the curves defined by Equation 16.24b and Equation 16.24c almost overlap although there are small differences between the curves defined by Equations 16.24a through 16.24c.

Differentiating Equations 16.24a through 16.24c twice with respect to x gives

$$\varphi_1''(x) = -\frac{8}{L^2} \tag{16.25a}$$

$$\varphi_2''(x) = -\frac{192x(L-x)}{5L^4} \tag{16.25b}$$

$$\varphi_3''(x) = -\frac{\pi^2}{L^2}\sin\frac{\pi x}{L} \tag{16.25c}$$

Figure 16.6c compares the shapes of the three second derivatives. The curves defined by Equations 16.25b and 16.25c are close to each other. However, there are significant differences between the curves defined by Equations 16.25a through 16.25c.

Case 1: This assumed shape satisfies $\varphi(0)=0$ and $\varphi(L)=0$. However, $\varphi''(0) \neq 0$ and $\varphi''(L) \neq 0$. Using Equations 16.20 and 16.23 gives

$$M_m = \bar{m}\int_0^L \left[\frac{4x}{L}\left(1 - \frac{x}{L}\right)\right]^2 dx = \frac{8\bar{m}L}{15}$$

$$K_m = EI\int_0^L \varphi''(x)^2 = EI\int_0^L \left(-\frac{8}{L^2}\right)^2 dx = \frac{64EI}{L^3}$$

$$\omega^2 = \frac{K_m}{M_m} = \frac{64EI}{L^3}\frac{15}{8\bar{m}L} = \frac{120EI}{\bar{m}L^4} \tag{16.26a}$$

Case 2: Equation 16.24b is the shape of the static deflection curve when the beam is subjected to a uniformly distributed load. This function satisfies the required boundary conditions and

$$M_m = \bar{m} \int_0^L \varphi(x)^2 \, dx = \frac{3968\bar{m}L}{7875}$$

$$K_m = EI \int_0^L \varphi''(x)^2 = EI \int_0^L \left(-\frac{192x(L-x)}{5L^4} \right)^2 dx = \frac{6144EI}{125L^3}$$

$$\omega^2 = \frac{K_m}{M_m} = \frac{6144EI}{125L^3} \frac{7875}{3968\bar{m}L} = \frac{3024EI}{31\bar{m}L^4} = \frac{97.55EI}{\bar{m}L^4} \qquad (16.26b)$$

Case 3: Equation 16.24c is the true shape of the fundamental mode of a simply supported beam and leads to

$$M_m = \bar{m} \int_0^L \varphi(x)^2 \, dx = \frac{\bar{m}L}{2}$$

$$K_m = EI \int_0^L \varphi''(x)^2 = EI \int_0^L \left(-\frac{\pi^2}{L^2} \sin\frac{\pi x}{L} \right)^2 dx = \frac{\pi^4 EI}{2L^3}$$

$$\omega^2 = \frac{K_m}{M_m} = \frac{\pi^4 EI}{2L^3} \frac{2}{\bar{m}L} = \frac{\pi^4 EI}{\bar{m}L^4} = \frac{97.41EI}{\bar{m}L^4} \qquad (16.26c)$$

Table 16.1 compares coefficients of M_m, K_m and ω^2 in Equation 16.26 and gives the relative errors (RE) against the exact values. The following can be seen from Table 16.1:

- Equation 16.26b produces a solution very close to the true solution given by Equation 16.26c.
- The solution obtained using Equation 16.26a is 23% larger than the true solution, which is because the assumed mode shape does not fully satisfy the boundary conditions.
- The differences between the modal stiffnesses are larger than the differences between the modal masses, as there are small differences between the three shape functions which are used for calculating the modal masses, but there are larger differences between the second derivatives of the three functions which are used to calculate the modal stiffnesses.
- Using the assumed mode shapes overestimates the modal stiffness, modal mass and natural frequency of the beam.

TABLE 16.1 Comparison of Coefficients of M_m, K_m and ω^2

	$K_m = a_1 EI/L^3$		$M_m = a_2 \bar{m}L$		$\omega^2 = a_3 EI/(\bar{m}L^4)$	
	a_1	RE (%)	a_2	RE (%)	a_3	RE (%)
Equation 16.26a	64	31.4	8/15	6.7	120	23.2
Equation 16.26b	6144/125	0.9	3968/7875	0.8	3024/31	0.1
Equation 16.26c	$\pi^4/2$	0	1/2	0	π^4	0

Conceptually, the use of assumed mode shapes rather than the true mode shapes is equivalent to applying additional external constraints on the structure to force the structure to deform in the assumed mode shape. These constraints effectively stiffen the structure, leading to an increased stiffness, and hence overestimate the true natural frequencies. This indicates that the lowest natural frequency obtained using a number of different assumed mode shapes would be the closest to the true value. For instance, the estimated natural frequency in Case 2 is smaller than that in Case 1; thus, Case 2 provides a better estimation than Case 1.

16.2.3 Multi-Degree-of-Freedom System

Consider the free vibration of a linear system that consists of r degrees of freedom. The equation of undamped free vibration of the system has the following form:

$$M\ddot{v} + Kv = 0 \tag{16.27}$$

where:
   M is the mass matrix
   K is the stiffness matrix
   v is the vector of displacements

The natural frequencies and mode shapes of the system can be obtained by solving the eigenvalue problem:

$$\left[K - \omega_i^2 M\right]\varphi_i = 0 \tag{16.28}$$

where:
   ω_i and φ_i are the natural frequency and shape of the ith mode ($i = 1, 2, \ldots, r$)
   ω_i^2 is also known as an eigenvalue or characteristic value

The N modes obtained from Equation 16.28 are independent, that is, no one mode can be expressed by a linear combination of the other modes. These modes satisfy the following orthogonality conditions:

$$\varphi_i^T K \varphi_j = 0 \quad \varphi_i^T M \varphi_j = 0 \quad i \neq j \tag{16.29a}$$

$$\varphi_i^T K \varphi_i = K_{m,i} \quad \varphi_i^T M \varphi_i = M_{m,i} \quad i = j \tag{16.29b}$$

where $K_{m,i}$ and $M_{m,i}$ are the modal stiffness and the modal mass of the ith mode and are positive, and related by

$$K_{m,i} = \omega_i^2 M_{m,i} \tag{16.30}$$

The summation of the response of several modes of a structure can be used to represent the structural response. As the N independent mode vectors form a base in the N-dimensional space, any other vector in the space can be expressed as a linear combination of the N mode vectors, that is,

$$v = z_1\varphi_1 + z_2\varphi_2 + \cdots + z_N\varphi_N = \sum_{i=1}^{N} Z_i\varphi_i \tag{16.31}$$

where z_i is the magnitude of the ith mode, defining the contribution of the mode to the total response.

In practice, it is frequently the case that the first few modes can reasonably represent the total dynamic response of structures. This leads to an effective simplification of analysis without significant loss of accuracy. Figure 16.7 shows the vibration of a cantilever column. Its deflected shape can be reasonably represented by the summation of the response of the first three modes. (The shapes of the first two modes of vibration of a cantilever are shown in Section 16.3.5.)

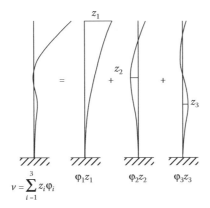

$$v = \sum_{i=1}^{3} z_i \varphi_i \qquad \varphi_1 z_1 \qquad \varphi_2 z_2 \qquad \varphi_3 z_3$$

FIGURE 16.7
Deflected shape as the sum of modal components. (From Clough, R. W. and Penzien, J. *Dynamics of Structures*, McGraw-Hill, New York, 1993.)

16.2.4 Relationship between Fundamental Natural Frequency and Maximum Displacement of a Beam

The fundamental natural frequency of a uniform beam can be expressed using Equations 16.23 and 16.26:

$$f = \frac{\omega}{2\pi} = \frac{b_1}{2\pi} \sqrt{\frac{EI}{\bar{m}L^4}} \tag{16.32}$$

where b_1 is a constant related to the boundary conditions.

The values of b_1 for beams with common boundary conditions are given in Table 16.2. For example, $b_1 = \pi^2$ for a simply supported beam. From Chapter 8, it is known that the maximum displacement of the uniform beam, listed in Table 16.2, subjected to self-weight can be expressed in a unified form as follows:

$$\Delta = d_1 \frac{\bar{m}gL^4}{EI} \tag{16.33}$$

where:

- g is gravity
- $\bar{m}$ is the distributed mass of the beam along its length L
- d_1 is a constant dependent on the boundary conditions

For example, $d_1 = 5/384$ for a simply supported beam. The constant d_1 for the beams with four common boundary conditions are given in Table 16.2.

TABLE 16.2	Coefficients for Single-Span Beams			
	Simply Supported Beam	Fixed-End Beam	Propped Cantilever Beam	Cantilever Beam
b_1	9.87	22.4	15.4	3.52
d_1	0.0130	0.00260	0.00542	0.125
A_1 (mm$^{1/2}$ s^{-1})	17.74	18.01	17.88	19.63
B_1 (mm s^{-2})	315	325	320	385

It can be noted that Equations 16.32 and 16.33 contain the same term EI, through which a relationship between the fundamental natural frequency, f, and the maximum displacement, Δ, due to the self-weight of a beam can be established. Equations 16.32 and 16.33 can be expressed as

$$EI = \frac{4\pi^2 \bar{m} L^4}{b_1^2} f^2 \quad EI = d_1 \frac{\bar{m} g L^4}{\Delta}$$

leading to

$$f = \frac{b_1}{2\pi} \sqrt{d_1 g} \sqrt{\frac{1}{\Delta}} = A_1 \sqrt{\frac{1}{\Delta}} \tag{16.34}$$

$$\Delta = \frac{b_1^2 d_1 g}{4\pi} \frac{1}{f^2} = B_1 \frac{1}{f^2} \tag{16.35}$$

As the constants b_1 and d_1 are given, the coefficients A_1 and B_1 can be calculated for the beams listed in Table 16.2 and have units of $\text{mm}^{1/2}\,\text{s}^{-2}$ and mm s^{-2}, respectively. The corresponding values of A_1 and B_1 are given in Table 16.2, using $g = 9810 \text{ mm s}^{-2}$. Thus, the displacement Δ is measured by millimetres in Equations 16.34 and 16.35. The fundamental natural frequency of a beam can be calculated using Table 16.2 and Equation 16.34 if the maximum displacement of the beam due to its self-weight is known. Alternatively, the maximum displacement of a beam can be estimated using Table 16.2 and Equation 16.35 if the fundamental natural frequency is available either from calculation or vibration measurement. Examples 16.3 and 16.4 illustrate some applications.

EXAMPLE 16.3

A cantilever beam supporting a floor has a self-weight of 20 kN m^{-1}. A static analysis of the beam shows the maximum displacement of 60 mm when subjected to live load of 30 kN m^{-1}. Calculate the fundamental natural frequency of the beam.

SOLUTION

The maximum displacement of the structure subjected to its self-weight should be $60 \times (20/30) = 40$ mm. For a cantilever beam, select $A_1 = 19.63$ from Table 16.2. The fundamental natural frequency of the cantilever beam is thus calculated as

$$f = A_1 \sqrt{\frac{1}{\Delta}} = 19.63 \sqrt{\frac{1}{40}} = 3.10 \text{ Hz} \tag{16.36}$$

EXAMPLE 16.4

A simply supported beam bridge has a self-weight of 30 kN m^{-1} and the fundamental natural frequency of the bridge is 4.5 Hz. Estimate the maximum displacement of the bridge for a live load of 150 kN m^{-1}.

SOLUTION

For a simply supported beam, select $B_1 = 315$ from Table 16.2. Using Equation 16.35 gives the maximum displacement of the bridge for self-weight only as

$$\Delta = B_1 \frac{1}{f^2} = 315 \frac{1}{4.5^2} = 15.6 \text{ mm}$$

The maximum displacement due to the live load will be

$$15.6 \times \frac{150}{30} = 78 \text{ mm}$$

It is interesting to note in Table 16.2 that the values of A_1 and B_1 for the four cases are close, in particular for the first three cases, indicating that A_1 and B_1 are not sensitive to boundary conditions. This observation is useful for practical application of Equations 16.34 and 16.35 as boundary conditions are often not either truly pinned or fixed. Actually, the effect of boundary conditions has been included in the value of Δ used to determine the fundamental natural frequency f, and in f for calculating the maximum displacement Δ. Often, $A_1 = 18.0$ is used to determine the natural frequency from a known displacement of a simple structure in engineering practice [5].

16.3 MODEL DEMONSTRATIONS

16.3.1 Free Vibration of a Pendulum System

This demonstration shows *the definitions of natural frequency and period of a simple system.*

Figure 16.8 shows a pendulum system consisting of five balls suspended by strings with equal length (known as a Newton's cradle). Push the balls in a direction parallel to the plane of the frames to apply an initial displacement and release the balls. The balls then move from side to side in free vibration at the natural frequency of the system.

Count the number of oscillations of the balls in 30 s. The number of oscillations divided by 30 is the natural frequency of the oscillations of the pendulum system in cycles per second or hertz. The inverse of the natural frequency is the period of the oscillations. For this system, 40 cycles of oscillation in 30 s were counted. Thus, the natural frequency of the system is 40/30 = 1.33 Hz and the period is 1/1.33 = 0.75 s.

The vertical distance between the supports of the strings and the connecting points on the balls is 140 mm, so the theoretical value of the natural frequency is

$$f = \frac{1}{2\pi}\sqrt{\frac{g}{l}} = \frac{1}{2\pi}\sqrt{\frac{9810}{140}} = 1.33 \text{ Hz} \tag{14.3}$$

FIGURE 16.8
Free vibration and natural frequency of a pendulum system.

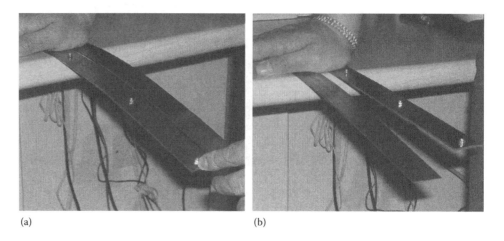

(a)   (b)

FIGURE 16.9
The decay of free vibration and the natural frequencies of members. (a) Applying the same initial displacement. (b) The double ruler stops vibration more quickly than the single ruler.

16.3.2 Vibration Decay and Natural Frequency

This set of models demonstrates that *the higher the natural frequency of a structure, the quicker the decay of its free vibrations*, which has been illustrated numerically in Example 16.1.

A single steel ruler and a pair of identical steel rulers, bolted firmly together, are placed side by side as cantilevers, as shown in Figure 16.9a. Give the ends of the two cantilevers the same initial displacement and release them suddenly at the same time (Figure 16.9a). Free vibrations of the rulers follow and it will be observed that the bolted double ruler stops vibrating much more rapidly than the single ruler (Figure 16.9b).

The rate of decay of free vibration is proportional to the product of the damping ratio and the natural frequency of the structure, as shown in Equation 16.15 and Example 16.1. The bolted double ruler has a higher natural frequency since its second moment of area is eight times that of the single ruler, making it eight times as stiff as the single ruler, while its mass is just double that of the single ruler. In fact, the fundamental natural frequency of the bolted double ruler is twice that of the single ruler. The damping ratios for the two cantilevers can be considered to be the same, although the bolted ruler will have higher damping. However, this demonstration verifies the concept that *the higher the natural frequency of a structure, the quicker the decay of its free vibration.*

16.3.3 Overcritically Damped System

This demonstration shows *the movement rather than the vibrations of an overcritically damped beam subject to an initial displacement.*

Figure 16.10 shows two cantilever metal strips representing two cantilever beams. The one on the left is a conventional metal strip and the one on the right uses two metal strips with a layer of damping material constrained between the two strips. Apply the same initial displacement on the free ends of the two cantilevers and release them at the same time. It is observed that the cantilever on the right returns to its original position slowly, without experiencing any vibration. This is the phenomenon of an overcritically damped system described in Section 16.2.1 and Figure 16.3.

FIGURE 16.10
Demonstration of an overcritically damped system.

16.3.4 Mode Shapes of a Discrete Model

This demonstration shows *the two mode shapes of a discrete two-degree-of-freedom (TDOF) system.*

Figure 16.11 shows a TDOF model that has two natural frequencies and two vibration modes. Apply initial horizontal displacements of the two masses in the same direction and then release them at the same time; the first mode of vibration is generated, as shown in Figure 16.11a. Apply initial horizontal displacements of the two masses in opposite directions and then release them at the same time; the second mode of vibration is generated, as shown in Figure 16.11b.

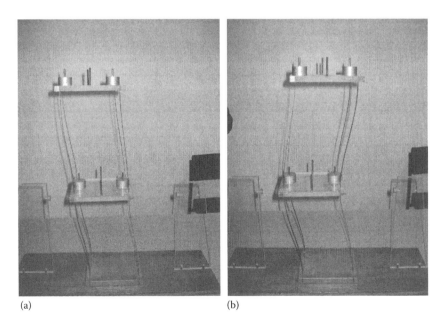

(a)   (b)

FIGURE 16.11
Mode shapes of a TDOF model. (a) The first mode of vibration. (b) The second mode of vibration.

(a)   (b)

FIGURE 16.12
Mode shapes of a cantilever. (a) The first mode of vibration. (b) The second mode of vibration.

16.3.5 Mode Shapes of a Continuous Model

This demonstration shows *the shapes of the first two modes of vibration of a continuous beam*.

Take a long plastic strip and hold one end of the strip as shown in Figure 16.12. Then move the hand forwards and backwards slowly, which generates the first mode of vibration as shown in Figure 16.12a. Do the same but increase the speed of the movement to excite the second mode of vibration as shown in Figure 16.12b. The two mode shapes of the strip are similar to those shown in Figure 16.7.

16.4 PRACTICAL EXAMPLES

16.4.1 Musical Box

A musical box is a device that produces music using mechanical vibration. Figure 16.13a shows one of many decorative music boxes which are readily available. The core of the music box is the unit shown in Figure 16.13b.

A spring is used to rotate a music tube, converting the potential energy stored in the spring into the kinetic energy which drives the rotation of the tube. Raised points on the tube displace

(a)   (b)

FIGURE 16.13
A musical box and its core. (The models are provided by Professor B. Zhuang, Zhejiang University, China.) (a) A musical box. (b) The core of the musical box.

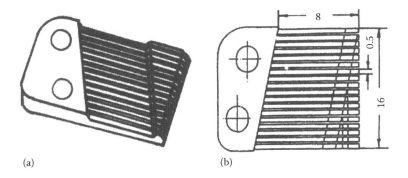

(a)   (b)

FIGURE 16.14
(a,b) Cantilever beams with different lengths and sections in the music box (mm). (From Zhuang, B., Zhu, Y., Bai, C., Cui, J. and Zhuang, B. *Design of Music Boxes and Vibration of Tuning Bars*, New Times Press, China, 1996.)

TABLE 16.3   Relations between Natural Frequency and Music Note

Natural frequency (Hz)	261.6	293.7	329.6	349.2	392	440	493.9
Music note	(Middle) C	D	E	F	G	A	B

Source: Zhuang, B., Zhu, Y., Bai, C., Cui, J. and Zhuang, B. *Design of Music Boxes and Vibration of Tuning Bars*, New Times Press, China, 1996. White, H. E. and White, D. H. *Physics and Music: The Science of Musical Sound,* Saunders College, Philadelphia, 1980.

cantilever metal bars causing them to vibrate and generate sound. Different geometries (lengths and cross sections) of the bars (see Figure 16.14) provide different natural frequencies of the bars, generating different music notes. The distribution of the raised points on the tube is designed to create a particular musical tune when the tube rotates. The music unit shown in Figure 16.13b has 18 metal bars generating 18 different musical notes.

A given musical note relates to a particular natural frequency of the vibrating body. Table 16.3 gives the relationships between some natural frequencies and music notes.

Figure 16.15 shows part of the keyboard of a piano. The main reference point on a piano is known as Middle C. This is the white note located approximately in the centre of the keyboard and immediately to the left of a pair of black keys. Striking the Middle C key, the sound generated corresponds to a frequency of 261.6 Hz. The next white key, D, to the right of the Middle C key, produces a sound corresponding to a frequency of 293.2 Hz.

FIGURE 16.15
The keyboard of a piano.

FIGURE 16.16
The test building. (Copyright of Building Research Establishment Ltd, reproduced by kind permission.)

16.4.2 Measurement of Fundamental Natural Frequency of a Building through Free Vibration Generated Using Vibrators

When the free vibrations of a structure can be measured as accelerations, velocities or displacements, the natural frequencies of the structure can be determined from their vibration time histories.

Figure 16.16 shows an eight-storey, steel-framed test building in the Cardington Laboratory of the Building Research Establishment Ltd. Different dynamic test methods were used to determine the natural frequencies of this building. One of the methods used was to record the free vibrations of the building [8].

Four vibrators mounted at the corners of the roof of the building were used to generate movement of the structure. Once movement was initiated, the vibrators were turned off and the resulting free vibrations of the structure were measured using accelerometers. Figure 16.17 shows the acceleration–time history of the decaying vibrations. The frequency of the oscillations, which is a natural frequency of the structure, was determined from the time history as the inverse of the time interval between two successive acceleration peaks.

16.4.3 Measurement of Natural Frequencies of a Stack through Vibration Generated by the Environment

Environmental effects such as air movements around a structure can also induce vibrations, though such vibration may be very small. These types of vibration may not be exactly free vibration as they are caused by disturbances due to external effects, such as wind. However, the concept of free vibration can still be used to identify the natural frequencies of a structure.

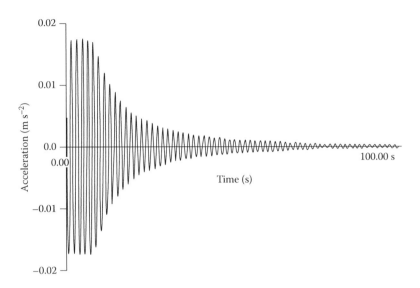

FIGURE 16.17
The free vibration record of the building.

FIGURE 16.18
Measuring the natural frequencies of a multiflare stack.

Figure 16.18 shows a 97.5 m tall multiflare stack which can be used to burn off excess gases. A laser test system was set up approximately 100 m from the stack to monitor the stack vibrations. Several velocity–time histories on selected measurement points of the stack were measured when wind was blowing.

The natural frequencies of 0.67 Hz and 0.73 Hz were measured which corresponded to the fundamental modes in the two orthogonal horizontal directions.

A similar application was to measure the fundamental natural frequency of a stadium roof remotely, as shown in Figure 16.19.

FIGURE 16.19
Measuring the fundamental natural frequency of a stadium roof.

PROBLEMS

1. An SDOF system has a natural frequency of 1 Hz. Consider its free vibration induced by an initial displacement of 1.0 cm and compare the time histories for the first 10 s when $\xi = 0.5$, 1.0 and 1.5, respectively.

2. A cantilever has a rigidity of EI, a span of L and a uniformly distributed mass $\bar{m}$. Estimate its modal mass, modal stiffness and fundamental natural frequency using the following two assumed mode shapes:

   a. $\varphi_1(x) = \dfrac{2x^2}{L^2} - \dfrac{4x^3}{3L^3} + \dfrac{x^4}{3L^4}$

   b. $\varphi_2(x) = \dfrac{3x^2}{2L^2} - \dfrac{x^3}{2L^3}$

   Compare and discuss the estimated results.

3. If the cantilever in Figure 16.2 has a fundamental natural frequency, f, and is shortened to $0.8\,L$ while the other parameters, EI and $\bar{m}$, remain the same, what is the new fundamental natural frequency?

4. Take a 1 m long plastic ruler, or similar, and hold the ruler vertically at its base, as shown in Figure 16.12. Move your hand forwards and backwards to generate vibration and observe the first and second modes of vibration of the ruler.

REFERENCES

1. Beards, C. F. *Structural Vibration: Analysis and Damping*, London: Arnold, 1996.
2. Clough, R. W. and Penzien, J. *Dynamics of Structures*, New York: McGraw-Hill, 1993.
3. Chopra, A. K. *Dynamics of Structures*, Upper Saddle River, NJ: Prentice Hall, 1995.
4. Morse, P. M. *Vibration and Sound*, New York: McGraw-Hill, 1948.
5. Smith A. L., Hicks S. J. and Devine P. J. *Design of Floors for Vibration: A New Approach*, London: The Steel Construction Institute, P354, 2007.

6. Zhuang, B., Zhu, Y., Bai, C., Cui, J. and Zhuang, B. *Design of Music Boxes and Vibration of Tuning Bars*, Beijing: New Times Press, 1996.
7. White, H. E. and White, D. H. *Physics and Music: The Science of Musical Sound*, Philadelphia, PA: Saunders College, 1980.
8. Ellis, B. R. and Ji, T. Dynamic testing and numerical modelling of the Cardington steel framed building from construction to completion. *The Structural Engineer*, 74, 186–192, 1996.

CHAPTER 17

CONTENTS

Resonance

17

17.1 DEFINITIONS AND CONCEPTS

Resonance is a phenomenon which occurs when the vibration of a system tends to reach its maximum magnitude. The frequency corresponding to resonance is known as the resonance frequency of the system. When the damping of the system is small, the resonance frequency is approximately equal to the natural frequency of the system.

- The resonance frequency is related to the damping ratio, the input (loading or ground motion) and the selected measurement parameter (relative or absolute movement, displacement, velocity or acceleration).
- Increasing damping will effectively reduce the response of the structure at resonance.

For a single-degree-of-freedom (SDOF) system subject to a harmonic input, such as a direct load or a base motion, there are three characteristics:

- The maximum dynamic displacement is close to the static displacement for a given load amplitude if the load frequency is less than a quarter of the natural frequency of the system.
- The maximum dynamic displacement is less than the static displacement if the load frequency is more than twice the natural frequency of the system.
- The maximum dynamic displacement is several times the static displacement if the load frequency is close to or matches the natural frequency of the system.

17.2 THEORETICAL BACKGROUND

Although the statements in Section 17.1 are abstracted from the study of an SDOF system subject to a harmonic input, they are applicable to many practical situations. This is because:

- The response of a structure can be expressed as a summation of the responses of several modes of vibration, and the response of each mode can be represented using an SDOF system.
- Some common forms of dynamic loading can be expressed as a summation of several harmonic components.

This section summarises the fundamental characteristics of the response of an SDOF system to a harmonic load. Many reference books, such as [1,2], deal with other types of dynamic load as well as with multi-degree-of-freedom systems.

17.2.1 SDOF System Subjected to a Harmonic Load

17.2.1.1 Equation of motion and its solution

Consider an SDOF system comprising a mass, m, attached to a spring of stiffness k, and a viscous damper of capacity c, as shown in Figure 17.1a, subjected to a harmonic load:

$$P(t) = P_0 \sin \omega_p t \tag{17.1}$$

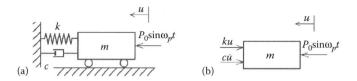

FIGURE 17.1
An SDOF system subject to a harmonic force. (a) An SDOF system subject to a harmonic load.
(b) A free-body diagram.

where:

P_0 is the force magnitude
ω_p is the angular frequency of the load

A free-body diagram of the system is shown in Figure 17.1b, from which the equation of motion can be obtained using Newton's second law:

$$m\ddot{u} = P_0\sin\omega_p t - c\dot{u} - ku \tag{17.2a}$$

or

$$m\ddot{u} + c\dot{u} + ku = P_0\sin\omega_p t \tag{17.2b}$$

The particular solution of Equation 17.2b can be written in the following form:

$$u(t) = A\sin\left(\omega_p t - \theta\right) \tag{17.3}$$

where A is a constant which is equal to the maximum amplitude of the motion that follows the force by θ, called the *phase lag*. Equation 17.3 indicates that the mass experiences simple harmonic motion at the load frequency ω_p.

Differentiating Equation 17.3 with respect to u once and twice and substituting u, $\dot{u}$ and $\ddot{u}$ into Equation 17.2b leads to

$$m A\omega_p^2\sin\left(\omega_p t - \theta + \pi\right) + cA\omega_p\sin\left(\omega_p t - \theta + \pi/2\right) + kA\sin\left(\omega_p t - \theta\right) = P_0\sin\omega_p t \tag{17.4}$$
$$\text{Inertial force} \qquad\qquad \text{Damping force} \qquad\qquad \text{Spring force} \quad \text{Applied force}$$

Equation 17.4 shows that

- The spring or elastic force lags the applied harmonic force by θ.
- The damping force precedes the spring force by $\pi/2$.
- The inertial force precedes the damping force by $\pi/2$ and has an opposite direction to the spring force.

A vector diagram of all the forces acting on the body is shown in Figure 17.2, from which it can be shown that

$$P_0^2 = (kA - m\omega_p^2 A)^2 + (c\omega_p A)^2 \tag{17.5}$$

The magnitude of the motion A can be obtained from Equation 17.5 as

$$A = \frac{P_0}{\sqrt{(k - m\omega_p^2)^2 + (c\,\omega_p)^2}}$$

$$= \frac{P_0}{k} \frac{1}{\sqrt{(1 - \beta^2)^2 + (2\xi\beta)^2}} \tag{17.6}$$

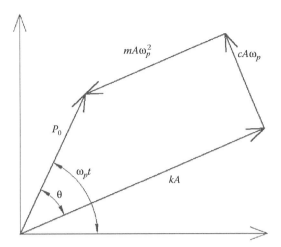

FIGURE 17.2
A diagram of force vectors.

where

$$\beta = \frac{\omega_p}{\omega} \qquad (17.7)$$

$$c = 2\xi m\,\omega \qquad (17.8)$$

The frequency ratio, β, is the ratio of the frequency of the load, ω_p, to the natural frequency of the system, ω. The phase lag between the applied force and the displacement can also be found from Figure 17.2 as

$$\tan\theta = \frac{cA\,\omega_p}{kA - mA\,\omega_p^2} = \frac{2\xi\omega_p/\omega}{1 - (\omega_p/\omega)^2} = \frac{2\xi\beta}{1 - \beta^2} \qquad (17.9)$$

Equations 17.3, 17.6 and 17.9 are the solutions of an SDOF system subject to a harmonic force in steady-state vibration. They are the same as those derived from the solution of the differential equation (Equation 17.2) [1,2]. The derivation and solution indicate that

- The displacement of the mass lags the force by θ.
- The velocity of the mass is ahead of its displacement by $\pi/2$.
- The acceleration of the mass leads its velocity by $\pi/2$ and the displacement by π.
- When $1 - \beta^2 = 0$ or $\theta = 90°$ in Equation 17.9, the force diagram in Figure 17.2 becomes a rectangle. The inertial force and the spring force balance each other and the external load is countered by the damping force. As the damping force is generally much less than the inertial and spring forces, the system response will be much bigger than when the other system forces dominate.

17.2.1.2 Dynamic magnification factor

Considering the static displacement of the mass as

$$\Delta = \frac{P_0}{k} \qquad (17.10)$$

Equation 17.6 can be written in the following form:

$$A = \Delta \times M_D \qquad (17.11)$$

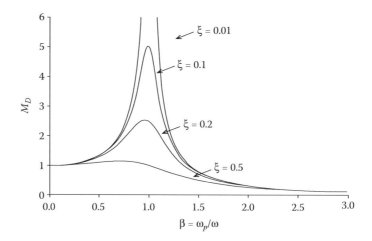

FIGURE 17.3
Variation of dynamic magnification factor against frequency ratio.

where

$$M_D = \frac{1}{\sqrt{(1-\beta^2)^2 + (2\xi\beta)^2}} \qquad (17.12)$$

Equation 17.11 indicates that *the maximum dynamic displacement (A) can be expressed as a product of the corresponding static displacement (Δ) and a dynamic magnification factor (M_D), which is thus a function of the frequency ratio and the damping ratio.* M_D is therefore related to the dynamic characteristics of the system subject to a harmonic force.

For several damping ratios ($\xi = 0.01$, 0.1, 0.2 and 0.5), M_D is shown in Figure 17.3 as a function of the frequency ratio (β). These curves can be interpreted as follows:

- When the load frequency is less than one-quarter the natural frequency of the system, the maximum dynamic displacement is close to the static displacement of the system.
- When the load frequency is more than twice the natural frequency, the maximum dynamic displacement is much lower than the static displacement of the system.
- When the load frequency is equal or close to the natural frequency, resonance will occur and the maximum dynamic displacement can be several times the static displacement of the system.

The significance of these observations is

- If the load moves slowly, or the frequency of the load is much lower than the natural frequency of a structure, it can be treated as a static problem.
- If the load frequency is significantly larger than the natural frequency of a structure, dynamic analysis may not be required.
- The situation when the load frequency is close to the natural frequency of the structure will generate a resonant response and this is a situation to avoid. Increasing system damping will, however, effectively reduce the response at resonance.

17.2.1.3 Phase lag

The particular solution of Equation 17.2b can be obtained by solving the differential equation and is in the following form:

$$u(t) = \frac{P_0}{k}\left[\frac{1}{(1-\beta^2)^2 + (2\xi\beta)^2}\right][(1-\beta^2)\sin\omega_p t - 2\xi\beta\cos\omega_p t] \qquad (17.13)$$

Let

$$\sin\theta = \frac{2\xi\beta}{\sqrt{(1-\beta^2)^2 + (2\xi\beta)^2}} \qquad (17.14a)$$

$$\cos\theta = \frac{1-\beta^2}{\sqrt{(1-\beta^2)^2 + (2\xi\beta)^2}} \qquad (17.14b)$$

Using the relationship:

$$\sin\omega_p t\cos\theta - \cos\omega_p t\sin\theta = \sin(\omega_p t - \theta) \qquad (17.15)$$

and substituting Equations 17.10 through 17.12 and 17.14 into Equation 17.13 leads to a concise expression:

$$u(t) = \Delta M_D \sin(\omega_p t - \theta) \qquad (17.16)$$

where θ and M_D have been given in Equations 17.9 and 17.12, respectively. Equation 17.16 is normally used instead of Equation 17.13 in calculations because it has a simpler form.

It should be noted that Equations 17.13 and 17.16 do not always give identical results. The difference between Equations 17.13 and 17.16 is due to the definitions of the angle θ in Equations 17.9 and 17.14. Since $2\xi\beta$ will always be positive, the variation of the phase lag should be between 0 and π in Equation 17.14, as shown in Figure 17.4a. However, the angle θ in Equation 17.16 defined by Equation 17.9 varies between $-\pi/2$ and $\pi/2$ shown in Figure 17.4b, which does not match that in the original equation (Equation 17.13), as shown in Figure 17.4a. This can be examined in detail, as shown in Table 17.1.

This difference is important in certain situations.

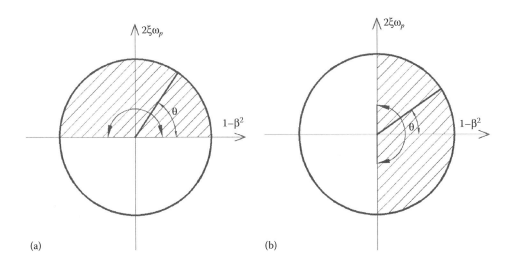

(a)   (b)

FIGURE 17.4
Definition of the range of the phase lag, θ. (a) $0 < \theta < \pi$ in the actual situation (Equation 17.14). (b) $-\pi/2 < \theta < \pi/2$ in Equation 17.9.

TABLE 17.1 Comparison of the Phase Lags Defined in Figure 17.4a (Equation 17.14) and Figure 17.4b (Equation 17.9)

Situation	θ in Figure 17.4a	θ in Figure 17.4b	Conclusion
When $1-\beta^2 > 0$	$0 < \theta < \pi/2$	$0 < \theta < \pi/2$	Equation 17.14 = Equation 17.9
When $1-\beta^2 < 0$	$\pi/2 < \theta < \pi$	$-\pi/2 < \theta < 0$	Equation 17.14 ≠ Equation 17.9

EXAMPLE 17.1

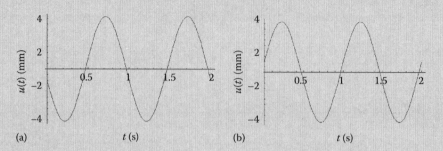

FIGURE 17.5
Harmonic vibration of a damped SDOF system. (a) Displacements calculated using Equation 17.13. (b) Displacements calculated using Equation 17.16.

An SDOF system subjected to a harmonic load has the following parameters:

$$P_0 = 1\,N; \quad k = 1\,kN/m; \quad \beta = 1.1; \quad \xi = 0.02; \quad \omega_p = 2\pi \text{ rad s}^{-1}$$

Calculate the response of the system when the phase lags are calculated using Equations 17.9 and 17.14, respectively.

SOLUTION

Substituting the data into Equation 17.14 gives

$$2\xi\beta = 0.044 \quad 1 - \beta^2 = -0.21 \quad \sqrt{(1-\beta^2)^2 + (2\xi\beta)^2} = 0.2146$$

$$\sin\theta = 0.2050 \quad \theta \text{ can be } 11.83° \text{ or } 168.17°$$

$$\cos\theta = -0.9787 \quad \theta \text{ can be } 168.17° \text{ or } -168.17°$$

The actual phase lag should satisfy both Equations 17.14a and 17.14b. Thus, it must be 168.17°.
When Equation 17.9 is used, it yields

$$\tan\theta = -0.2095 \quad \text{and} \quad \theta = -11.83°$$

The phase difference between −11.83° and 168.17° is exactly 180°. The response curves, generated using Equations 17.13 and 17.16, are given in Figure 17.5 showing the 180° phase difference.

To this incompatibility, a supplementary condition must be introduced and Equation 17.9 should be written in the following form:

$$\theta = \begin{cases} \tan^{-1}\left[\dfrac{2\xi\beta}{1-\beta^2}\right] + \pi & \text{if} \quad 1-\beta^2 < 0 \\[3mm] \tan^{-1}\left[\dfrac{2\xi\beta}{1-\beta^2}\right] & \text{if} \quad 1-\beta^2 > 0 \end{cases} \tag{17.17}$$

Then, the response calculated using Equations 17.16 and 17.17 is the same as that using Equations 17.13 and 17.14.

The supplementary condition (Equation 17.17) is necessary and straightforward when Equation 17.16 is used, but this has not been emphasised elsewhere. The possible error in calculating

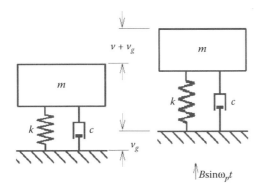

FIGURE 17.6
An SDOF system subject to vertical ground movements.

phase lag does not affect the magnitude of the response when an SDOF system is subjected to a single harmonic load; however, when several harmonics, such as those for human jumping loads, need to be considered, the errors in the calculation of phase lags will affect the magnitudes and pattern of the loads and consequently the response of the structure subjected to these loads [3].

17.2.2 SDOF System Subject to a Harmonic Support Movement

Consider an SDOF system subjected to vertical harmonic ground or support movements. The ground motion is defined as follows:

$$v_g = B\sin\omega_p t \qquad (17.18)$$

where:
B is the magnitude of the ground movement
ω_p is the frequency of the ground movement

The equation of motion of the SDOF system is

$$m(\ddot{v}_g + \ddot{v}) + c\dot{v} + kv = 0 \qquad (17.19)$$

where v is the movement of the mass relative to the ground. Substituting Equation 17.18 into Equation 17.19 gives

$$m\ddot{v} + c\dot{v} + kv = m\omega_p^2 B\sin\omega_p t \qquad (17.20)$$

Comparing Equations 17.20 and 17.2b, it can be seen that they are identical when $P_0 = Bm\omega_p^2$. Thus, the solution of Equation 17.20 has the same form as that of Equation 17.3 and the magnitude of the vibration using Equation 17.6 is

$$A = \frac{Bm\omega_p^2}{\sqrt{(k - m\omega_p^2)^2 + (c\omega_p)^2}}$$

$$= \frac{Bm\omega_p^2}{m\omega^2} \frac{1}{\sqrt{(1 - \omega_p^2/\omega^2)^2 + (2\xi\,\omega_p/\omega)^2}} = \frac{B\beta^2}{\sqrt{(1 - \beta^2)^2 + (2\xi\beta)^2}} \qquad (17.21)$$

Thus, the solution of Equation 17.20 is

$$v(t) = BM_R \sin(\omega_p t - \theta) \qquad (17.22)$$

where the phase lag θ has been given in Equation 17.17 and M_R is the dynamic magnification factor for the relative movement of the system to ground and is defined as

$$M_R = \frac{\beta^2}{\sqrt{(1-\beta^2)^2 + (2\xi\beta)^2}} \qquad (17.23)$$

If the absolute motion, $v_a = v_g + v$, of the mass is considered, Equation 17.19 can be rewritten as

$$m\ddot{v}_a + c\dot{v}_a + kv_a = c\dot{v}_g + kv_g$$

$$= cB\omega_p \cos\omega_p t + kB\sin\omega_p t$$

$$= B\sqrt{k^2 + (c\omega_p)^2} \, \sin(\omega_p t + \phi) \qquad (17.24)$$

where

$$\phi = \tan^{-1}\left[\frac{c\omega_p}{k}\right] = \tan^{-1}\left[\frac{2\xi\omega_p}{\omega}\right] \qquad (17.25)$$

As both $2\xi\omega_p$ and ω are positive, φ varies between 0 and $\pi/2$. Thus, similar to Equation 17.22, the solution of Equation 17.24 can be written as

$$v_a(t) = BM_A\sin(\omega_p t + \varphi - \theta) \qquad (17.26)$$

where M_A is the dynamic magnification factor for the absolute movement of the system and is expressed as

$$M_A = \frac{\sqrt{1 + (2\xi\beta)^2}}{\sqrt{(1-\beta^2)^2 + (2\xi\beta)^2}} \qquad (17.27)$$

The phase lag in Equation 17.26 has been defined in Equation 17.17.

Comparing the ground motion, Equation 17.18, and the absolute movement of the system, Equation 17.26, it can be noted that the magnification factor M_A in Equation 17.27 is also called the *motion transmissibility*, that is the ratio of the amplitude of the absolute vibration of the system to the amplitude of the ground motion. Similar to Figure 17.3 for the magnification factor when the system is subjected to a harmonic load, Figures 17.7 and 17.8 show the dynamic magnification factors for the relative and absolute displacement, respectively, for four damping values, 0.01, 0.1, 0.2 and 0.5 when the system is subjected to harmonic ground motion.

The same qualitative observations as those from Figure 17.3 can be obtained from Figure 17.8.

17.2.3 Resonance Frequency

It can be observed from Figures 17.3, 17.7 and 17.8 that the resonance frequency is different to and maybe larger or smaller than the natural frequency of a system. This difference is more pronounced for higher damping ratios. The resonance frequency can be obtained by differentiating the magnification factors with respect to the frequency ratio and equating the functions to zero. For the magnification factors given in Equations 17.12, 17.23 and 17.27:

$$\frac{\partial M_D}{\partial \beta} = \frac{\partial}{\partial \beta}\left(\frac{1}{\sqrt{(1-\beta^2)^2 + (2\xi\beta)^2}}\right) = 0 \qquad (17.28a)$$

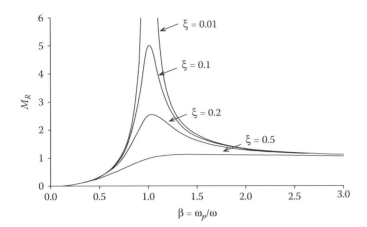

FIGURE 17.7
The magnification factor for the relative movement of the system.

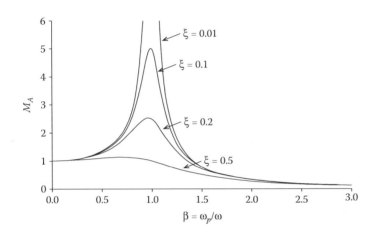

FIGURE 17.8
The magnification factor for the absolute movement (motion transmissibility) of the system.

$$\frac{\partial M_R}{\partial \beta} = \frac{\partial}{\partial \beta}\left(\frac{\beta^2}{\sqrt{(1-\beta^2)^2+(2\xi\beta)^2}}\right) = 0 \qquad (17.28b)$$

$$\frac{\partial M_A}{\partial \beta} = \frac{\partial}{\partial \beta}\left(\frac{\sqrt{1+(2\xi\beta)^2}}{\sqrt{(1-\beta^2)^2+(2\xi\beta)^2}}\right) = 0 \qquad (17.28c)$$

The solution of Equation 17.28 gives the relationship between the resonance and natural frequencies of an SDOF system as follows:

When subjected to a harmonic load:

$$f_R = f\sqrt{1-2\xi^2} \qquad (17.29a)$$

When subjected to a harmonic ground motion and considering relative movements:

$$f_{R,R} = \frac{f}{\sqrt{1-2\xi^2}} \qquad (17.29b)$$

TABLE 17.2   Effect of Damping Ratio on the Resonance Frequency

	ξ	0.01	0.05	0.1	0.2	0.3	0.4	0.5
Equation 17.29a	f_R/f	0.9999	0.9975	0.9899	0.9592	0.9055	0.8246	0.7071
Equation 17.29b	$f_{R,R}/f$	1.0001	1.0025	1.0102	1.0426	1.1043	1.2127	1.4142
Equation 17.29c	$f_{R,A}/f$	0.9999	0.9975	0.9903	0.9647	0.9301	0.8927	0.8556

When subjected to a harmonic ground motion and considering absolute movements:

$$f_{R,A} = f\sqrt{\frac{\sqrt{1+8\xi^2}-1}{4\xi^2}} \qquad (17.29c)$$

Thus, the differences between the natural frequency and the three resonance frequencies can be quantified using Equation 17.29. Table 17.2 lists the ratios of the resonance frequency to the natural frequency for different damping ratios for the three expressions.

It can be observed from Table 17.2 that

- The resonance frequency is related to the damping ratio. When the damping ratio is less than 0.1, there are negligible differences between the natural frequency and resonance frequency.
- The resonance frequency is also related to the input loading or ground motion and to the type of displacement (relative or absolute movement).

Structures in civil engineering normally have damping ratios smaller than 10% ($\xi = 0.10$), often much smaller, and therefore there is usually no need to distinguish between the resonance frequency and natural frequency in the vibration of structures.

17.3 MODEL DEMONSTRATIONS

To see a resonant response requires an input device that can generate either harmonic base motion, such as a shaking table, or harmonic load, such as using a vibrator, together with a simple structure that has a natural frequency within the frequency range of the input.

17.3.1 Dynamic Response of an SDOF System Subject to Harmonic Support Movements

This demonstration shows *the observations obtained from Figures 17.3 and 17.8, that is, the relationship between the response of an SDOF system and the ratio of the frequency of input to the natural frequency of the system.*

A mass attached to a spring forms a simple SDOF system, as shown in Figure 17.9. Hold the end ring of the spring and move it up and down harmonically to simulate harmonic support movement of the SDOF system. This corresponds to the model shown in Figure 17.6. The movement of the mass can be described using Equation 17.24, and the ratio of the maximum movement of the mass to that of the hand is shown in Figure 17.8.

The dynamic phenomena shown in Figure 17.8 can be demonstrated qualitatively as follows:

- The person holding the string first moves his or her hand slowly up and down, creating a situation where the frequency of the support (hand) movement is much smaller than the natural frequency of the SDOF system. It will be observed that the amplitude of movement of the mass is almost the same as that of the support (hand), when the frequency ratio is less than 0.25.
- When the hand moves up and down quickly, it creates a situation where the frequency of the hand movement is larger than the natural frequency of the system. It can be seen that the hand movements are much larger than the movements of the mass, when the frequency ratio is larger than 2.0.

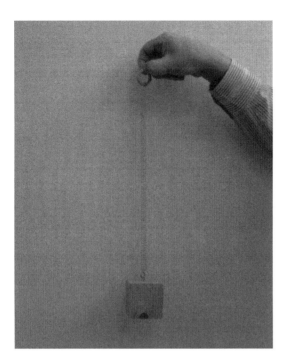

FIGURE 17.9
The model of an SDOF system.

- Finally, when the hand moves up and down at a frequency close to the natural frequency of the system, a situation is created in which resonance develops. It is observed that the movements of the mass are much larger than the hand movements, when the frequency ratio is near to unity.

17.3.2 Effect of Resonance

This demonstration shows that *a cantilever experiences significant vibration at resonance.*

Figure 17.10a shows a medical shaker that can generate vibrations at a frequency of 5 or 10 Hz in three perpendicular directions. A steel ruler and a longer and thicker steel ruler are mounted on

(a)

(b)

FIGURE 17.10
Demonstration of resonance. (a) A medical shaker and two cantilevers. (b) Simulated shaking table test.

a wooden plate that can be firmly placed on the shaker, as shown in Figure 17.10b. Both rulers are selected to have their fundamental natural frequencies slightly less than 5 Hz.

Switch on the shaker at the frequency of 10 Hz and it will be seen that the two rulers vibrate with very small amplitudes in comparison with the movement of the shaker.

Change the vibration frequency from 10 to 5 Hz and it will then be observed that both rulers vibrate with significant amplitudes as the fundamental frequencies of the rulers are close to the shaking frequency.

17.4 PRACTICAL EXAMPLES

In most situations, resonance should be avoided in civil engineering structures. This requires knowledge of the natural frequencies of the structure and the frequencies of the dynamic loading applied to it.

17.4.1 London Millennium Footbridge

The London Millennium Footbridge is the first new pedestrian bridge crossing over the Thames in Central London for more than a century. The bridge is located between Peter's Hill on the North Bank leading to Saint Paul's Cathedral and the new Tate Modern Art Gallery on the South Bank. The bridge has three spans over a total length of 325 m. The lengths of the three spans are 81 m for the north span, 144 m for the main span and 100 m for the south span.

The London Millennium Footbridge opened on 10 June 2000 (Figure 17.11). It was estimated that between 80,000 and 100,000 people crossed the bridge during the opening day, with a maximum of 2,000 people on the bridge at any one time. During the opening day, unexpected lateral movements, or 'wobbling', occurred when people walked across the bridge. Two days later, the bridge was officially closed to investigate the causes of the vibration. After investigation and modification, the bridge reopened in December 2001 [4]. Since then, millions of people have visited the bridge and there has been no recurrence of the 'wobbling' problems.

As constructed, the bridge had lateral natural frequencies as follows:

- South span: First lateral natural frequency was about 0.8 Hz
- Central span: First two lateral natural frequencies were about 0.5 and 0.95 Hz
- North span: First lateral natural frequency was about 1 Hz

It was reported that excessive vibrations of the bridge in the lateral direction did not occur continuously but built up when a large number of pedestrians were on the south and central spans of the bridge and then died down if the number of people on the bridge decreased or if the people stopped walking [4].

(a)

(b)

FIGURE 17.11
(a,b) The London Millennium Footbridge.

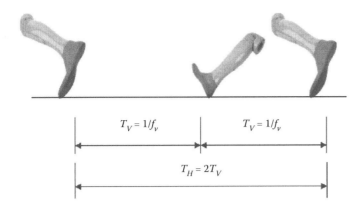

FIGURE 17.12
Periods of walking forces.

Resonance is related to forcing frequencies. Walking is a periodic movement on a flat surface in which two feet move alternately from one position to another and do not leave the surface simultaneously. When people walk in a normal way, walking has its own frequency. This is illustrated in Figure 17.12.

When people walk, they produce vertical loading on the walking surface, such as floors and footbridges. The time from one vertical load generated by one foot to the next vertical load induced by the other foot is the period of the walking load in the vertical direction, denoted as T_V. People walking generate not only vertical forces, but also lateral forces. The force generated laterally by the right foot is normally in the opposite direction to the force induced by the left foot. Therefore, the time required for reproducing the same pattern of force can be counted from the left (or right) foot to the next step of the left (or right) foot. In other words, the period of lateral forces (T_H) is just twice the period of the vertical force induced by people walking, or the frequency of the lateral walking loads is a half of that of the vertical walking loads [5].

Figure 17.13 shows the distribution of walking frequencies obtained from 400 people walking on two footbridges in Manchester [6]. The horizontal axis indicates the walking frequency in the vertical direction, while the vertical axis shows the number of observations of particular frequencies. It can be seen from Figure 17.13 that most of the 400 frequencies of people walking are between 1.6 and 2 Hz in the vertical direction. As the walking frequency in the lateral direction is just half that in the vertical direction, the corresponding frequencies of people walking in the lateral direction are between 0.8 and 1 Hz.

It can be noted that the frequencies of walking loads in the lateral directions were close to the lateral natural frequencies of the south and central spans of the London Millennium Footbridge where excessive vibrations occurred. It was also observed that people walking in large groups

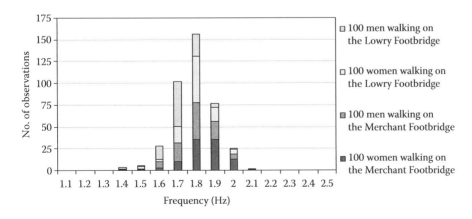

FIGURE 17.13
Distribution of frequency of walking loads in the vertical direction. (From Pachi, A. and Ji, T. *Struct Eng.*, 83, 36–40, 2005. With Permission.)

tended to synchronise their walking paces. When the footbridge started wobbling, more people would walk at the frequency of the wobbling, which enhanced the synchronisation. This synchronisation magnified the effect of the lateral footfall forces on the footbridge. The wobbling of the footbridge was caused by resonance induced by the walking loads.

17.4.2 Avoidance of Resonance: Design of Structures Used for Pop Concerts

British Standard BS 6399: Part 1, Loading for Buildings [7], introduced in September 1996, included a new section on synchronised dance loading. It stated that any structure that might be subjected to this form of loading should be designed in one of two ways: to withstand the anticipated dynamic loads or to avoid significant resonance effects. The first method requires dynamic analysis to assess the structural response to the loading in order to calculate a safety margin. The second method requires that the structure should be designed to be sufficiently stiff so that the lowest relevant natural frequency of the structure is above the range of load frequencies considered. The second approach is simpler and only requires the calculation of natural frequencies. However, both approaches need knowledge of the load frequencies.

Dance-type activities, such as keep-fit exercises, aerobics and audience movements at pop concerts, are more common now than ever before. These activities are likely to be held on grandstands, dance floors and in sports centres. Therefore, the use or the design of these structures or both, should consider the effect of the human-induced dance-type loads. Figure 17.14 shows a pop concert where people moved, jumped, bobbed and swayed, in time to the music.

Dance is movement with rhythmic steps and actions, usually to accompanying music. Efforts have been made to study dance-type loads since a prediction of the response of a structure subject to this loading requires an understanding of the loading. Dance-type loads are functions of the type of dance activity, the density and distribution of the dancers, the frequency of the music, load factors and the dynamic crowd effect.

There are many different types of dancing and a wide range of beat frequencies for dance music; however, dance frequencies tend to be in the range of 1.5–3.5 Hz for individuals and 1.5–2.8 Hz for groups of people.

Figure 17.15 shows an autospectrum obtained from accelerometers monitoring vertical motions of a grandstand during a pop concert [8]. The vertical axis indicates the normalised acceleration squared per hertz, while the horizontal axis shows frequency. It can be seen that the frequencies

FIGURE 17.14
At a pop concert.

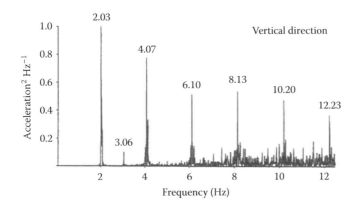

FIGURE 17.15
Structural response to the music beat frequency at a pop concert. (From Littler, J. D. *Proceedings of the First International Conference on Stadia 2000*, 123–134, 1998.)

corresponding to the peaks of responses are the beat frequency of the music and integer multiples of the beat frequency. This phenomenon has been observed in a number of measurements taken during pop concerts and can be explained theoretically. As spectators at pop concerts move with the music beat, the frequency of dance-type loads can be determined from the beat frequency of music played on these occasions. The beat frequencies of 210 modern songs have been determined [9].

The 210 songs consisted of 30 songs from each of the 1960s, the 1970s and the 1980s, and 120 songs from the 1990s. The 1990s music was further classified into four main types, dance, indie, pop and rock, with 30 songs for each group. All the songs selected had been popular in their time and thus provided a good sample of popular music. The frequency distribution of the 210 songs surveyed is given in Figure 17.16. It can be seen that the majority of songs (202 out of 210) are in the frequency range of 1–2.8 Hz.

Human loading, such as jumping, bobbing and walking, contains several harmonic components with the load frequency, two times the frequency, three times the frequency and so on. However, only the first two or three components are significant and need to be considered in design. For example, if the load or music frequency is not larger than 2.8 Hz and the first three load components are considered for a floor subject to jumping-type loading, the highest frequency in the load to be considered will not be larger than $2.8 \times 3 = 8.4$ Hz. This corresponds to the threshold value of natural frequency in BS6399: Part 1 of 8.4 Hz in the vertical direction, for which no dynamic analysis of the structure is required [7].

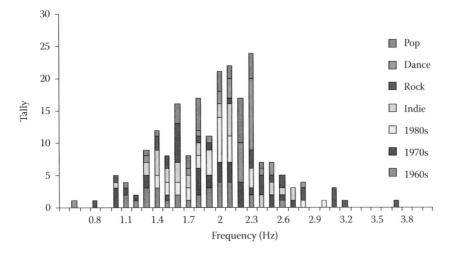

FIGURE 17.16
Frequency distribution of the 210 modern songs. (From Ginty, D., Derwent, J.M. and Ji, T. *J Struct Eng.*, 79, 27-31, 2001.)

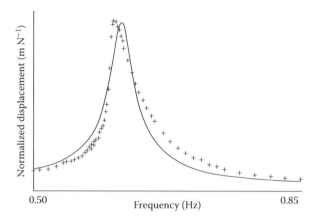

FIGURE 17.17
Frequency spectrum in one main direction of the building shown in Figure 17.16. (From Ellis, B. R. and Ji, T. *Struct Eng.*, 74, 186–192, 1996.)

17.4.3 Measurement of Resonance Frequency of a Building

The concept of resonance can be used to identify the natural frequencies of a structure. A harmonic force with an appropriate amplitude is applied to the structure and the corresponding maximum structural response is then recorded. The procedure is repeated a number of times with different forcing frequencies. The frequency corresponding to the largest value of these maximum responses is the resonant frequency of the structure. As civil engineering structures usually have damping ratios far less than 10%, the resonance frequency measured from the forced vibration is actually the natural frequency of the structure, as shown in Table 17.2.

Among a number of experiments carried out on the test building shown in Figure 16.16 were forced vibration tests with vibration generators used to shake the structure in a controlled manner at frequencies within the range 0.3–20 Hz [10]. Four vibration generators were placed at the four corners of the roof of the building and the building response was monitored using accelerometers aligned to measure motion in appropriate directions. The response of the building was sampled using optimised filtering, amplification and curve fitting and then normalised by converting the measured accelerations to equivalent displacements which were then divided by the applied forces. The maximum normalised displacement corresponding to a particular load frequency was plotted as a cross in Figure 17.17. This process was repeated for a number of load frequencies to produce many crosses in Figure 17.17, which are linked by a best-fit one-degree-of-freedom curve. The frequency corresponding to the largest response is a resonance frequency of the structure. For this building the resonant frequency was 0.617 Hz in one main direction. This technique has been widely used in structural engineering, mechanical engineering and other areas.

The building has many degrees of freedom. However, it behaves like an idealised SDOF system when it vibrates at its particular mode, which is evidenced by a close match between the crosses (real behaviour) and the solid line (SDOF model) in Figure 17.17. In addition, the measured resonance frequency can be considered as the natural frequency of the building in the related direction.

17.4.4 An Entertaining Resonance Phenomenon

Figure 17.18a shows a replica bronze washbowl which was used in ancient China. Four fish are cast into the bottom of the washbowl in an anticlockwise direction around the centre of the base, as shown in Figure 17.18b. The washbowl is symmetric about the plane that passes through the centre of the bowl and is parallel to the two handles. Now, such bowls are used for entertainment, as the water in the washbowl can be made to spurt up to 300 mm into the air around the locations

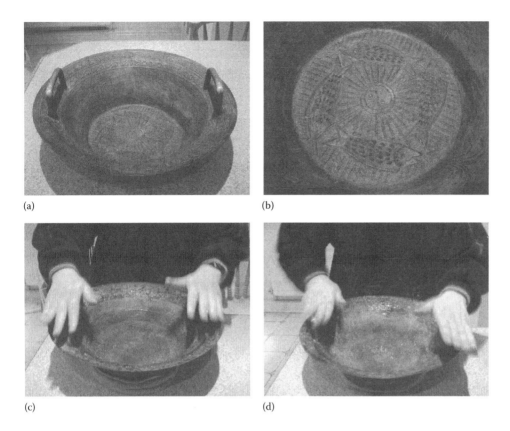

(a)

(b)

(c)

(d)

FIGURE 17.18
An entertaining resonance phenomenon. (a) Replica bronze washbowl. (b) Four fishes cast into the bottom of the bowl. (c) Ripple on water surface. (d) Water jumping out vertically.

of the fishes' mouths when two hands rub the handles on the washbowl. This effect can be produced as follows:

1. The washbowl is filled half-full with water and placed on a wet kitchen towel set in the shape of a ring with a radius of about 80 mm.
2. Two hands are cleaned using soap and water and the palms are then used to rub the handles of the washbowl, as shown in Figure 17.18c. The two palms move forwards and backwards alternatively and periodically, creating a pair of antisymmetric periodical forces applied to the symmetrically located handles of the washbowl.
3. Sound will generate from between the palms and the handles and ripples will form on the water surface (Figure 17.18c).
4. Continuing the hand movements will cause drops of water to jump out almost vertically from locations around the mouths of the fishes. Figure 17.18d shows the situation when the resonance occurs.

Only small forces are applied to the two handles of the washbowl in the vertical and forward–backward directions. As the hands move periodically, resonance occurs when the frequency of hand movement matches one of the natural frequencies of the washbowl–water system. The evidence of the resonance is sound and 'jumping water'.

PROBLEMS

1. Explain conceptually why the response of a structure becomes very large at resonance.
2. Derive the magnification factor for acceleration of an SDOF system subjected to a harmonic load. Plot the magnification factor – frequency ratio curves with four damping ratios of 0.01, 0.1, 0.2 and 0.5, respectively – and determine the resonance frequency for acceleration.
3. Consider an SDOF system that has a stiffness of $1\,kN\,m^{-1}$ and a mass of 10 kg. A periodical load, $P(t) = 20\sin2\pi t + 30\sin4\pi t$ (N), is applied on the system. Calculate the responses of the system in which the phase lag is determined using Equations 17.9 and 17.17, respectively, and plot the results, similar to Figure 17.5, for comparison.

REFERENCES

1. Beards, C. F. *Structural Vibration: Analysis and Damping*, London: Arnold, 1996.
2. Clough, R. W. and Penzien, J. *Dynamics of Structures*, New York: McGraw-Hill, 1993.
3. Ji, T. and Wang, D. A supplementary condition for calculating periodical vibration, *Journal of Sound and Vibration*, 241, 920–924, 2001.
4. Dallard, P., Fitzpatrick, A. J., Flint, A., Le Bourva, S., Low, A., Ridsdill-Smith, R. M. and Willford, M. The London Millennium Footbridge, *The Structural Engineer*, 79, 17–33, 2001.
5. Ellis, B. R. Serviceability evaluation of floor vibration induced by walking loads, *The Structural Engineer*, 79, 30–36, 2001.
6. Pachi, A. and Ji, T. Frequency and velocity of walking people, *The Structural Engineer*, 83, 36–40, 2005.
7. BSI. *BS 6399: Part 1: Loading for Buildings*, London, 1996.
8. Littler, J. D. Full-scale testing of large cantilever grandstands to determine their dynamic response. *Proceedings of the First International Conference on Stadia 2000*, 123–134, 1998.
9. Ginty, D., Derwent, J. M. and Ji, T. The frequency ranges of dance-type loads, *Journal of Structural Engineer*, 79, 27–31, 2001.
10. Ellis, B. R. and Ji, T. Dynamic testing and numerical modelling of the Cardington steel framed building from construction to completion. *The Structural Engineer*, 74, 186–192, 1996.

CHAPTER 18

CONTENTS

Damping in Structures

18

18.1 CONCEPTS

- The larger the damping ratio, ξ, the larger the ratio of successive peak displacements in free vibration (u_n/u_{n+1}), and the quicker the decay of oscillations.
- The damping ratio can be determined from measurements using the vibration theory of a single-degree-of-freedom system.
- The damping ratio is a measure of the amount of damping in a structure which can effectively reduce structural vibration at resonance.
- The higher the amplitude of free vibration of a structure, the larger will be the critical damping ratio and the smaller will be the natural frequency.

18.2 THEORETICAL BACKGROUND

The process by which vibrations decrease in amplitude with time is called **damping**. Damping dissipates the energy of a vibrating system and can do this in a number of ways. There are several different types of damping: friction, hysteretic and viscous. For structural systems, the damping encountered is best described using a viscous model in which the damping force is proportional to its velocity. The damping ratio ξ is expressed as a fraction of critical damping and is often written as a percentage (i.e. 1% critical).

The value of the damping ratio is required for predicting structural responses induced by different forms of dynamic loading. When structural vibrations are too large, artificial damping devices can be used to increase the damping and therefore reduce the levels of the vibrations. For example, dampers were installed in the London Millennium Footbridge to reduce the large lateral vibration of the bridge that occurred as a result of people walking across the bridge.

18.2.1 Evaluation of Viscous-Damping Ratio from Free Vibration Tests

Consider the equation of motion of a single-degree-of-freedom system with viscous damping:

$$m\ddot{u} + c\dot{u} + ku = 0 \tag{18.1}$$

The solution of Equation 18.1 has been given in Equation 16.15 as follows:

$$u(t) = \left[\frac{\dot{u}(0) + u(0)\xi\omega}{\omega_D} \sin\omega_D t + u(0)\cos\omega_D t \right] e^{-\xi\omega t} \tag{16.15}$$

where ω and ω_D are the angular frequencies of the undamped and damped systems, respectively.

EXAMPLE 18.1

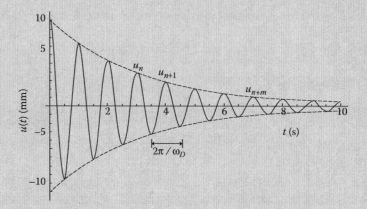

FIGURE 18.1
Damped free vibration.

Consider free vibrations of an undercritically damped system that has the following properties:

$$f = \frac{\omega}{2\pi} = 1.0\,\text{Hz}, \quad \xi = 0.05, \quad u(0) = 10\,\text{mm} \quad \text{and} \quad \dot{u}(0) = 0$$

SOLUTION

Substituting the data into Equation 16.15, the displacements of the system can be represented in graphical form, as shown in Figure 18.1.

It can be seen that

- The damped system oscillates about its neutral position with a constant angular frequency ω_D.
- The oscillation decays exponentially due to the damping.

Consider any two successive positive peaks such as u_n and u_{n+1} which occur at time $n(2\pi/\omega_D)$ and $(n+1)(2\pi/\omega_D)$, respectively. Equation 16.15 can be used to obtain the ratio of the two successive peak values as

$$\frac{u_n}{u_{n+1}} = \exp\left(\frac{2\pi\xi\omega}{\omega_D}\right) \tag{18.2}$$

Taking the natural logarithm of both sides of the equation and substituting $\omega_D = \omega\sqrt{1-\xi^2}$ into Equation 18.2, the logarithmic decrement of damping, δ, is obtained since ξ is normally small for structural systems:

$$\delta = \ln\frac{u_n}{u_{n+1}} = \frac{2\pi\xi}{\sqrt{1-\xi^2}} \doteq 2\pi\xi \quad \text{or} \quad \xi = \frac{\delta}{2\pi} = \frac{1}{2\pi}\ln\frac{u_n}{u_{n+1}} \tag{18.3}$$

Equation 18.3 indicates that

The larger the damping ratio ξ, the larger the ratio of u_n/u_{n+1} and thus the quicker the decay of the oscillation.

Considering two positive peaks from Figure 18.1 which are not adjacent, say u_n and u_{n+m}, Equation 18.3 becomes

$$\delta = \ln \frac{u_n}{u_{n+m}} = \frac{2\pi m \xi}{\sqrt{1-\xi^2}} = 2\pi m \xi \quad \text{or} \quad \xi = \frac{\delta}{2m\pi} \tag{18.4}$$

Thus, the damping ratio can be evaluated from Equation 18.3 or Equation 18.4 by measuring any two positive peak displacements and the number of cycles between them.

If the damping is perfectly viscous, Equations 18.3 and 18.4 will give the same result. However, this may not be true in all practical situations as will be shown in Section 18.4.

The advantages of using free vibration test responses to obtain values of damping ratio are

- The requirements for equipment and instrumentation are minimal in comparison with forced vibration test methods.
- Only relative displacement amplitudes need to be measured.
- The initial vibration can be generated by any convenient method, such as an initial displacement, an impulse or a sudden change of motion.

18.2.2 Evaluation of Viscous-Damping Ratio from Forced Vibration Tests

The dynamic magnification factor of a single-degree-of-freedom system subject to a harmonic load is shown in Section 17.2, that is

$$M_D = \frac{A}{\Delta_{st}} = \frac{1}{\sqrt{(1-f_p^2/f^2)+(2\xi f_p/f)^2}} \tag{17.12}$$

where:
f_p is the frequency of the load
f is the natural frequency of the system

At resonance of a structure with a low damping ratio, the frequency of oscillation is approximately equal to the natural frequency of the structure, that is, $f_p = f$ and Equation 18.12 becomes

$$M_D = \frac{A}{\Delta_{st}} = \frac{1}{2\xi} \tag{18.5}$$

Thus, if both the static displacement and the maximum dynamic displacement of a single-degree-of-freedom system (or a particular mode of a structure) at resonance can be determined experimentally, the critical damping ratio can be calculated from Equation 18.5. More accurate methods for determining the damping ratio from forced vibration tests can be found in [1–3].

18.3 MODEL DEMONSTRATIONS

18.3.1 Observing the Effect of Damping in Free Vibrations

This set of models demonstrates *the effect of damping provided by oil in free vibration, which can be observed by eye.*

Take a steel ruler and form a cantilever as shown in Figure 16.9a. Give the tip of the ruler an initial displacement and release it. The steel ruler will perform many cycles of oscillation before it becomes stationary, indicating that the damping ratio associated with steel alone is low.

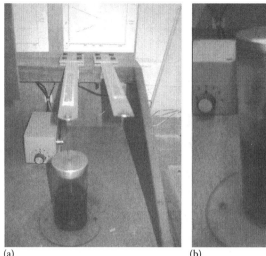

(a)   (b)

FIGURE 18.2
(a,b) Effect of the damping provided by oil.

Figure 18.2a shows two identical steel strips acting as cantilevers. The only difference between the two cantilevers is that connected to the free end of the cantilever on the left is a vertical metal bar which in turn is attached to a disc immersed in oil. Figure 18.2b shows the disc by lifting up the free end of the strip. Thus, the effect of the oil and the device can be examined.

Press the free ends of the two cantilevers down by the same amount and then release them suddenly. It will be observed that the cantilever on the left only vibrates for a small number of cycles before stopping while the cantilever on the right oscillates through many cycles demonstrating the effect of viscous damping provided by the oil.

18.3.2 Hearing the Effect of Damping in Free Vibrations

This set of models shows *the effect of damping provided by rubber bands in free vibration, which can be heard by ears.*

The sound heard from the free vibrations of a taut string, such as a violin string, links two different physical phenomena, sound transmission and the string vibrations, which both can be described using the same differential equation of motion. Thus, hearing a sound can be related to observing free vibrations, in this case those of a taut string.

Figure 18.3 shows two identical steel bars, one is a bare bar and the other has rubber bands wrapped around it. The effect of the damping added by the rubber can be demonstrated as follows:

- Suspend the bare bar and give it a knock at its lower end using the other metal bar, as shown in Figure 18.3a. A sound will be generated from the bar for several seconds as it reverberates.
- Suspend the wrapped bar and give it a similar knock on the exposed metal part (Figure 18.3b). This time only a brief dull sound is heard as the rubber wrapping dissipates much of the energy of the vibration.

18.4 PRACTICAL EXAMPLES

18.4.1 Damping Ratio Obtained from Free Vibration Tests

The true damping characteristics of typical structural systems are normally very complex and difficult to define with few structures actually behaving as ideal single-degree-of-freedom systems.

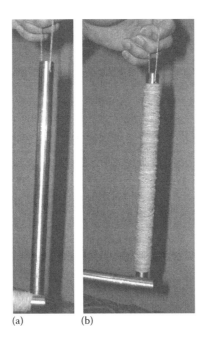

FIGURE 18.3
(a,b) Effect of damping on sound transmission.

Notwithstanding this, the earlier discussion of single-degree-of-freedom systems (Chapters 16 and 17) can be useful when considering more complex practical structures.

A free vibration test was conducted on a full-sized eight-storey test building (Figure 16.17) to identify the damping ratio of its fundamental mode [4]. In order to amplify the displacements of the structure, the building was shaken by a set of vibrators mounted at the four corners on the roof of the building, at the fundamental frequency of the building. After the vibrators were suddenly stopped, the ensuing free vibrations of the roof of the building were measured. The decay of the vibrations in one of the two main directions of the building is shown in Figure 16.17. As the excitation caused vibrations effectively only in the fundamental mode, the contributions of other modes of vibration to the decaying response of the structure were negligible.

The natural frequency and damping ratio of the response of the structure can be determined from the records shown in Figure 16.17. Five, continuous 10 s samples of vibrations were extracted from the response and a curve-fitting technique was used to produce smooth curves from which the natural frequency and damping ratio could be determined. One such smoothed curve, superimposed on the measured curve, is shown in Figure 18.4. The response frequency and damping ratio values

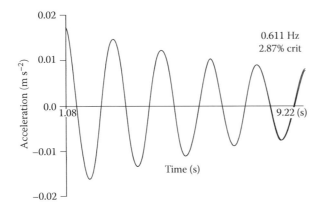

FIGURE 18.4
Extraction of natural frequency and damping ratio from free vibration records. (From Ellis, B. R. and Ji, T. *The Structural Engineer,* 74, 186–192, 1996.)

TABLE 18.1 Natural Frequency and Damping Ratio
Determined from Various Sections of Decay

Relative Amplitude	Natural Frequency (Hz)	Damping Ratio (%)
1.00	0.611	2.87
0.366	0.636	1.81
0.181	0.645	1.28
0.106	0.647	1.02
0.062	0.656	0.85

extracted from the five samples are given in Table 18.1 and related to the amplitude of vibration at the start of each sample [4].

From Table 18.1 it can be observed that

- The higher the amplitude of vibration, the smaller the natural frequency and the larger the damping ratio.
- The natural frequency for the relative amplitude of 1.00 is 6.86% lower than that for the relative amplitude of 0.062, while the damping ratio at the relative amplitude of 1.00 is 238% higher than that at the relative amplitude of 0.062.

When the building vibrated with small amplitudes, the relative movements between joints and other connections in the structure were small, involving frictional forces doing less work, and leading to lower damping ratios than were found when amplitudes were larger with associated larger relative joint movements and friction-related work. These variations have been observed in many different types of structure.

18.4.2 Damping Ratio Obtained from Forced Vibration Tests

The forced vibration tests of the framed steel building for obtaining its resonance frequency are described in Section 18.4.3. The frequency spectrum obtained from the experiment and curve fitting are shown in Figure 17.17.

The natural frequency and damping ratio obtained from the forced vibration tests were 0.617 Hz and 2.25%, respectively. It can be noted that the measured values have a characteristic negative skew compared to the best-fit curve. This is typical of this type of measurement and shows one aspect of nonlinear behaviour as observed in free vibration tests of buildings and other structures. It can be observed from Table 18.1 that the measurements from the forced vibration tests agree favourably with those obtained from the free vibration tests for relative amplitudes between 0.366 and 1.000. In general, forced vibration tests normally provide larger forces than free vibration tests. Therefore, forced vibration tests frequently produce smaller values for natural frequencies and larger values for damping ratios than those obtained from free vibration tests.

18.4.3 Damping Ratios for Floor Structures

It has been shown in Sections 18.4.1 and 18.4.2 that damping is not a factor which can always be determined accurately. However, damping reflects how a system is constructed as well as its constituent materials. For example, a steel framework supporting precast concrete floor beams will have a much larger damping value than that of the steel frame alone or the beams by themselves, the joints between the two materials being a source of energy dissipation. Also, a bolted steel framework will have much higher damping than a similar welded framework. Thus, damping of a structure is greater than that of the elements from which it is constructed; which suggests that damping is not a quantity readily compatible with finite elements. Instead, it is often sensible to estimate the damping value for each mode of vibration based upon measurements taken on similar structures.

TABLE 18.2 Measured Characteristics of the Principal Mode of Vibration of a Single Panel for a Range of Floors

Floor type	Natural Frequency (Hz)	Damping (% crit)
Steel beams with composite floor	6.39	0.61
Slimfloor	12.03	0.67
In situ concrete	11.89	0.86
Steel beams with composite floor	5.31	0.87
Steel beams with composite floor	5.43	1.34
Steel girders with prestressed concrete planks	4.95	1.35
Deep profiled concrete slab	7.43	1.56
Steel beams with composite floor	8.50	1.65
Steel beams with composite floor	6.70	1.75
Composite steel/concrete beams with precast hollow floor units	7.91	1.81
Steel beams with precast concrete planks and false floor	6.96	1.90
Steel beams with composite floor	5.26	2.72
Steel beams with composite floor plus furnishings	9.26	4.45
Sprung wooden dance floor	7.62	4.54
Beam and pot floor	11.93	5.08

Source: Ellis, B. R., Ji, T., El-Dardiry, E. and Zheng, T. *The Structural Engineer*, 88, 18–26, 2010.

The previous paragraphs, and indeed most literature, suggest that damping is a constant factor, but measurements show that damping does vary with amplitude of vibration, the damping generally increasing with increasing amplitude of vibration. So, damping values should be measured at amplitudes of vibration of a similar level to those under consideration.

A number of measurements have been made on various types of floor (at vibration amplitudes similar to those induced by people walking) and these are given in Table 18.2, which lists the floors in order of increasing damping so the range of damping values can be appreciated. It can be observed from the table that there is quite a wide range of damping values.

The floors with the lower damping values have one thing in common: they have a continuous floor system. The floors built up with precast concrete panels tend to have higher damping values due to the joints which provide an extra damping mechanism. The furnished floors also tend to have higher damping values, but it is questionable whether damping from furnishings can be relied on and it is often the least furnished areas of multipanel floors that encounter problems. The two highly damped floors are both quite old and of a type of construction with many joints.

18.4.4 Damping Ratios for Buildings

As indicated in the previous section, it is not really possible to calculate damping values accurately, as they depend on how structures are assembled and on somewhat nebulous items like the quality of construction. Usually, a viscoelastic model is used to represent damping, and this is because observed behaviour aligns with this model, at least for a small range of amplitudes of motion, rather than any identified damping mechanism. Hence, if a value for damping is required, it is sensible to see what values have been measured on similar structures for similar amplitudes of vibration. However, it is wise to see how the damping has been measured, because there are many ways of determining it from experiment and some can have large potential errors [6].

Damping measurements were taken from forced vibration tests on a number of buildings in the United Kingdom [6], and a range of damping ratios, from 1.0% to 2.5% critical, was apparent for the translational modes of buildings where soil-structure interaction was insignificant, with the exception of 0.5% critical for one very tall building. This observation that a very tall building appears to have lower levels of damping may be because any forced testing can only generate low amplitude vibration, so this may be, in part, a function of nonlinear behaviour. The damping values for torsional modes also seem to be slightly higher than those for the translational modes.

(a)   (b)

FIGURE 18.5
(a,b) Damping devices installed on the London Millennium Footbridge.

18.4.5 Reducing Footbridge Vibrations Induced by Walking

A total of 37 viscous dampers were installed on the London Millennium Footbridge in order to reduce the lateral vibrations of the bridge which occurred when people walked across the bridge. The majority of these dampers are situated beneath the bridge deck, supported by transverse members (Figure 18.5a). One end of each viscous damper is connected to the apex of a steel V brace, known as a chevron. The apex of the chevron is supported on roller bearings that provide vertical support but allow sliding in the other directions. The other ends of the chevron are fixed to the neighbouring transverse members [7].

Viscous dampers were also installed in the planes between the cables and the deck at the piers (Figure 18.5b) to provide damping of the lateral and lateral–torsional modes of vibration.

18.4.6 Reducing Floor Vibration Induced by Walking

Damping can be introduced into concrete floors by sandwiching a layer of high damping material between the structural concrete floor and a protective concrete topping. This acts in a similar manner to the demonstration model described in Section 16.3.3.

During the bending of a floor induced by footfall vibrations, energy is dissipated through the shear deformation produced in the damping material by the relative deformations of the two concrete layers (Figure 18.6). The technique was originally used to damp out vibrations of aircraft fuselage panels when the resonant frequencies of the panels were over 200 Hz.

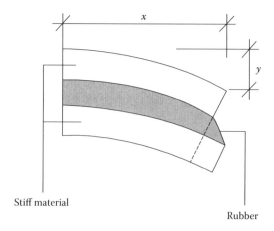

Stiff material

Rubber

FIGURE 18.6
Deformation of bonded layers.

FIGURE 18.7
The test floor.

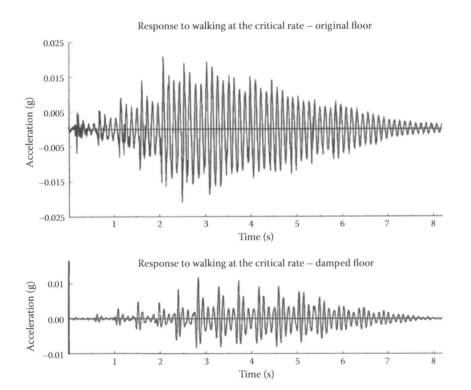

FIGURE 18.8
Floor vibration induced by walking loads, with and without the damping layer.

A floor panel, 6 × 9 m, in the steel-frame test building at the BRE Cardington Laboratory, was selected for testing with and without damping layers (Figure 18.7). A variety of comparative tests were conducted including heel-drop tests, forced vibration tests and walking tests at different speeds.

Figure 18.8 shows the comparison of acceleration–time histories induced by the same individual walking on the floor panel without and with the damping layer. The benefit of the constrained damping layer in reducing the vibration of the floor is obvious.

PROBLEMS

1. The damping ratio in a structure is likely to be affected by the methods for processing the measurements. Carry out the following exercises:
   a. Digitise the acceleration and time curve shown in Figure 18.4.
   b. Use Equation 18.3 to determine the damping ratio using the first two peak values and the third and fourth peak values respectively.
   c. Abstract the natural frequency and damping ratio from the digitised data.
2. A chimney has a fundamental natural frequency of 1.3 Hz and a damping ratio of 0.4% and a building has a fundamental natural frequency of 1.0 Hz and a damping ratio of 0.5%. If the same initial disturbance is applied to each of the two structures to induce free vibration in their fundamental modes, does the vibration of the building decay more quickly than that of the chimney? Give reasons to justify your answer.

REFERENCES

1. Beards, C. F. *Structural Vibration: Analysis and Damping*, London: Arnold, 1996.
2. Clough, R. W. and Penzien, J. *Dynamics of Structures*, New York: McGraw-Hill, 1993.
3. Chopra, A. K. *Dynamics of Structures*, Upper Saddle River, NJ: Prentice Hall, 1995.
4. Ellis, B. R. and Ji, T. Dynamic testing and numerical modelling of the Cardington steel framed building from construction to completion, *The Structural Engineer*, 74, 186–192, 1996.
5. Ellis, B. R., Ji, T., El-Dardiry, E. and Zheng, T. Determining the dynamic characteristics of multi-panel floors, *The Structural Engineer*, 88, 18–26, 2010.
6. Ellis, B. R. Full-scale measurements of the dynamic characteristics of tall buildings in the UK, *Journal for Wind Engineering and Industrial Aerodynamics*, 59, 365–382, 1996.
7. Dallard, P., Fitzpatrick, A. J., Flint, A., Le Bourva, S., Low, A., Ridsdill-Smith, R. M. and Willford, M. The London Millennium Footbridge, *The Structural Engineer*, 79, 17–33, 2001.

CONTENTS

CHAPTER 19

Vibration Reduction

19.1 DEFINITIONS AND CONCEPTS

The dynamic vibration absorber (DVA) or tuned mass damper (TMD) is a device which can reduce the amplitude of vibration of a structure through interactive effects. A TMD consists of a mass, connected by means of an elastic element and a damping element to the structure.

- The amplitude of vibration of a structure at resonance can be effectively reduced through slightly increasing or reducing the natural frequency of the structure, thus avoiding resonance from an input at a given frequency.
- The amplitude of vibration of a structure can be reduced through base isolation which changes the natural frequencies of the system.
- The amplitude of vibration of a particular mode of a structure can be effectively reduced using a TMD.

19.2 THEORETICAL BACKGROUND

Reducing vibration levels induced by different kinds of dynamic loads is a major requirement in the design of some civil engineering structures. The loads include those induced by wind, earthquakes, machines and human activities. For example, some structures prone to wind loads are those which possess a relatively large dimension, either in height or in length, such as tall buildings and long-span roofs.

Common methods of reducing structural vibration include base isolation, passive energy dissipation and active or semi-active control.

Base isolation: An isolation system is typically placed at the foundation of a structure, which alters the natural frequencies of the structure and avoids resonance from the principal frequency of the excitation force. In earthquake resistance design, an isolation system will deform and absorb some of the earthquake input energy before the energy can be transmitted into the structure.

Passive energy dissipation: Dampers installed in a structure can absorb some of the input energy from dynamic loading or alter the natural frequencies of the structure or both, thereby reducing structural vibrations. There are several types of damper used in engineering practice, for example viscoelastic dampers, friction dampers, TMDs and tuned liquid dampers (TLDs).

Semi-active and active control: The motion of a structure is controlled or modified by means of a control system through providing external forces which oppose the action of the input.

In a similar manner to the previous chapters, limited theoretical background is provided to explain the demonstration models and practical examples illustrated in this chapter. More detailed information on this topic can be found from [1–3].

Altering the structural stiffness, increasing the damping of a structure or placing a TMD, are the relatively simple and effective measures to reduce structural vibrations.

19.2.1 Change of Dynamic Properties of Systems

Equation 17.11 showed that the dynamic response of a single-degree-of-freedom (SDOF) system is the product of the static displacement and the dynamic magnification factor which is defined in Equation 17.12. This has been shown graphically in Figure 17.3 with different damping ratios in the frequency ratio range between 0.0 and 3.0. In Figure 19.1, the same figure is shown but with a range of between 0.9 and 1.1, with $\xi = 0.01$, 0.02, 0.03 and 0.05, for examining the response behaviour

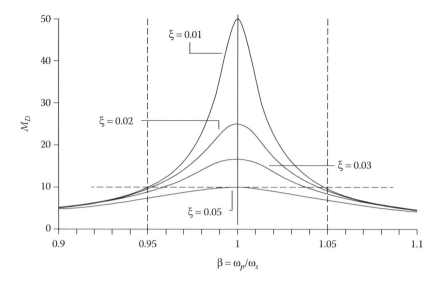

FIGURE 19.1
Magnification factor as a function of the force frequency ratio.

TABLE 19.1 Magnification Factors M_D Close to Resonance

	$\xi = 0.01$	$\xi = 0.02$	$\xi = 0.03$	$\xi = 0.05$
$\beta = 0.95$	**10.1**	9.56	8.85	7.35
$\beta = 1.0$	50.0	25.0	16.7	**10.0**
$\beta = 1.05$	**9.56**	9.03	8.31	6.81

around resonance. Table 19.1 compares the magnification factors of the SDOF system when the frequency ratios are 0.95, 1.0 and 1.05, respectively, using Equation 17.12. The following can be observed from Figure 19.1 and Table 19.1.

- Increasing the damping ratio can effectively reduce the vibrations at resonance; simply doubling the damping ratio reduces the peak response by a factor of approximately two.
- A slight change of the natural frequency of the SDOF system away from the resonance frequency can effectively reduce the dynamic response. For instance, when the system has a damping ratio of 0.01, the resonance response can be reduced by about 80% by increasing or reducing the natural frequency by 5%. The same level of vibration reduction requires increasing the damping ratio from 0.01 to 0.05. The three related values for comparison are highlighted in boldface in Table 19.1.

Altering the natural frequency of a structure or increasing the corresponding damping ratio or both can reduce vibration. The concepts behind the two measures are different. Altering the natural frequency of a structure aims to avoid resonance and, for an input at the previous frequency, the structure will experience lower levels of vibration. Increasing damping will increase the energy dissipation in a structure, and although the structure will still experience resonance, the response is reduced through energy dissipation.

Increasing structural stiffness to increase the natural frequency of the structure may not affect the damping ratio or damping mechanism of the structure. The practical case given in Section 10.4.4 is a good example, where the floor was stiffened with the profiled external tendons and the fundamental natural frequency of the floor was increased slightly to avoid resonance. It was thought that the added tendons would not change the damping mechanism of the floor. It is important to note that structural alterations to increase stiffness may also result in increased mass, thus limiting changes to the natural frequencies.

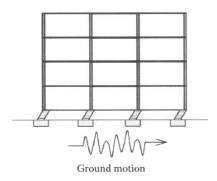

FIGURE 19.2
Base isolation of a building.

Reducing structural stiffness reduces the natural frequency of a structure and this approach may be used to avoid resonance. Base isolation systems have been used successfully for the design of earthquake-resistant buildings. An earthquake will shake a building. Placing a base isolation system between the building and the foundation/ground can reduce the level of earthquake forces transmitted to the building. The base isolation system reduces the fundamental natural frequency of the building to a value lower than the predominant frequencies of ground motion to avoid resonance. The first vibration mode of the isolated building is primarily deformation of the isolated system, and the building itself appears rigid, as illustrated in Figure 19.2. This type of isolation system may absorb a relatively small amount of earthquake energy to suppress any possible resonance at the isolation frequency, but its main function is to change the natural frequency of the building and deform itself. This type of isolation works when the system is linear and elastic, even when it is lightly damped.

Increasing the damping ratio of a structure by incorporating viscoelastic dampers into the structure may also provide additional stiffness along with a dissipative mechanism. Thus, the vibration reduction of the structure due to the added damping materials, or dampers, could be the combined effect of the increased damping and the increased stiffness. Examples are the London Millennium Footbridge, where viscous dampers were fitted in conjunction with the use of a continuous bracing system as discussed in Section 18.4.4, and the test floor panel incorporated with a layer of rubber together with an additional layer of concrete, as discussed in Section 18.4.5.

19.2.2 Tuned Mass Dampers

The DVA, or TMD, is a device which can reduce the amplitude of vibration of a system through interactive effects. A TMD consists of a mass, connected by means of an elastic element and a damping element to the system. The principle of a TMD reducing the vibration of a primary structure is to transfer some structural vibration energy from the structure to the TMD and to split the resonance peak into two less-significant resonances. Thus, it reduces the vibration of the primary structure while the TMD itself may vibrate significantly.

Figure 19.3b shows an SDOF system subjected to a harmonic load and its response has been shown in Chapter 17. A TMD can be placed on the system, forming a two-degree-of-freedom (TDOF) system, as shown in Figure 19.3c. The equations of motion of the new system can be described as

$$\begin{bmatrix} M_s & 0 \\ 0 & M_T \end{bmatrix} \begin{Bmatrix} \ddot{u}_S \\ \ddot{u}_T \end{Bmatrix} + \begin{bmatrix} C_s + C_T & -C_T \\ -C_T & C_T \end{bmatrix} \begin{Bmatrix} \dot{u}_S \\ \dot{u}_T \end{Bmatrix} + \begin{bmatrix} K_S + K_T & -K_T \\ -K_T & K_T \end{bmatrix} \begin{Bmatrix} u_S \\ u_T \end{Bmatrix} = \begin{Bmatrix} P_0 \sin\omega_p t \\ 0 \end{Bmatrix} \quad (19.1)$$

The displacements of the SDOF structure system and the TMD are denoted by u_S and u_T, respectively; the damping coefficient and stiffness are denoted by C_T and K_T for the TMD and C_S and K_S for the structure; and P_0 is the magnitude of the external force applied to the structure system with a frequency ω_p.

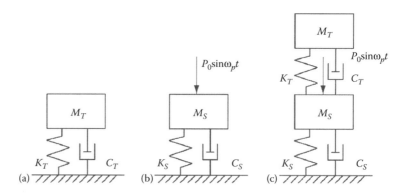

FIGURE 19.3
(a) A tuned mass damper (TMD). (b) An SDOF structure subjected to a harmonic load. (c) The TMD is placed on the SDOF structure.

For an appreciation of the efficiency of a TMD in reducing structural vibration, consider a simplified situation where the structural damping is neglected, that is, $C_S = 0$. Thus, the response of the structural system is infinite at resonance without a TMD. Similar to the dynamic magnification factor defined in Chapter 17 for an SDOF system, the dynamic magnification factor for an undamped structural system with a TMD is [3]:

$$R = \sqrt{\frac{(\gamma^2 - \beta^2)^2 + (2\xi_T\gamma\beta)^2}{[(\gamma^2 - \beta^2)^2(1 - \beta^2)^2 - \alpha\gamma^2\beta^2]^2 + (2\xi_T\gamma\beta)^2(1 - \beta^2 - \alpha\beta^2)^2}} \qquad (19.2)$$

where:

$\beta = \omega_p/\omega_S$	is the forcing frequency ratio
$\gamma = \omega_T/\omega_S$	is the natural frequency ratio
$\omega_S = \sqrt{K_S/M_S}$	is the natural frequency of the structural system
$\omega_T = \sqrt{K_T/M_T}$	is the natural frequency of the TMD system
$\xi_T = C_T/(2M_T\omega_T)$	is the damping ratio of the TMD system
α	is the mass ratio of the TMD to the structure

It can be seen from Equation 19.2 that the magnification factor is a function of the four variables, α, ξ_T, β and γ. Plots of the magnification factor R as a function of the frequency ratio β are given in Figure 19.4a for $\gamma = 1$ and $\alpha = 0.05$, with three ξ_T values (0.0, 0.03 and 0.1); and in Figure 19.4b for $\gamma = 1$ and $\xi_T = 0.03$, with three α values (0.01, 0.05 and 1.0); and in Figure 19.4c for $\alpha = 0.05$ and $\xi_T = 0.03$, with three γ values (0.9, 1.0 and 1.1).

From Figure 19.4 it can be observed that

- When $\xi_T = 0$ and $\beta = \gamma$, the structural mass is stationary, that is, the applied load and the force generated by the TMD cancel each other (Equation 19.2 and Figure 19.4a).
- The damping in the TMD is important for reducing the vibration at the resonance frequencies of the new TDOF system (Figure 19.4a).
- The larger the mass ratio, the wider the spread of the two new natural frequencies of the TDOF system, and the smaller the response at $\beta = 1$ (Figure 19.4b).
- A TMD is more effective at reducing the vibration of the structure at the natural frequency of the TMD than at that of the structure (Figure 19.4c).
- A TMD can effectively reduce the resonant vibration of a structure with relatively low damping.

It is noted that large movements of a TMD should be considered. In practice, several types of TMD have been developed to accommodate different requirements [1].

In a TMD, a solid (concrete or metal) block often acts as the mass, but in tall buildings a tank filled with water serves the same purpose; this may be considered to be a TLD. Liquids are used to

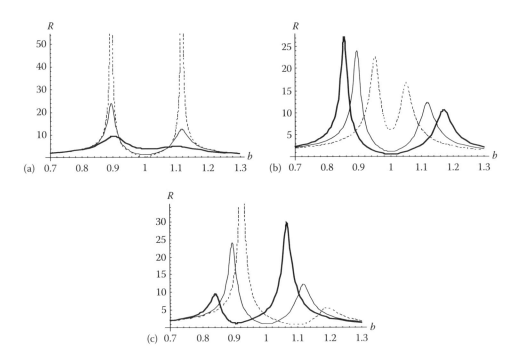

FIGURE 19.4
Dynamic magnification factor as function of β. (a) $\gamma=1$, $\alpha=0.05$ with $\xi_T=0.0$ (dashed line), 0.03 (solid line) and 0.1 (dark solid line). (b) $\gamma=1$, $\xi_T=0.03$ with $\alpha=0.01$ (dashed line), 0.05 (solid line) and 0.1 (dark solid line). (c) $\alpha=0.05$, $\xi_T=0.03$ with $\gamma=0.9$ (dark solid line), 1.0 (solid line) and 1.1 (dashed line).

provide not only all the necessary characteristics of the TMD system but also the damping through a sloshing action. The mathematical description of TLD response is difficult, but structural implementation is often quite simple.

19.3 MODEL DEMONSTRATIONS

19.3.1 Tuned Mass Damper

This demonstration *compares free vibrations of an SDOF system and a similar system with a TMD,* and *shows the effect of the TMD in free vibration.*

Figure 19.5 shows an SDOF system consisting of a mass and a spring attached to the left hand of a cross arm. A similar SDOF system is attached to the right end of the cross arm and a smaller SDOF system is suspended from it. The smaller SDOF system, which is used as a TMD, has the same natural frequency as that of the main SDOF system. In other words, the ratio of the stiffnesses of the two springs is the same as the ratio of the two masses.

Displace the two main SDOF systems downward vertically by the same amount and then release them simultaneously. It can be seen that the amplitude of vibration of the SDOF system on the left is greater than that of the one on the right. The TMD vibrates more than the mass to which it is attached. The vibration of the TMD applies a harmonic force to the SDOF system which suppresses the movements of the main mass.

The effect of a TMD is most significant if a harmonic load is applied to the system with a frequency close to the natural frequency of the system.

19.3.2 Tuned Liquid Damper

This demonstration *compares free vibration of two SDOF systems, one with and one without water,* and *shows the effect of a turned liquid damper in reducing free vibration.*

FIGURE 19.5
Comparison of vibration with and without a tuned mass damper.

A TLD is basically a tank of liquid that can be tuned to slosh at the same frequency as the natural frequency of the structure to which it is attached.

Two circular tanks are attached to the top of two identical flexible frames, as shown in Figure 19.6. The tank on the right is filled with some coloured water. The effect of water sloshing on the vibration of the supporting structure can be demonstrated as follows:

1. Displace the tops of the two frames laterally by similar amounts.
2. Release the two frames simultaneously and observe free vibrations of the two frames. The amplitude of vibration of the frame on the right decays much more quickly than that of the frame on the left.

The difference is due to the water sloshing in the tank and dissipating the energy of the system.

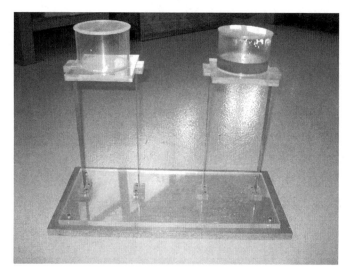

FIGURE 19.6
Comparison of vibration with and without the effect of liquid.

FIGURE 19.7
Altering structural frequency using base isolation.

19.3.3 Vibration Isolation

This demonstration *compares forced vibrations of two glasses containing similar amounts of water, one with and one without a plastic foam support*, and *shows the effect of base isolation*.

A medical shaker can be used as a shaking table to generate harmonic base movements in three perpendicular directions. A glass is fixed directly to a wooden board while a similar glass is glued to a layer of plastic foam which is mounted on the wooden board, as shown in Figure 19.7.

Fill the two glasses with similar amounts of water. When the shaker moves at a preset frequency, it can be seen that the water in the glass on the plastic foam moves less than that in the other glass. The difference is due to the effect of the plastic foam, which isolates the base motion and also produces a natural frequency of the glass–water–foam system which is lower than that of the glass–water system and is thus farther away from the vibration frequency.

19.3.4 Pendulum Tuned Mass Damper

This demonstration shows *the effect of a pendulum TMD on reducing vibration at resonance, which simulates the pendulum used to reduce lateral vibration of the 509-m-high building of Taipei 101.*

A large pendulum TMD is used in Taipei 101 to reduce the lateral vibration of the building induced by wind or (possibly) earthquakes. The principle of the vibration reduction can be demonstrated using a medical shaking table that can generate horizontal harmonic motion, two similar rulers that represent slender tall buildings, and a string and a golf ball that model the pendulum.

Figure 19.8 shows the set-up for the demonstration. The frequency of the movement of the shaker can be adjusted manually between 0 and 4 Hz. The two rulers (one with and one without a pendulum) are placed vertically on the shaking table and clamped at their lower parts, using two horizontally placed members on the shakers. Two identical small rulers are placed horizontally at the top of the two rulers, for hanging the pendulum. It can be seen from Figure 19.8 that the only difference between the two 'structures' is that one has an attached pendulum – a string and a golf ball. The length of the string is determined so that the pendulum has a similar natural frequency to

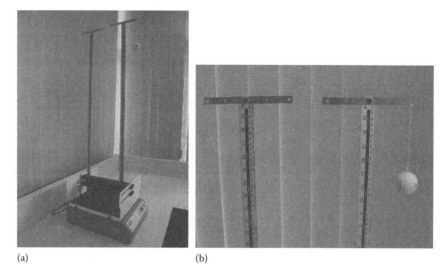

(a)   (b)

FIGURE 19.8
A medical shaker and the test structures. (a) The test set-up. (b) The difference between the two structures.

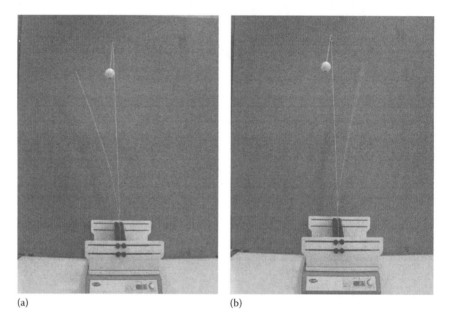

(a)   (b)

FIGURE 19.9
Comparison of the lateral vibration of the two rulers, with and without a pendulum, at two different times.

the cantilever ruler. The fundamental natural frequency of the cantilever is experimentally determined, using the shaker to determine when resonance occurs.

The efficiency of the pendulum TMD is demonstrated when the shaker frequency is close to the fundamental natural frequency of the ruler, as shown in Figure 19.9. It can be observed that

- The ruler without a pendulum vibrates significantly and its vibration magnitude is several times that of the ruler with the pendulum.
- For the ruler with the pendulum, the pendulum vibrates more than the ruler.

By gradually increasing the frequency of the shaker, between 0 and 4 Hz, which covers the natural frequencies of the ruler–pendulum system, the responses of the two systems subjected to base motion can be observed.

19.4 PRACTICAL EXAMPLES

19.4.1 Tyres Used for Vibration Isolation

Figure 19.10 shows two tyres placed between the ground and a generator in a rural area of a developing country. The presence of the tyres reduced the natural frequency of the generator and moved it away from the operating frequency. Although the operators of the generator may not have known much about vibration theory, they knew from experience that the presence of the tyres could reduce the level of vibration.

19.4.2 London Eye

The British Airways London Eye is one of the largest observation wheels in the world (Figure 19.11a). It was built by the River Thames in the heart of London, near to the Palace of Westminster. The wheel has a height of 135 m and carries 32 capsules which can hold up to 800 people in total.

In order to reduce vibrations in the direction perpendicular to the plane of the wheel due to wind loads, 64 TMDs were installed in steel tubes which are uniformly distributed around the wheel. One of these tubes is indicated in Figure 19.11b. The relation between the TMD and the tube is illustrated in Figure 19.12. A mass and a spring in the tube were designed with a natural frequency close to the natural frequency of the wheel in the direction perpendicular to the plane of the wheel. Plenty of room is available in the tube to allow large movements of the TMD. No excessive vibration of the wheel has been observed in this direction since it was erected [4,5].

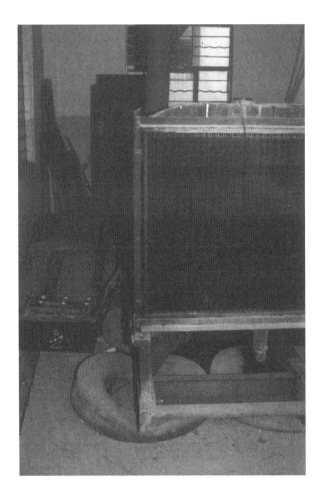

FIGURE 19.10
Tyres used for base isolation. (Courtesy of Professor B. Zhuang, Zhejiang University, China.)

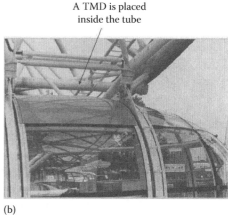

A TMD is placed inside the tube

(a)   (b)

FIGURE 19.11
The London Eye. (a) Front view. (b) A tuned mass damper placed inside a tube.

FIGURE 19.12
A tuned mass damper in a tube of the London Eye.

19.4.3 London Millennium Footbridge

A total of 26 pairs of vertical TMDs were installed over the three spans of the London Millennium Footbridge [6]. These comprise masses of between 1 t and 3 t supported on compression springs. The TMDs are situated on the top of the transverse arms beneath the deck. One pair of TMDs is shown in Figure 19.13.

FIGURE 19.13
Two tuned mass dampers under the deck of the London Millennium Footbridge.

These TMDs become effective if the footbridge experiences relatively large vertical vibrations, though such vibrations were not the primary source of the well-publicised problems when the bridge first opened.

PROBLEMS

1. Assume that the generator shown in Figure 19.10 has a vertical natural frequency of 10 Hz and a mass of 2000 kg. If the tyres have a vertical stiffness of 100,000 N m^{-1} and are placed underneath the generator, what is the natural frequency of the tyre-generator system (the mass of the tyres can be ignored in this analysis)?
2. Make models similar to those shown in Figure 19.6 and conduct the following comparative experiments:
   a. Free vibration test: one with and one without water.
   b. Free vibration test: one with and one without a solid mass on the top.
   c. Free vibration test: one with the solid mass on the top and one with water that has a similar weight to the solid mass.
   d. Record the observations and compare the observations for the three sets of test.

REFERENCES

1. Soong, T. T. and Dargush, G. F. *Passive Dissipation Systems in Structural Vibration*, New York: John Wiley, 1997.
2. Soong, T. T. *Active Structural Control: Theory and Practice*, Harlow: Longman Scientific & Technical, 1990.
3. Korenev, B. G. and Reznikov, L. M. *Dynamic Vibration Absorbers*, New York: John Wiley, 1993.
4. Rattenbury, K. *The Essential Eye*, London: HarperCollins, 2006.
5. Ji, T. Yu, X. Zheng, T. and Ellis, B. R. *Vibration Measurement of the London Eye*, Technical report to Jacobs, 2007.
6. Dallard, P., Fitzpatrick, A. J., Flint, A., Le Bourva, S., Low, A., Ridsdill-Smith, R. M. and Willford, M. The London Millennium Footbridge, *The Structural Engineer*, 79, 17–33, 2001.

CHAPTER 20

CONTENTS

Human Body Models in Structural Vibration

20

20.1 CONCEPTS

- A stationary person, that is, one who is sitting or standing on a structure, acts as a mass-spring-damper rather than as an inert mass in both vertical and lateral structural vibration.
- A walking or jumping person acts solely as a dynamic load on a structure and thereby induces structural vibration.
- A bouncing person, that is, one who is in constant contact with a structure, acts as both a dynamic load and a mass-spring-damper on a structure in vertical vibration.

20.2 THEORETICAL BACKGROUND

20.2.1 Brief Introduction to Human–Structure Interaction

How people interact with their environment is a topical issue and one of increasing importance. One form of physical interaction which is understood poorly, even by professionals, is concerned with human response to structural vibration. This is important, for example when determining how dance floors, footbridges and grandstands respond to moving crowds and for determining how stationary people are affected by vibration in their working environment. Human–structure interaction provides a new topic that *describes the independent human system and the structural system working as a whole and studies structural vibration when people are involved, and human body response to structural movements.*

When a structure is built on soft soil, the interaction between the soil and the structure may be considered; when a structure is in water, such as an offshore platform, the interaction between the structure and the surrounding fluid may be considered. Similarly, when a structure is loaded with people, the interaction between people and structure may need to be considered. An interesting question is why has this not been considered earlier? There are two reasons for this.

1. The human body is traditionally considered as an inert mass in structural vibration. For example, Figure 20.1 is a question taken from a well-known textbook on engineering mechanics [1] where a girl is modelled as an inert mass in the calculation of the natural frequency of the human–beam system.
2. The human mass is small in comparison with the masses of many structures and in this situation its effect is negligible. Thus, there have been no requirements from practice for considering such effects as human–structure interaction.

Dynamic measurements were taken on the North Stand at Twickenham when it was empty and when it was full of spectators (Section 20.4.1). The observations on the stand suggested a new concept that *the stationary human whole body acts as a mass-spring-damper rather than an inert mass in structural vibrations.* The phenomenon was reproduced in the laboratory (Section 20.3.1). It was confirmed that *the stationary human whole body did not act as an inert mass but appeared to act as a mass-spring-damper system in structural vibrations* [2].

Today, many structures are lighter and have longer spans than earlier similar types of construction and as a consequence the effect of human bodies has become more important. Newly emerging

Problem 8/17

FIGURE 20.1
A girl standing on a beam. (Permission of John Wiley) (From Meriam, J. L. and Kraige, L. G. *Engineering Mechanics: Dynamics*, 4th edn, Vol. 2, John Wiley, New York, 1998. Reproduced with permission.)

problems are the human-induced vibrations of grandstands and footbridges, where crowds of people are involved, and the human perception of structural vibration induced by human actions.

The study of human–structure interaction is concerned with both structural dynamics and body biomechanics [3,4]. The former belongs to engineering while the latter is part of science. Figure 20.2 describes the study of structural dynamics. The structure may range from a simple beam to a complex building, or from a car to an aeroplane. The relationships between input, output and the model of a structure can normally be described by governing equations and the solution of the equations is the output. In the diagram, the input, output and the structure can be quantified or at least quantified statistically. In addition, the input and output normally have a one-to-one relationship, that is, one input produces one output.

If a similar diagram is required to describe the basic studies in biomechanics of the human body, it may be represented as shown in Figure 20.3.

The objective of the study of human response to vibration is to establish relationships between cause and effect [5]. However, there are no governing equations available to describe the relationships between causes, people and effects. This might be because one cause may generate a range of effects or different causes may induce the same effect or both. In addition, the effects, relating to comfort, interference and perception of vibration may be descriptive and difficult to quantify.

As human–structure interaction is a relatively new topic and the problems in practice have only recently emerged, there is only limited information available on the topic. Yet, understanding

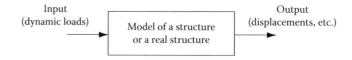

FIGURE 20.2
The basic study in structural dynamics.

FIGURE 20.3
The basic study in body biomechanics.

the human whole-body models in structural vibration and the dynamic properties of the whole body subject to low amplitude vibration are critical in the development of this topic. The concept of human–structure interaction has been partly considered in a design code and a design guide [6,7].

In future, it is probable that structures will have longer spans and be lighter, and that human expectation of the quality of life and the working environment will be greater. Therefore, engineers will need an improved understanding of human–structure interaction to tackle problems where structural serviceability or human comfort or both are concerned.

20.2.2 Identification of Human Body Models in Structural Vibration

Human whole-body models in structural vibration can be qualitatively identified through experimental methods by placing a person on a vibrating structure.

A structure is considered as a single-degree-of-freedom (SDOF) system, as shown in Figure 20.4b, which has a natural frequency of $\omega_S = \sqrt{K_S / M_S}$, where K_S and M_S are the stiffness and mass of the SDOF system. If a human body acts as an inert mass, M_H, on the SDOF structure system (Figure 20.4c), the natural frequency of the human–structure system becomes

$$\omega_{HS} = \sqrt{\frac{K_S}{M_S + M_H}} < \sqrt{\frac{K_S}{M_S}} = \omega_S \tag{20.1}$$

Therefore, the natural frequency of the human–structure system would be less than that of the structure system.

If a human whole body is considered as a mass-spring-damper system, with coefficients M_{H1}, K_{H1} and C_{H1} which are consistent with the vibration of the first mode of the body, as shown in Figure 20.5a, its natural frequency is

$$\omega_H = \sqrt{\frac{K_{H1}}{M_{H1}}} \tag{20.2}$$

Placing the SDOF human body model (Figure 20.5a) onto the SDOF structure system (Figure 20.5b) forms a two-degree-of-freedom (TDOF) system (Figure 20.5c). The equation of motion of the human–structure system is the same as that for an SDOF structure system with a tuned mass damper attached.

The natural frequencies of the combined human–structure system (Figure 20.5c) can be obtained by solving the corresponding eigenvalue problem of the following equations of motion:

$$\begin{bmatrix} M_s & 0 \\ 0 & M_{H1} \end{bmatrix} \begin{Bmatrix} \ddot{u}_S \\ \ddot{u}_H \end{Bmatrix} + \begin{bmatrix} C_S + C_{H1} & -C_{H1} \\ -C_{H1} & C_{H1} \end{bmatrix} \begin{Bmatrix} \dot{u}_S \\ \dot{u}_H \end{Bmatrix} + \begin{bmatrix} K_S + K_{H1} & -K_{H1} \\ -K_{H1} & K_{H1} \end{bmatrix} \begin{Bmatrix} u_S \\ u_H \end{Bmatrix} = \begin{Bmatrix} 0 \\ 0 \end{Bmatrix} \tag{20.3}$$

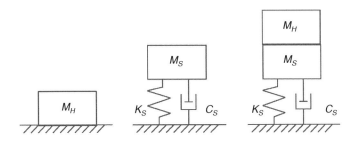

FIGURE 20.4
A human–structure model when the body acts as an inert mass. (a) Model of a body. (b) Model of a structure. (c) Model of a human–structure system.

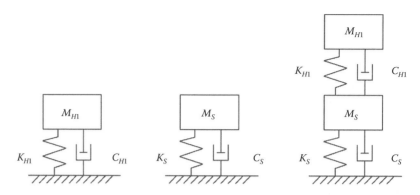

FIGURE 20.5
A human–structure model when the body acts as a mass-spring-damper. (a) Model of a body.
(b) Model of a structure. (c) Model of a human–structure system.

By neglecting the damping terms, a harmonic solution that satisfies Equation 20.3 can be determined as

$$\begin{Bmatrix} u_S \\ u_H \end{Bmatrix} = \begin{Bmatrix} A_S \\ A_H \end{Bmatrix} \sin(\omega t + \phi) \tag{20.4}$$

The symbols A_S and A_H in Equation 20.4 represent the amplitudes of the vibration of the system. Substituting Equation 20.4 into Equation 20.3 gives the following equations:

$$\begin{bmatrix} K_S + K_{H1} - \omega^2 M_S & -K_{H1} \\ -K_{H1} & K_{H1} - \omega^2 M_{H1} \end{bmatrix} \begin{Bmatrix} A_S \\ A_H \end{Bmatrix} = \begin{Bmatrix} 0 \\ 0 \end{Bmatrix} \tag{20.5}$$

For convenience, the following terms are defined

$$\alpha = \frac{M_{H1}}{M_S} \qquad \gamma = \frac{\omega_H}{\omega_S}$$

where α and γ are called the modal mass ratio and the frequency ratio of the SDOF human body system to the SDOF structure system, and have positive values.

Solving Equation 20.5 gives the natural frequencies of the TDOF system ω_1 and ω_2 represented using ω_S, ω_H and α:

$$\omega_1^2 = \frac{1}{2}\left[\omega_S^2 + \alpha\omega_H^2 + \omega_H^2 - \sqrt{(\omega_S^2 + \alpha\omega_H^2 + \omega_H^2)^2 - 4\omega_S^2\omega_H^2} \right] \tag{20.6a}$$

$$\omega_2^2 = \frac{1}{2}\left[\omega_S^2 + \alpha\omega_H^2 + \omega_H^2 + \sqrt{(\omega_S^2 + \alpha\omega_H^2 + \omega_H^2)^2 - 4\omega_S^2\omega_H^2} \right] \tag{20.6b}$$

Thus, frequency relationships between the human–structure system (ω_1, ω_2) and the independent human and structure systems, (ω_S, ω_H), can be found as follows:

Relationship 1:

$$\omega_1^2 + \omega_2^2 = \omega_S^2 + (1 + \alpha)\omega_H^2 > \omega_S^2 + \omega_H^2 \tag{20.7}$$

This relationship indicates that *the sum of the squares of the natural frequencies of the combined human–structure system is larger than that of the corresponding human and structure systems.* Equation 20.7 is obtained by adding Equation 20.6a to Equation 20.6b.

Relationship 2:

$$\omega_1\omega_2 = \omega_S\omega_H \tag{20.8}$$

This relationship indicates that *the product of natural frequencies of the human–structure system equals that of the corresponding human and structure systems.* Equation 20.8 can be derived by multiplying Equation 20.6a and Equation 20.6b.

Relationship 3:

$$\omega_1 < (\omega_{S,}\, \omega_H) < \omega_2 \tag{20.9}$$

This relationship indicates that *the natural frequencies of the human and structure systems are always between those of the human–structure system.* Equation 20.9 can be determined as follows:

$$\frac{\omega_2}{\omega_S} = \sqrt{\frac{1.0 + \alpha\gamma^2 + \gamma^2 + \sqrt{(1.0 + \alpha\gamma^2 + \gamma^2)^2 - 4\gamma^2}}{2.0}}$$

$$> \sqrt{\frac{1.0 + \alpha\gamma^2 + \gamma^2 + \sqrt{(1.0 + \gamma^2)^2 - 4\gamma^2}}{2.0}}$$

$$= \sqrt{\frac{1.0 + \alpha\gamma^2 + \gamma^2 + \sqrt{(1.0 - \gamma^2)^2}}{2.0}} = \sqrt{1 + \alpha\gamma^2/2} > 1.0 \tag{20.10}$$

Substituting Equation 20.10 into Relationship 2 (Equation 20.8) leads to Relationship 3.

The three relationships are valid without any limitation on the values of the natural frequencies of the human and structure systems and valid for any system that can be represented as a TDOF system in which damping is not considered.

Equations 20.1 and 20.9 provide qualitative relationships to identify human body models in structural vibration obtained through experiments.

Further information on human–structure interaction and on human body models in structural vibration can be found in [8–10].

20.3 DEMONSTRATION TESTS

20.3.1 Body Model of a Standing Person in the Vertical Direction

The tests demonstrate that *a standing person acts as a mass-spring-damper rather than an inert mass in vertical structural vibration, while a walking or jumping person acts solely as loading on structures.*

Figure 20.6a shows a simply supported, reinforced concrete beam with an accelerometer placed at the centre of the beam. Striking the middle of the beam vertically with a rubber hammer caused vertical vibrations of the beam. The acceleration–time history recorded from the beam and the

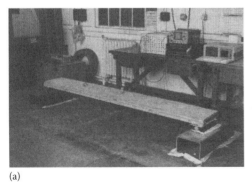

(a)   (b)

FIGURE 20.6
Test set-up for identifying human body models in vertical structural vibration. (a) An empty beam. (b) A person standing on the beam.

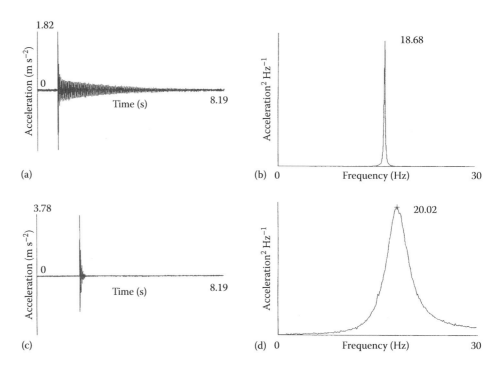

FIGURE 20.7
Measurements of the identification tests in vertical directions. (a) Free vibration of the beam alone. (b) Response spectrum of the beam alone. (c) Free vibration of a human–beam system. (d) Response spectrum of the human–beam system. (From Ellis, B. R. and Ji, T. *Structures and Buildings, the Proceedings of Civil Engineers*, 122, 1–9, 1997.)

frequency spectrum abstracted from the record are shown in Figure 20.7a and b. It can be observed that the simply supported beam has a vertical natural frequency of 18.7 Hz and the system has a very small damping ratio shown by the decay of the free vibrations which lasts more than 8 s.

A person then stood on the centre of the beam, as shown in Figure 20.6b, and a human–structure system was created. A rubber hammer was again used to induce vibrations. Figure 20.7c and d show the acceleration–time history and frequency spectrum of the beam with the person. Comparing the two sets of measurements in Figure 20.7 shows that

- The measured vertical natural frequency of the beam with the person was 20.0 Hz, which is larger than that of the beam alone (Figure 20.7b and d). This observation coincides with Relationship 3, that is, $\omega_S < \omega_2$ (Equation 20.9).
- The beam with the person possesses a much larger damping ratio than the beam alone, as the free vibration of the beam with the person decays very quickly (Figure 20.7c). This is also evident from Figure 20.7d, as the peak in the spectrum of the beam with the person has a much wider bandwidth than that of the beam alone (Figure 20.7b).

Further tests were conducted to identify the qualitative effects of human bodies in structural vibration, including tests on the beam with a dead weight and the beam occupied by a person who moved, both jumping and walking. The measured frequencies for these cases are listed in Table 20.1, which shows

- The dead weights, which were placed centrally on the beam for two tests, reduced the natural frequency as expected. This can be accurately predicted using Equation 20.1.
- The test with the person standing on the beam showed an increase in the measured natural frequency. This observation cannot be explained using the inert mass model or Equation 20.1. Thus, it is clear that the standing human body does not act as an inert mass in structural vibration.
- The measured frequency for the vibrations when the person sat on a stool on the beam was also higher than that of the beam alone.
- Significant damping contributions from the human whole body were observed for both standing and sitting positions, as can be appreciated from Figure 20.7 for the standing person.

TABLE 20.1 Natural Frequencies Observed on the Beam

Description of Experiments	Measured Natural Frequency (Hz)
Bare beam (Figure 20.6a)	18.7
Beam plus a mass of 45.4 kg (100 lb)	15.8
Beam plus a mass of 90.8 kg (200 lb)	13.9
Beam with T. Ji standing (Figure 20.6b)	20.0
Beam with T. Ji sitting on a high stool	19.0
Beam with T. Ji jumping on spot	18.7
Beam with T. Ji walking on spot	18.7

Source: Ellis, B. R. and Ji, T. *Structures and Buildings, the Proceedings of Civil Engineers*, 122, 1–9, 1997.

- Jumping and walking also provided interesting results in that they did not affect either the natural frequency or damping. The unchanged system characteristics would appear to be because the moving human body is not vibrating with the beam.

Two concepts can be identified from these tests:

- A stationary person, for example sitting or standing, acts as a mass-spring-damper rather than as an inert mass in structural vibration.
- A walking or jumping person acts solely as a load on a structure.

When a structure is very flexible and its vertical motion becomes significant, it is very difficult to jump at the natural frequency of the structure [8]. This forms another type of human–structure interaction.

20.3.2 Body Model of a Standing Person in the Lateral Direction

The tests demonstrate that *a standing person acts as a mass-spring-damper rather than an inert mass in lateral structural vibration* [9].

Figure 20.8a shows an SDOF rig for both vertical and horizontal directions used for identification tests of human body models in structural vibration. The test rig consists of two circular top plates bolted together, three identical springs supporting the plates and a thick base plate. The test procedure is simple and is the same as that conducted in Section 20.3.1. The free vibration test of the test rig alone was first conducted using a rubber hammer to generate an impact on the rig in the lateral direction. Then, a person stood on the test rig and an impact was applied in the lateral direction parallel to the shoulder of the test person, as shown in Figure 20.8b. Figure 20.9 shows the

(a)
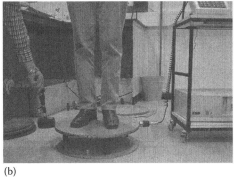
(b)

FIGURE 20.8
Test set-up for identifying human body models in lateral structural vibration. (a) An SDOF test rig. (b) A person standing on the test rig.

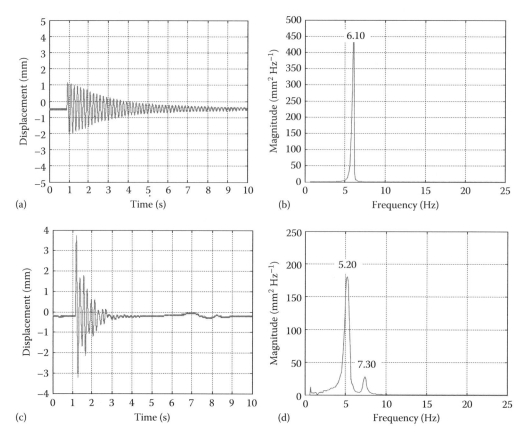

FIGURE 20.9
Measurements of the identification tests in lateral directions. (a) Free vibration of the test rig alone. (b) Response spectrum of the test rig alone. (c) Free vibration of a human–rig system. (d) Response spectrum of the human–rig system.

displacement time histories and the corresponding spectra of the test rig alone and the human-occupied test rig in the lateral directions. It can be noted from Figure 20.9 that

- The standing body contributes significant damping to the test rig in the lateral direction (Figure 20.9a and c).
- There is one single resonance frequency recorded on the test rig alone (Figure 20.9b) but two resonance frequencies are observed from the human–structure system in the lateral direction (Figure 20.9d).
- The single resonance frequency of the test rig alone is between the two resonance frequencies of the human-occupied test rig (Relationship 3 in Section 20.2.2).
- Human whole-body damping in the lateral directions is large, but less than that in the vertical direction, as shown by the vibration time history of the human–structure system in the lateral directions, which experiences more cycles of free vibration than that in the vertical directions (Figure 20.7c), although the test structures are different.

The experimental results of the identification tests conducted in the lateral directions clearly indicate that *a standing human body acts in a similar manner to a mass-spring-damper rather than an inert mass in lateral structural vibration.*

Further identification tests have been conducted on the same test rig and on a simply supported concrete beam with a bouncing person who maintains contact with the structure. It is observed that [10]

- A bouncing person acts as both loading and a mass-spring-damper on structures in vertical structural vibration.
- The interaction between a bouncing person and the test rig is less significant than that between a standing person and the test rig.

20.4 PRACTICAL EXAMPLES

20.4.1 Effect of Stationary Spectators on a Grandstand

Measurements were taken to determine the dynamic behaviour of the North Stand (Figure 20.10) at the Rugby Football Union ground at Twickenham. The grandstand has three tiers and two of them are cantilevered. Dynamic tests were performed on the roof and cantilevered tiers of the stand, with further measurements of the response of the middle cantilevered tier to dynamic loads induced by spectators during a rugby match [2].

Spectra for the empty and full grandstand are given in Figure 20.11. Figure 20.11a shows a clearly defined fundamental mode of vibration for the empty structure. Instead of the expected reduction in the natural frequency of the stand as the crowd assembled, the presence of the spectators appeared to result in the single natural frequency changing into two natural frequencies (Figure 20.11b). Figure 20.11 shows that the dynamic characteristics of the grandstand changed significantly when a crowd was involved and that the structure and the crowd interacted. This pattern was also noted in two other locations of the stand where measurements were taken.

Comparing the spectra for the empty stand and the fully occupied stand, three significant phenomena are apparent:

- An additional natural frequency was observed in the occupied stand.
- The natural frequency of the empty stand was between the two natural frequencies of the occupied stand.
- The damping increased significantly when people were present.

Considering human bodies simply as masses cannot explain these observations. The observations suggest that the crowd acted as a mass-spring-damper rather than just as a mass. When the

FIGURE 20.10
The North Stand, Twickenham.

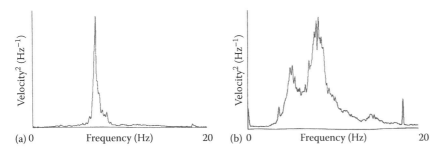

FIGURE 20.11
Response spectra of the north stand, Twickenham. (a) Without spectators. (b) With spectators.

crowd is modelled as an SDOF system, the structure and the crowd form a TDOF system. Based on this model, the observations can be explained. These observations complement the laboratory tests described in Section 20.3.

20.4.2 Calculation of Natural Frequencies of a Grandstand

It is common for a grandstand, either permanent or temporary, to be full of spectators during a sports event or a pop concert, as shown in Figure 20.12. In this situation, the human mass can be of the same order as the mass of the structure. When calculating the natural frequencies of a grandstand for design purposes, it is necessary to consider how the human mass should be represented. It is recommended in BS6399: Part 1: Loading for Buildings [6] and in a document produced by the Institution of Structural Engineers [7] that empty structures should be used for calculating their natural frequencies, that is, the human mass should not be included. This is because the worst situation for design consideration is when people move rather than when people are stationary. Therefore, engineers will not underestimate the natural frequencies of grandstands through adding the mass of a crowd to the structural mass.

20.4.3 Dynamic Response of a Structure Used at Pop Concerts

During a pop concert, it is rare that everyone moves in the same way following the music. Some people will move enthusiastically, jumping or bouncing, some will sway and some will remain stationary, either sitting or standing. Those who are stationary will provide significant damping to the structure, as observed from the previous demonstrations, and thus will alter the dynamic characteristics of the structure and effectively damp the level of vibration induced by the movements of others. This, in part, explains why the predicted structural vibrations induced by human movements, where stationary people are not considered, are often much larger than vibrations measured on-site.

The effect of stationary people on structural vibration has been considered in a design document [7] for predicting the response of grandstands used for pop concerts.

20.4.4 Indirect Measurement of Fundamental Natural Frequency of a Standing Person

If a stationary person, or a crowd, should be modelled as an SDOF system in structural vibration, what are the natural frequency, damping ratio and the modal mass of the human SDOF system?

The natural frequency of a human body cannot be obtained directly using traditional methods and tools of structural dynamics, such as sensors (accelerometers) which cannot be conveniently

FIGURE 20.12
A grandstand full of spectators.

mounted on a human body. However, a method has been developed to estimate the natural frequency of a standing person by experiment without touching the person whilst still using the methods of structural dynamics.

A simple formula can be derived based on the three frequency relationships derived in Section 20.2.2 using the measurements of the natural frequency of the empty structure and the resonance frequency/frequencies of the human–structure system, as demonstrated in Section 20.3. For example, the demonstration given in Section 20.3.2 shows the natural frequency of the test rig alone of 6.10 Hz and the two resonance frequencies of the human–rig system of 5.20 and 7.30 Hz, respectively, in Figure 20.9. Using Equation 20.8 gives an estimated natural frequency of the standing person of 6.22 Hz in the lateral direction. It should be noted that this is an approximation because

- The measurements in Figure 20.9d are the resonance frequencies which include the effect of human body damping, while Equation 20.8 does not consider any effect of damping.
- The simple mass-spring-damper model is good enough for identifying the models of a human body in structural vibration qualitatively, though this body model is developed on fixed ground rather than on a vibrating structure.

A better human body model in vertical vibration has been developed [11] based on human–structure interaction. The fundamental natural frequency and damping ratio of a standing human body can be determined through matching between the measured and calculated frequency responses using the updated human body model.

20.4.5 Indirect Measurement of Fundamental Natural Frequency of a Chicken

Six hundred million chickens are consumed each year in the United Kingdom. It has been observed sometimes during transportation that healthy chickens become ill and are unable to stand. Possible causes for this have been found to arise from the effect of resonance associated with their transportation.

In order to prevent the resonance, in which the natural frequency of the truck matches the body natural frequency of chickens, the natural frequency of a typical chicken needs to be identified. When studying body biomechanics to determine the characteristics of a human body, a subject is asked to sit or to stand on a shaking table for a short period of vibration. However, this technique cannot be applied to studying the natural frequency of a chicken, as the chicken will fly off when the shaking table moves.

The method developed for indirectly measuring human body natural frequencies could be used to obtain the natural frequency of a chicken. Figure 20.13 shows a chicken perched on a wooden beam. A slight impact on the beam can generate vibration of the beam and the chicken, which will

FIGURE 20.13
Indirect measurement of the natural frequency of a chicken. (Courtesy of Dr. J. Randall.)

not cause anxiety in the chicken. When the frequencies of the bare beam and the beam with the chicken are measured, the natural frequency of the chicken can be estimated from the two measurements and a simple equation.

PROBLEMS

1. You may feel vibration when sitting in a chair on a large floor, such as in an airport terminal, or standing on a footbridge, while people pass by. However, you would not feel the vibration when you are walking rather than sitting in the same area. What is the reason for this?

2. Use an iPhone or iPad to download a free application such as xSensor. Then conduct the following experiment:
   a. Select a relatively flexible structure, such as a steel staircase or a light footbridge.
   b. Place your iPhone or iPad at the centre of the structure and switch on the application for measuring accelerations.
   c. Jump repeatedly several times at a distance of about 300 mm from your device and read or record the acceleration.
   d. Ask one of your friends to stand over or nearby the device and repeat the procedure set out in (c).
   e. You will note that the acceleration produced in (d) is less than that in (c).
   f. Explain the reason for the difference.

REFERENCES

1. Meriam, J. L. and Kraige, L. G. *Engineering Mechanics: Dynamics*, 4th edn, Vol. 2, New York: John Wiley, 1998.
2. Ellis, B. R. and Ji, T. Human–structure interaction in vertical vibrations, *Structures and Buildings, the Proceedings of the Institution of Civil Engineers*, 122, 1–9, 1997.
3. Ji, T. On the combination of structural dynamics and biodynamics methods in the study of human–structure interaction. *The 35th United Kingdom Group Meeting on Human Response to Vibration*, Southampton, UK, 13–15 September, 2000.
4. Ji, T. Understanding the interactions between people and structures, *The Structural Engineer*, 81, 12–13, 2003.
5. Griffin, M. J. *Handbook of Human Vibration*, London: Academic Press, 1990.
6. BSI. *BS 6399: Part 1: Loading for Buildings*, London: BSI, 1996.
7. Institution of Structural Engineers. *Dynamic Performance Requirements for Permanent Grandstands Subject to Crowd Action—Recommendations for Management, Design and Assessment*, London: Institution of Structural Engineers, 2008.
8. Harrison, R. E., Yao, S., Wright, J. R., Pavic, A. and Reynolds, P. Human jumping and bobbing forces on flexible structures-effect of structural properties, *Journal of Engineering Mechanics*, 134(8): 663–675, 2008.
9. Duarte, E. and Ji, T. Measurement of human–structure interaction in vertical and lateral directions: A standing body, *ISMA International Conference on Noise and Vibration Engineering*, Leuven, Belgium, 2006.
10. Duarte, E. and Ji, T. Action of individual bouncing on structures. *Journal of Structural Engineering*, 135, 818–827, 2009.
11. Ji, T., Zhou, D. and Zhang, Q. Models of a standing human body in vertical vibration. *Structures and Buildings, the Proceedings of the Institution of Civil Engineers*, 166, 367–378, 2013.

PART III

SYNTHESIS

CHAPTER 21

CONTENTS

Static and Modal Stiffnesses

21

21.1 GENERAL COMMENTS ON STIFFNESS

Stiffness is an important physical quantity and is a characteristic of a structure that is used widely. The stiffness of a structure is generally understood to be its ability to resist deformation. Structural stiffness describes the capacity of a structure to resist deformations induced by applied loads. If a discrete model of a structure is considered, the structural stiffness can be completely described by its stiffness matrix. However, it is difficult to sense how stiff a structure is from a stiffness matrix, so in engineering practice a single value of stiffness is more useful.

Stiffness is an important concept used in engineering to relate force to structural response. However, there are different definitions of stiffness for different applications. One is obtained from a static analysis where the displacement at the critical point of a structure is calculated when a unit load is applied at that point in the required direction. Then, the inverse of the displacement is defined as the static stiffness of the structure and this can be easily calculated (Section 9.2). However, the calculated value may be significantly different from the real value. This may be due to differences between the actual structure and the mathematical model of the structure used for analysis [1].

Another form of stiffness is evaluated using structural dynamics methods and is termed *modal stiffness*. This can be determined by either dynamic analysis or dynamic test and relates to one vibration mode, its mode shape, modal mass and natural frequency. One advantage of using dynamic methods on real structures is that the natural frequencies can be measured directly and quickly.

In structural analysis and design, static, dynamic and eigenvalue analyses are normally conducted independently. Therefore, the relationship between the static and modal stiffnesses of a structure has not been explored. Numerical models will provide detailed information including static and modal stiffnesses, but experimentally, static and dynamic measurements are quite different, and often one type of measurement is far easier to take than the other. For example, measuring the static lateral stiffness of a tall building is nearly impossible but dynamic measurements are routine. A different example is where the deflection of a bridge can be measured remotely using a laser as a vehicle of known load crosses, but dynamic stiffness measurements will require closure of the bridge. Therefore, knowing the relationship between the two stiffnesses may lead to further applications in practice.

One simple but useful application in engineering is that the fundamental natural frequency of a beam can be determined using the maximum static displacement of a simply supported uniform beam due to its self-weight, that is, $f = 18\sqrt{1/\Delta}$ (Section 16.2.4), where f is the fundamental natural frequency of the structure and Δ is the maximum static displacement induced by its self-weight and is measured in millimetres. The two physical quantities are defined by two basic equations independently, but they can be linked using the same rigidity of the cross section (EI) presented in the two formulae. Thus, for relatively simple structures, the fundamental natural frequency can be directly determined using the maximum static displacement without conducting an eigenvalue analysis, and vice versa.

As the static stiffness and modal stiffness of the fundamental mode of a structure are defined independently and differently, the values calculated from the two definitions are different. However, as the two values are calculated on the basis of the same stiffness matrix, there should be a relationship between them [2]. This chapter derives the relationship between the two stiffnesses and explores the application of such a relationship in practice. The next section provides the definitions of the static stiffness and modal stiffness of a structure. Section 21.3 derives the relationship between the two stiffnesses and demonstrates that the modal stiffness is always larger than the static stiffness. Verification of the relationship is given in Section 21.4 using analytical, numerical and experimental examples. Section 21.5 shows three applications of the relationship in real structures.

21.2 DEFINITIONS OF THE STATIC STIFFNESS AND MODAL STIFFNESS

21.2.1 Static Stiffness

The static stiffness of a structure is defined as the inverse of the displacement at the critical point where a unit force is applied. This is expressed in Equations 9.4 and 9.5.

Consider a structure that is modelled by m elements and n nodes with each node having d degrees of freedom. The static equilibrium equation, containing $n \times d$ unknowns, is expressed as

$$[K]\{U\} = \{P\} \tag{21.1}$$

where:

$\{U\}$ and $\{P\}$ are the displacement and load vectors, respectively

$[K]$ is the stiffness matrix that includes the effect of the boundary conditions for linear systems

Based on the definition of the static stiffness, a single unit load is applied at the critical node c (Chapter 9) and in a particular direction l:

$$\{P\} = \{0, 0, \dots 1, \dots, 0, 0\}^T \tag{21.2}$$

Substituting Equation 21.2 into Equation 21.1 and solving leads to the determination of the displacements and thus the maximum displacement u_{cl}. Therefore, according to Equation 9.5 the static stiffness of the structure in the loading direction is

$$K_S = \frac{1}{u_{cl}} \tag{21.3}$$

The static stiffness has a further physical meaning. Equation 21.1 can be expressed as

$$
\begin{Bmatrix} u_1 \\ \vdots \\ u_{cl} \\ \vdots \\ u_n \end{Bmatrix} = [K]^{-1}\{P\} = [\delta]\{P\} =
\begin{Bmatrix} \delta_{1,1} & \cdots & \delta_{1,cl} & \cdots & \delta_{1,n} \\ \vdots & \ddots & \vdots & \ddots & \vdots \\ \delta_{cl,1} & \cdots & \underline{\delta_{cl,cl}} & \cdots & \delta_{cl,n} \\ \vdots & \ddots & \vdots & \ddots & \vdots \\ \delta_{n,1} & \cdots & \delta_{n,cl} & \cdots & \delta_{n,n} \end{Bmatrix}
\begin{Bmatrix} 0 \\ \vdots \\ 1 \\ \vdots \\ 0 \end{Bmatrix} =
\begin{Bmatrix} \delta_{1,cl} \\ \vdots \\ \delta_{cl,cl} \\ \vdots \\ \delta_{n,cl} \end{Bmatrix} \tag{21.4}
$$

where:

$[\delta]$ is the flexibility matrix of the structure and is the inverse of the stiffness matrix

$\delta_{cl,cl}$ is the diagonal element at row cl and column cl in the flexibility matrix

Equation 21.4 shows that

$$u_{cl} = \delta_{cl,cl} \tag{21.5}$$

Equations 21.3 and 21.5 lead to

$$K_S = \frac{1}{\delta_{cl,cl}} \tag{21.6}$$

Equation 21.6 indicates that the static stiffness of a structure in the clth degree-of-freedom is the inverse of the largest diagonal element in the l direction in the flexibility matrix [3]. This definition (Equation 21.6) of the static stiffness of a structure is independent of the location of the load. When evaluating the static stiffness of a structure, it may be convenient to use Equation 21.3 rather than Equation 21.6.

The location of the critical point is easy to identify for many structures, for example at the end of a cantilever, the centre of a floor or the top of a frame.

21.2.2 Modal Stiffness

There are several ways to determine the modal stiffnesses of a structure which are related to the vibration modes of the structure. Three definitions, or methods of calculation, of modal stiffness are described next.

21.2.2.1 Numerical method

The eigenvalue analysis of the free vibrations of a structure can be expressed as

$$([K] - \omega^2 [M])\{\phi\} = \{0\} \tag{21.7}$$

where:
    $\{\phi\}$ is the eigenvector or mode shape
    $[M]$ is the mass matrix and ω^2 is the eigenvalue

Solving Equation 21.7 gives the eigenvalues and eigenvectors of the structure which describe the circular natural frequencies and mode shapes of the structure. The modal mass and modal stiffness of the structure are defined as

$$\{\phi\}_i^T [M] \{\phi\}_j = \begin{cases} M_{m,i} & i = j \\ 0 & i \neq j \end{cases} \tag{21.8}$$

$$\{\phi\}_i^T [K] \{\phi\}_j = \begin{cases} K_{m,i} & i = j \\ 0 & i \neq j \end{cases} \tag{21.9}$$

where $M_{m,i}$ and $K_{m,i}$ are the modal mass and modal stiffness of the ith mode, respectively. The ith vector ϕ_i is normalised so that its largest component is equal to 1.0.

21.2.2.2 Relationship between modal stiffness and modal mass

The modal stiffness and modal mass of the ith mode have the following relationship:

$$K_{m,i} = \omega_i^2 M_{m,i} \tag{21.10}$$

where ω_i is the circular natural frequency of the ith mode, and can be either calculated or measured.

The modal mass $M_{m,i}$ can be calculated using Equation 21.8, where the vibration mode can be either calculated or estimated. For example, the deformed shape of a simple structure induced by its self-weight is a good approximation to one of the mode shapes.

21.2.2.3 Energy method

The $n \times d = r$ independent vectors in $[\Phi]$ constitute the base of an r-dimensional space. Any vector in the space can be expressed as a linear combination of the components of $[\Phi]$. Thus, the static displacement vector $\{U\}$ can be expressed as a linear combination of the base vectors:

$$\{U\} = \sum_{i=1}^{n \times d} z_i \{\phi_i\} = [\Phi]\{Z\} \tag{21.11}$$

where z_i is a generalised coordinate, indicating the contribution to the response from the ith mode.

The total strain energy of a structural system written in a discrete form is [4]

$$
\begin{aligned}
E &= \tfrac{1}{2}\{U\}^T[K]\{U\} \\
&= \tfrac{1}{2}\{Z\}^T[\Phi]^T[K][\Phi]\{Z\} \\
&= \tfrac{1}{2}\{Z\}^T[K_m]\{Z\} = \tfrac{1}{2}\sum_{i=1}^{r} K_{m,i} z_i^2
\end{aligned}
\tag{21.12}
$$

Equation 21.12 indicates that the *total strain energy of a structure can be expressed as the sum of the energy in each mode*. The modal stiffness of the ith mode can be obtained from Equation 21.12 by differentiating the total strain energy E twice in respect to the generalised coordinate z_i, that is [4]:

$$
K_{m,i} = \frac{\partial^2 E}{\partial z_i^2}
\tag{21.13}
$$

Equation 21.13 is useful, in particular when using generalised coordinates, as the strain energy of a structure system can often be easily determined.

Equations 21.9, 21.10 and 21.13 give alternative definitions/expressions of the modal stiffness of a structure.

21.3 RELATIONSHIP BETWEEN STATIC AND MODAL STIFFNESSES OF A STRUCTURE

Substituting Equation 21.11 into the static equilibrium Equation 21.1 and premultiplying by $[\Phi]^T$ gives

$$
[\Phi]^T[K][\Phi]\{Z\} = [\Phi]^T\{P\}
$$

or

$$
\begin{bmatrix} K_{m,1} & & & \\ & K_{m,2} & & \\ & & \cdots & \\ & & & K_{m,r} \end{bmatrix}
\begin{Bmatrix} z_1 \\ z_2 \\ \cdots \\ z_r \end{Bmatrix}
=
\begin{Bmatrix} \{\phi_1\}^T\{P\} \\ \{\phi_2\}^T\{P\} \\ \cdots \\ \{\phi_r\}^T\{P\} \end{Bmatrix}
=
\begin{Bmatrix} \phi_{cl,1} \\ \phi_{cl,2} \\ \cdots \\ \phi_{cl,r} \end{Bmatrix}
\tag{21.14}
$$

that is:

$$
K_{m,i} z_i = \{\phi_i\}^T\{P\} = \phi_{cl,i} \rightarrow i = 1, 2, \ldots, r
\tag{21.15}
$$

Using Equations 21.11 and 21.15, the structural response at the critical point and in the load direction is,

$$
u_{cl} = \sum_{i=1}^{r} \phi_{cl,i} z_i = \sum_{i=1}^{r} \phi_{cl,i} \times \frac{\phi_{cl,i}}{K_{m,i}} = \sum_{i=1}^{r} \frac{\phi_{cl,i}^2}{K_{m,i}}
\tag{21.16}
$$

Substituting Equation 21.16 into Equation 21.3 gives

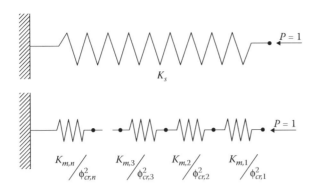

FIGURE 21.1
Graphical presentation of the relationship between static stiffness and modal stiffness.

$$\frac{1}{K_S} = \sum_{i=1}^{r} \frac{\phi_{cl,i}^2}{K_{m,i}} = \frac{\phi_{cl,1}^2}{K_{m,1}} + \frac{\phi_{cl,2}^2}{K_{m,2}} + \frac{\phi_{cl,3}^2}{K_{m,3}} + \cdots + \frac{\phi_{cl,r}^2}{K_{m,r}} \tag{21.17}$$

Equation 21.17 provides the relationship between the static stiffness and the modal stiffnesses of a structure. This relationship is applicable to any linear system as there are no other limitations applied in the derivation.

The physical meaning of Equation 21.17 is that *the displacement of a single spring with stiffness K_S induced by a unit load is equal to that of a series of r springs with a stiffness of $K_{m,i} / \phi_{cl,i}^2$ ($i = 1, 2, 3, \ldots , r$) subjected to the same load.* Figure 21.1 shows the two equivalent spring systems, which have the same displacement when the same load is applied.

As $\phi_{cl,i}^2 \geq 0$ and $K_{m,i} > 0$ ($i = 1, 2, 3, \ldots , r \times d$), all terms on the right side of Equation 21.17 are positive. Thus, a deduction from Equation 21.17 is

$$\frac{1}{K_S} > \frac{\phi_{cl,1}^2}{K_{m,1}} \quad \text{or} \quad K_S < \frac{K_{m,1}}{\phi_{cl,1}^2} \tag{21.18}$$

If $\phi_{cl,1} = 1$, that is, the point where the fundamental mode has the maximum value (the maximum mode value point) is the critical point, this leads to

$$K_{m,1} > K_S \tag{21.19}$$

Equation 21.19 indicates that *the modal stiffness of the fundamental mode is always larger than the static stiffness of a linear system if the critical point and the maximum mode value point are at the same node.* For some simple structures, the critical point and the maximum mode value point for the fundamental mode are at the same node. Similar to the location of the critical point, the location of the maximum mode value point can be judged without calculation for relatively simple structures. For example, the critical point and the maximum mode value point are at the same position for a cantilever beam and a simply supported plate.

21.4 VERIFICATION

Analytical, experimental and numerical methods are used for different structures to verify Equations 21.17 and 21.19, that is,

- The static displacement at the critical point of a structure can be represented using a summation of modal displacements.
- The modal stiffness of the fundamental mode is always larger than the static stiffness of a structure.

21.4.1 Analytical Verification

21.4.1.1 Simply supported beam

The critical point of a simply supported beam is at the centre of the beam when a vertical unit load is applied at the point. The maximum displacement induced by the load is $L^3/48EI$. Based on the definition given in Equation 20.3, the static stiffness for the simply supported beam is

$$K_S = \frac{48\,EI}{L^3} \tag{21.20}$$

The natural frequency of the ith mode of the beam is [4]

$$\omega_i = i^2\pi^2\sqrt{\frac{EI}{\bar{m}L^4}} \quad \text{or} \quad f_i = \frac{i^2\pi}{2L^2}\sqrt{\frac{EI}{\bar{m}}} \tag{21.21}$$

The shape of the ith mode of the beam is

$$\phi_i(x) = \sin\left(\frac{i\pi x}{L}\right) \tag{21.22}$$

The modal displacements at the centre of the beam, $x = L/2$, are

$$\phi_{cl,i} = \phi_i\left(\frac{L}{2}\right) = \begin{cases} 1 & \text{when } i = 1, 3, 5, \dots \\ 0 & \text{when } i = 2, 4, 6, \dots \end{cases} \tag{21.23}$$

The modal masses of the beam are

$$M_{m,i} = \int_0^L \bar{m}(x)\phi_i^2(x)dx = \frac{\bar{m}L}{2} \quad i = 1, 2, 3, \dots \tag{21.24}$$

The modal stiffnesses of the beam can be calculated using Equation 21.10 as follows:

$$K_{m,i} = \omega_i^2 M_{m,i} = \frac{i^4\pi^4 EI}{\bar{m}L^4}\frac{\bar{m}L}{2} = \frac{i^4\pi^4 EI}{2L^3} \quad i = 1, 2, 3, \dots \tag{21.25}$$

Substituting the static stiffness (Equation 21.20), the modal stiffness of the ith mode (Equation 21.25) and the mode shape value (Equation 21.23) into Equation 21.17 leads to

$$\frac{L^3}{48EI} = \frac{2L^3(1)^2}{1^4\pi^4 EI} + \frac{2L^3(1)^2}{3^4\pi^4 EI} + \frac{2L^3(1)^2}{5^4\pi^4 EI} + \cdots$$

$$= \frac{2}{\pi^4}\frac{L^3}{EI}\left[\frac{1}{1^4} + \frac{1}{3^4} + \frac{1}{5^4} + \cdots\right] \tag{21.26}$$

Removing the common terms on the two sides of Equation 21.26 gives the following equation:

$$\pi^4 = 96\left(1 + \frac{1}{3^4} + \frac{1}{5^4} + \frac{1}{7^4} + \cdots\right) = 96\sum_{i=1,3,5,\dots}^{\infty}\frac{1}{i^4} \tag{21.27}$$

Equation 21.27 is a known application of the Fourier series [5], which demonstrates that the relationship (Equation 21.17) between the static stiffness and modal stiffness of a structure holds. It is of interest to note that the identical equation (Equation 21.27) is obtained from structural mechanics methods using the relationship between the two stiffnesses of a system (Equation 21.17).

TABLE 21.1	Relationship between Static and Modal Displacements of a Simply Supported Beam	
No	**Modes Considered**	**Ratio, % Δ_m/Δ_s**
1.	1	98.553
2.	1+3	99.771
3.	1+3+5	99.928
4.	1+3+5+7	99.969
5.	1+3+5+7+9	99.984

To examine the accuracy of the approximation to the maximum static displacement (the item on the left side of Equation 21.26) using the modal displacements (the items on the right side of Equation 21.26) when the first few items are considered, this equation can be rewritten as

$$\frac{\Delta_m}{\Delta_s} = \frac{96}{\pi^4}\left(1 + \frac{1}{3^4} + \frac{1}{5^4} + \frac{1}{7^4} + \cdots\right)$$

Table 21.1 shows the ratio of the modal displacement to the static displacement. The results show that the response calculated from the first mode of a simply supported beam takes nearly 99% of the total response and in this case the series converges quickly.

The ratio of the modal stiffness of the fundamental mode (Equation 21.25) to the static stiffness (Equation 21.20) is

$$\frac{K_{m,1}}{K_S} = \frac{\pi^4 EI}{2L^3} \times \frac{L^3}{48EI} = \frac{\pi^4}{96} \approx 1.015 \tag{21.28}$$

This result agrees with the conclusion given in Equation 21.19 that the modal stiffness of the fundamental mode is larger than, but very close to, the static stiffness for a simply supported beam.

21.4.1.2 Simply supported rectangular plate

The shape of the ijth mode of a simply supported rectangular plate is [4]

$$\varphi_{ij}(x,y) = \sin\frac{i\pi x}{a} \sin\frac{j\pi y}{b} \tag{21.29}$$

where a and b are the dimensions of the plate.

The square of the value for the ijth mode at the critical point of the plate is

$$\phi_{ij}^2\left(\frac{a}{2},\frac{b}{2}\right) = \left(\sin\frac{i\pi}{2}\sin\frac{j\pi}{2}\right)^2 = \begin{cases} 1 & \begin{array}{l} i=1,3,5,\ldots \\ j=1,3,5,\ldots \end{array} \\ \\ 0 & \begin{array}{l} i=2,4,6,\ldots \\ j=2,4,6,\ldots \end{array} \end{cases} \tag{21.30}$$

The strain energy of the ijth mode is

$$E_{ij} = \frac{\pi^4 ab D}{8} z_{ij}^2\left(\frac{i^2}{a^2} + \frac{j^2}{b^2}\right)^2 \tag{21.31}$$

The modal stiffness for the ijth mode of the plate, using Equation 21.13, is

$$K_{m,ij} = \frac{\partial^2 E_{ij}}{\partial z^2} = \frac{\pi^4 abD}{4}\left(\frac{i^2}{a^2} + \frac{j^2}{b^2}\right)^2 \tag{21.32}$$

The static displacement subject to a unit load applied at any point (x,y) of the plate is [6]

$$\frac{1}{K(x,y)} = \Delta(x,y) = \sum_{i=1}^{\infty}\sum_{j=1}^{\infty} \frac{\left(\sin\dfrac{i\pi x}{a}\sin\dfrac{j\pi y}{b}\right)^2}{\dfrac{\pi^4 abD}{4}\left(\dfrac{i^2}{a^2} + \dfrac{j^2}{b^2}\right)^2} = \sum_{i=1}^{\infty}\sum_{j=1}^{\infty} \frac{\phi_{ij}^2(x,y)}{K_{m,ij}} \tag{21.33}$$

It is observed that Equation 21.33 is a direct application of Equation 21.17. In other words, the static displacement can be expressed as the summation of the response of each mode. When $x=a/2$; $y=b/2$, the displacement is

$$\Delta\left(\frac{a}{2},\frac{b}{2}\right) = \frac{1}{K_S} = \sum_{j=1,3,5,\ldots}^{\infty}\sum_{j=1,3,5,\ldots}^{\infty} \frac{1}{K_{m,ij}} \tag{21.34}$$

Equations 21.33 and 21.34 indicate that both the critical point and the maximum mode value point of the odd-numbered modes are at the centre of the plate. The ratio of the modal stiffness of the fundamental mode to the static stiffness of the plate is

$$\frac{K_{m,1}}{K_S} = \frac{\pi^4 abD}{4}\left(\frac{1}{a^2} + \frac{1}{b^2}\right)^2 \sum_{i=1}^{\infty}\sum_{j=1}^{\infty} \frac{1}{\dfrac{\pi^4 abD}{4}\left(\dfrac{i^2}{a^2} + \dfrac{j^2}{b^2}\right)^2} = \left(1+\frac{a^2}{b^2}\right)^2 \sum_{i=1}^{\infty}\sum_{j=1}^{\infty} \frac{1}{\left(i^2 + \dfrac{a^2 j^2}{b^2}\right)^2} \tag{21.35}$$

The ratios of the two stiffnesses are given in Table 21.2, when several typical side ratios of the rectangular plates are considered.

The ratio of the modal stiffness of the fundamental mode to the static stiffness of the plate increases as the side ratio increases. Once again it shows that the modal stiffness of the fundamental mode is larger than the static stiffness, as predicted in Equation 21.19.

21.4.2 Experimental Verification

Errors are likely to occur in the theoretical modelling of structures due to the differences between the actual structure and the model of the structure. The errors can be due to incorrect or inaccurate input data, inaccurate description of boundary conditions, partitions, joints between beams and columns and connections between partitions and structural elements.

A reinforced concrete beam with dimensions of $0.45 \times 0.08 \times 3.20$ m was set up as a simply supported beam for testing, as shown in Figure 21.2. Young's modulus was taken as 30×10^9 N m^{-2}, a conventional value from books.

Simply supported conditions are normally used in calculations, but the actual supports may not be ideal. Therefore, two simple support conditions were examined in the experiment. The two supporting conditions are shown in Figure 21.3a and b, respectively. The difference between the two is that one has an upper plate placed on a roller whereas the other does not. The two cases are named

TABLE 21.2 Comparison between Modal Stiffness and Static Stiffness for Different Aspect Ratios of a Rectangular Plate

a/b	1.0	1.5	2.0	2.5	3.0
$K_{m,1}/K_S$	1.1237	1.1603	1.2442	1.3589	1.4940

FIGURE 21.2
A bare reinforced concrete beam.

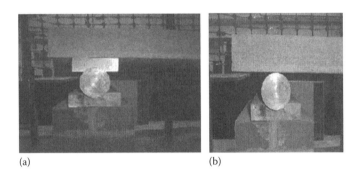

(a)   (b)

FIGURE 21.3
Two support conditions for a reinforced concrete beam. (a) Supports for the UM1 beam.
(b) Supports for the UM2 beam.

UM1 and UM2. Both static and dynamic impact tests were conducted on the two beams. It can be judged that UM2 is closer to the model of a simply supported beam.

21.4.2.1 Static tests

A static load was placed at the centre of the beam and the static displacement was measured. The static stiffness can be determined using Equation 21.20. The measured static stiffnesses of the UM1 and UM2 beams are summarised in Table 21.3.

TABLE 21.3 Comparison between Measured and Calculated Results for the UM Beams

Methods	Frequency, f		Modal Stiffness, K_m		Static Stiffness, K_s		Ratio %
	Hz	f/f_{cal} (%)	N/m	$K_m/K_{m,cal}$ (%)	N/m	$K_s/K_{s,cal}$ (%)	K_m/K_s
Calculation	14.44	100	10.60×10^5	100	10.45×10^5	100	102
UM1 Beam							
Impact test	9.625	66.7	4.708×10^5	44.4	–	–	109
Static test	–	–	–	–	4.318×10^5	41.3	–
UM2 Beam	–	–	–	–	–	–	–
Impact test	9.375	64.9	4.467×10^5	42.1	–	–	106
Static test	–	–	–	–	4.220×10^5	40.3	–

21.4.2.2 Impact tests

Impact tests were conducted on the beams and the acceleration–time histories recorded at the centre of the beams. The fundamental natural frequencies can be determined from the measurements. The modal stiffnesses of the fundamental modes can be calculated using Equation 21.10, and the measured frequencies. The modal mass of a simply supported beam is calculated using $M_{m,1} = \bar{m}(L/2)$, where $\bar{m}$ is the mass per unit length and L is the length of the beam between the two supports.

21.4.2.3 Calculations

The fundamental natural frequency of the idealised UM beam using Equation 21.21 is

$$f_{c,1} = \frac{\pi}{2L^2}\sqrt{\frac{EI}{m}} = \frac{\pi}{2 \times 2.980^2}\sqrt{\frac{(30 \times 10^9) \times (1/12 \times 0.45 \times 0.08^3)}{0.45 \times 0.08 \times 2400}} = 14.44\,\text{Hz}$$

The first mode modal stiffness of the UM beam, using Equation 21.25 is

$$K_{m,1} = \frac{\pi^4 EI}{2L^3} = \frac{\pi^4 \times (30 \times 10^9)(1/12 \times 0.45 \times 0.08^3)}{2 \times 2.980^3} = 1.060 \times 10^6\,\text{N/m}$$

The static stiffness of the UM beam worked out using Equation 21.20 is

$$K_S = \frac{48EI}{L^3} = \frac{48 \times (30 \times 10^9) \times (1/12 \times 0.45 \times 0.08^3)}{2.980^3} = 1.045 \times 10^6\,\text{N/m}$$

It is useful to compare the results from different sources and examine the differences between calculations and measurements. The measured and calculated values are summarised in Table 21.3. The ratio of the modal stiffness to the static stiffness is calculated based on each evaluation method. From the results in Table 21.3 it can be seen that

- The calculated natural frequency and stiffness are significantly larger than the measured ones. This may be because the elasticity modulus used in calculations was larger than the actual one and there were some slight cracks on the bottom of the beam.
- The modal stiffness is larger than the static stiffness from both calculation and measurement, as predicted by Equation 21.19.

Using the measured fundamental frequency to calculate the modal stiffness can provide much more reliable results than pure calculations, as the measured fundamental frequency reflects the actual behaviour of the structure and includes the factors that affect the stiffness, such as boundary conditions, elasticity modulus, dimensions and possible minor cracks.

21.4.3 Numerical Verification

A seven-storey reinforced concrete building was constructed, consisting of seven storeys of 3.75 m each in height, giving a total height of 26.25 m, three bays of 7.50 m constituting a width of 22.50 m and four bays of 7.50 m making a length of 30.00 m. The building floor plan with 12 panels is shown in Figure 21.4.

Flat thin shell elements (QSI4) were used to model the floor slabs. These elements take into account both membrane and flexural deformation. Three-dimensional thick beam elements (BMS3) were used to represent columns. An element of this type can resist axial stretching, bending and transverse shear force.

The data used in the modelling were

- Thickness of the floors: 250 mm
- Cross section of the columns: 400 × 250 mm (outer) and 400 × 400 mm (inner)

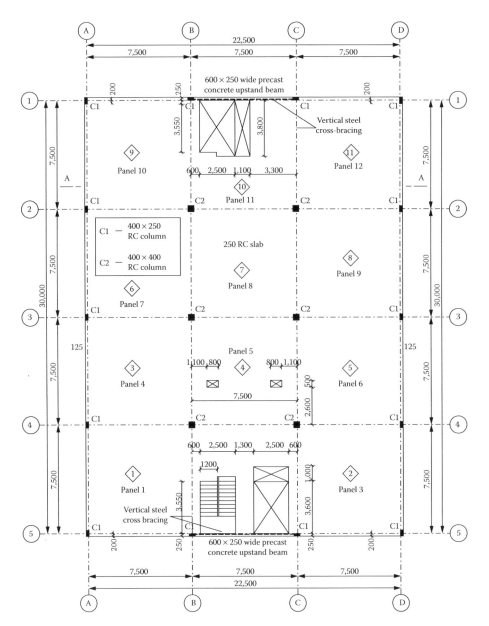

FIGURE 21.4
Typical floor plan of a seven-story concrete building. (From El-Dardiry, E., Wahyuni, E., Ji, T. and Ellis, B. R. *Journal of Computers and Structures*, 80, 2145–2156, 2002.)

- Young's modulus of the concrete: 35.5 GPa
- Poisson ratio of the concrete: 0.2
- Material density of the concrete: 2400 kg m^{-3}

The floor-column computer model consists of the flat slab floor and the full-length columns that connect to the floors between upper and lower storeys. This model gives an appropriate representation of the actual structure. The first 12 measured and calculated natural frequencies of the floor were compared and had a good agreement [7].

A static finite element (FE) analysis of the floor model was carried out and a uniform mesh was used. When a unit vertical load was applied at the centre of Panel 3, the maximum displacement was 2.476×10^{-8} m. Thus, the static stiffness, based on Equation 21.3, is $K_S = 1/2.476 \times 10^{-8} = 4.039 \times 10^{7}$ Nm^{-1}. An eigenvalue analysis was conducted and the natural frequencies of the first 300 modes obtained. Figure 21.5 shows the static deformation and the mode shapes of the first five modes [8].

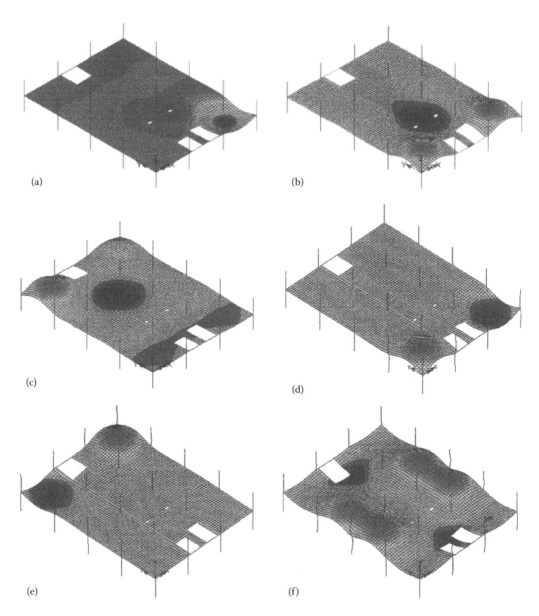

FIGURE 21.5
Static displacements and shapes of the first five modes of the floor. (a) Static deformed shape.
(b) First mode shape. (c) Second mode shape. (d) Third mode shape. (e) Fourth mode shape.
(f) Fifth mode shape.

The modal mass of the ith mode of the floor is

$$M_i = \iint m\phi_i^2(x,y)\,dxdy = \sum_{s=1}^{px}\sum_{t=1}^{py} m_{s,t}\phi_i^2(x_s,y_t) = m_u \sum_{s=1}^{px}\sum_{t=1}^{py}\phi_i^2(x_s,y_t) \qquad (21.36)$$

where:

px and py are the number of nodes in the x and y direction

m_u is the mass of a typical element in the uniform floor mesh

Thus, the modal mass of the ith mode is the product of the typical element mass and the sum of the mode nodal displacement squared. The modal stiffness can be determined using Equation 21.10.

Table 21.4 provides selected information from the 300 modes. The first column shows the order of mode, the second column the natural frequency, the third the modal mass and the fourth

TABLE 21.4   Natural Frequency, Modal Stiffness and Modal Displacement for Selected Modes

Mode i	Natural Frequency f_i (Hz)	Modal Mass $M_{m,i}$(kg) $\times 10^4$	Modal Stiffness $K_{m,i}$(Nm^{-1}) $\times 10^7$	Mode Shape Value at the Critical Point $\phi_{cr,i}$	Modal Displ. at the Critical Point (m) $\times 10^{-10}$	Total Modal Displ. (m) $u_{cr,i} = \sum\limits_{k=1}^{i} \dfrac{\phi_{cr,i}^2}{K_{m,i}}$ $\times 10^{-8}$	Ratio $u_{cr,i}/u_s \times$ 10 (%)
1	8.344	3.513	9.660	1.000	103.6	1.036	41.8
2	8.395	2.628	7.310	−0.126	2.161	1.057	42.7
3	8.617	3.003	8.800	−0.935	99.39	2.051	82.8
4	8.698	2.396	7.160	−0.017	0.0391	2.052	82.9
5	9.065	8.613	27.94	−0.444	7.063	2.122	85.7
6	9.282	8.132	27.68	−0.142	0.7259	2.129	86.0
20	17.94	3.315	42.10	0.182	0.7867	2.234	90.2
40	31.21	3.949	152.0	−0.452	1.347	2.283	92.2
90	64.75	3.062	506.7	−0.310	0.1894	2.416	97.6
160	104.9	1.728	750.6	0.224	0.0667	2.439	98.5
241	145.9	1.199	1007	−0.404	0.1622	2.465	99.5
300	176.0	1.619	1980	0.279	0.0392	2.473	99.9
Static displacement subjected to a unit vertical load at the centre of Panel 3						2.476	100

the modal stiffness. The mode shape value and modal displacement at the critical point or the loading point are given in the fifth and sixth columns. The seventh column shows a sum of the modal response at the loading point up to the ith mode and the final column gives the ratio of the modal displacement to the static displacement. The last row of Table 21.4 shows the static displacement at the critical point. From Table 21.4 it can be observed that

- The static stiffness at the loading point, 4.039×10^7 Nm^{-1}, is smaller than any modal stiffness of the 300 modes. The smallest modal stiffness is 7.160×10^7 Nm^{-1}, that is 1.773 times the static stiffness.
- The modal stiffness is related to both natural frequency and modal mass. It does not state that the modal stiffness of a higher mode is larger than that of a lower mode. It can be seen from Column 4 that the modal stiffness of the fourth mode is smaller than that of the first three modes and the modal stiffness of the second mode is smaller than that of the fundamental mode. The fourth mode shows less movement than the other modes and leads to a smaller modal mass.
- The first and third modes contribute dominant modal displacements (column 6 in Table 21.4 and mode shapes in Figure 21.5). Normally, higher modes have smaller contributions to the modal displacement than lower modes. However, this is not always true; for example, in this structure the displacement contributed from the fortieth mode is much larger or larger than that from the fourth, sixth and twentieth modes.

21.5 APPLICATIONS

In the past, measurements of resonance frequencies have been used for checking mathematical models, and in many cases this is good practice. However, in some situations measurements of static and modal stiffnesses provide extra valuable information. From the previous sections it can be seen that the static stiffness and fundamental modal stiffness are close for simple, single-span structures but diverge for more complex systems. Thus, these measurements, in isolation, can only provide an indication of the complexity of the structure. However, far more can be obtained in certain situations, as will be seen in the following examples.

Measurements have been taken on many different types of structure, albeit on some structures only some of the measurements could be made. To illustrate two of the three different applications, consider experiments using a composite floor in the eight-storey, steel-framed experimental building at Cardington. The full-size building is described in detail in [1].

TABLE 21.5 Comparison of the Measured and Predicted Results

Parameters	Measured	Continuous Model Value	Continuous Model Error (%)	Semi-Rigid Connection Model Value	Semi-Rigid Connection Model Error (%)
Frequency (Hz)	8.49	8.68	+2.24	8.39	−1.18
Modal stiffness (MNm^{-1})	20.8	54.8	+162	20.7	−0.48
Static stiffness (MNm^{-1})	9.32	11.31	+21.4	9.38	−1.00

21.5.1 Use of Stiffness Measurements of a Composite Floor

One of the experimental studies was carried out using the full-size, eight-storey, steel-framed experimental building at Cardington, UK, where the continuity of multipanel floors in the building was examined. The principal objective of the work was to establish a correct model of a multipanel composite floor. A critical observation made during the work was minor cracks noted above some supporting beams, suggesting that the continuity in the composite floor was not perfect. Initial, detailed numerical modelling of the floor provided natural frequencies which were not too dissimilar to the measured resonance frequencies (a difference of 2.24%), but the calculated stiffnesses were quite different to those measured (a difference of 162%). If the stiffness measurements had not been taken, it might have been believed that the FE model was good enough to represent the floor structure. Realising that the cracks in the floor indicated imperfect continuity, this was modelled using springs. Given the measured frequency, and the static and modal stiffnesses, the model could be tuned to correspond to the measurements [1]. The values for the continuous model and for the tuned model and measurements are given in Table 21.5.

The results in Table 21.5 show that the modal stiffness is significantly larger than the static stiffness. From the measurements, the ratio of modal to static stiffness was 2.23, whereas the ratio was 4.85 for the continuous FE model. Similar differences have been noted in other structures, including floors, cantilever grandstands supported by raker beams, multispan bridges and concrete and buttress dams, where the structural continuity in multispan/multipanel structures is not perfect.

21.5.2 Displacement of a Structure Subjected to Rhythmic Human Loads

Rhythmic human loads, such as those induced by walking, bouncing and jumping, can be described using the following equation [9]:

$$P(t) = G_v \left[1 + \sum_{n=1}^{3} r_{n,v} \sin(2\pi n f_p t + \phi_n) \right] \tag{21.37}$$

where:

G_v is the modal (contributing) weight of v persons
$r_{n,v}$ is the nth Fourier (load) coefficient induced by v persons
f_p is the beat (walking, bouncing or jumping) frequency
ϕ_n is the phase lag of the nth harmonic load component

The human load (Equation 21.37) can be interpreted as a summation of the static weight and the dynamic force components. When the response can be approximated by one single mode, the displacement of the structure subjected to the above load should be expressed as

$$\Delta(t) = \Delta_S + \Delta_D(t) = \frac{G_v}{K_S} + \frac{G_v}{K_{m,1}} \sum_{n=1}^{3} r_{n,v} D_n \sin(2\pi n f_p t - \theta_n + \phi_n) \tag{21.38}$$

where:

D_n is the nth dynamic magnification factor

θ_n is the phase lag between the nth response and the nth harmonic load component

It can be seen in Equation 21.38 that both static stiffness and modal stiffness are used, relating to static and dynamic displacements, respectively. It is important to distinguish between the two stiffnesses in Equation 21.38, as it has been noted in Section 21.5.1 that the modal stiffness $K_{m,1}$ is 2.23 times the static stiffness K_S for the floor structure in Figure 13.11 and Table 21.5. The structures in which human loads should be considered in design include footbridges, floors and grandstands.

21.5.3 Measuring Static Stiffness and Loads on Structures

Equation 21.3 shows that the static stiffness is the ratio of the applied unit load to the resulting displacement at the critical point of a structure. For a linear system, the displacement is proportional to the applied load. This equation provides a basis for different practical applications, such as determining the static stiffness of a structure when the load and the resulting displacement can be measured, or determining the load capacity when the static stiffness and the resulting displacement can be measured.

21.5.3.1 Determining the static stiffness of a footbridge

Figure 21.6 shows the measured vertical displacement as a person walked across a footbridge, to the centre and then stood still. The measurements were taken using a laser system, with a mirror directing the beam vertically under the centre of the bridge. As the person's weight is known, the displacement at the centre of the bridge (1.22 mm) can be used to determine the static stiffness of the bridge using Equation 21.3. In addition, the fundamental frequency and damping of the bridge can also be determined from the decay of vibration.

21.5.3.2 Checking the proof load of a bridge

Figure 21.7 shows the measured central displacement of the main span of a bridge as a large container lorry crossed. The centre of the bridge rose when the lorry was on the side span, but sagged when it was on the main span. Knowledge of the length of the main span would allow the velocity of the lorry to be determined. In this example, the load duration was far longer than the fundamental period of the structure, so there was no dynamic amplification to consider.

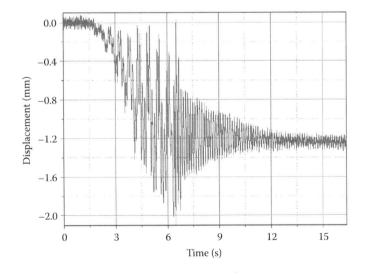

FIGURE 21.6
Measured central displacement when a person walked to the centre of a footbridge.

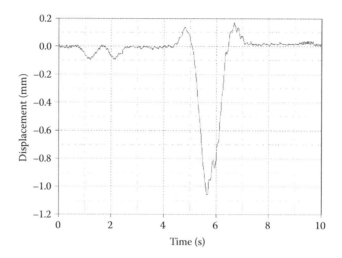

FIGURE 21.7
Measured central displacement of a concrete bridge as a lorry passed.

Displacement measurement is somewhat meaningless unless the weight of the vehicle is known; however, there are occasions when very heavy loads are moved and here various axle loads arc known. So, if the bridge was monitored when such a load crossed, then the measurements would be equivalent to a load test, but without the usual requirement to control the traffic. This could be useful for checking design calculations, or repeated measurements every few years, maybe as part of a regular assessment, to check for changes to the bridge.

These two examples consider bridges, but similar concepts have been applied to floors [10], barriers [11], stairs [12], grandstands [13] and even whole buildings [14].

21.6 DISCUSSION

The definitions of the modal mass and modal stiffness are given in Equations 21.8 and 21.9 as follows:

$$\{\phi\}_i^T [M]\{\phi\}_j = \begin{cases} M_{m,i} & i = j \\ 0 & i \neq j \end{cases} \tag{21.8}$$

$$\{\phi\}_i^T [K]\{\phi\}_j = \begin{cases} K_{m,i} & i = j \\ 0 & i \neq j \end{cases} \tag{21.9}$$

As the ith mode $\{\phi\}_i$ is a shape that does not have unique values, the modal mass and modal stiffness can have any value. However, this does not affect the value of the corresponding natural frequency. In practice, two ways are used to calculate the modal mass and modal stiffness: (1) the maximum value in the mode is normalised to one; (2) the mode shape is normalised so that the modal mass has a unit value. The first method is adopted in this chapter.

From a comparison with the static stiffness, as illustrated in Figure 21.1, the modal stiffness for the ith spring (mode) should be $K_{m,i} / \phi_{cr,i}^2$ rather than $K_{m,i}$, as there may be a significant difference in values between the two. $K_{m,i} / \phi_{cr,i}^2$ *means the modal stiffness when the ith mode shape is normalised so that the mode value at the critical point becomes a unit, that is,* $\phi_{cr,i} = 1$. For the simply supported beam example in Section 21.4.1.1 and for the simply supported plate example in Section 21.4.1.2, $\phi_{cr,i} = 1$ for all the modes, thus $K_{m,i} / \phi_{cr,i}^2 = K_{m,i}$. Otherwise $K_{m,i}$ means the modal stiffness of the ith mode at the point where the mode shape value is the maximum in the mode and is unity. As a deduction, the modal stiffnesses for different modes, calculated by Equation 21.9 using the first method, are the modal stiffnesses at different points where the maximum mode values are equal to unity.

TABLE 21.6 Modal Mass and Modal Stiffness at Different Points

Mode *i*	Natural Frequency f_i (Hz)	Modal Mass $M_{m,i} \times$ 10^4(kg)	Modal Stiffness $K_{m,i} \times$ 10^7(Nm^{-1})	Mode Shape Value at the Critical Point $\phi_{cr,i}$	Modal Mass $M_{m,i}/\phi_{cr,i}^2 \times$ 10^4(kg)	Modal Stiffness $K_{m,i}/\phi_{cr,i}^2 \times$ 10^7(Nm^{-1})
1	8.344	3.513	9.660	1.000	3.513	9.66
2	8.395	2.628	7.310	−0.126	165.5	460.4
3	8.617	3.003	8.800	−0.935	3.435	10.07
4	8.698	2.396	7.160	−0.017	8291	24,780
5	9.065	8.613	27.94	−0.444	43.69	141.7
6	9.282	8.132	27.68	−0.142	403.3	1372
20	17.94	3.315	42.10	0.182	100.1	1271
40	31.21	3.949	152.0	−0.452	19.33	743.9
90	64.75	3.062	506.7	−0.310	31.86	5263
160	104.9	1.728	750.6	0.224	34.44	14,950
241	145.9	1.199	1007	−0.404	7.346	6170
300	176.0	1.619	1980	0.279	20.80	25,440

Using the floor example in Section 21.4.3, the modal mass and modal stiffness at different points can be compared quantitatively in Table 21.6. The values in the third and fourth columns in Table 21.6 are the modal mass and modal stiffness in which the maximum mode value is normalised to unity. These maximum mode value points are at different locations in the floor. The fifth column gives the mode shape value at the critical point for the selected modes, whilst the sixth and seven columns show the modal mass and modal stiffness normalised at the critical point for the related modes. There are obvious differences between the two sets of modal masses and modal stiffnesses.

From the discussion, it can be seen that

1. The modal stiffness, based on Equation 21.9, can have many different values depending on which point the unit value in the mode is selected.
2. When a modal stiffness is mentioned, it should be clearly stated which particular point is concerned rather than as a modal stiffness in general.

These points are important when the measured and calculated modal stiffnesses are compared. When the modal stiffness or modal mass is evaluated from a measurement point, the calculated modal mass or modal stiffness should be evaluated at the same point where the mode shape value is taken as unity. Otherwise, errors will occur, as the evaluations are conducted at different locations.

For buildings, sensors are often placed on roofs and the modal masses or modal stiffnesses for the first few modes can be evaluated through a forced vibration test. If the modal stiffness, calculated using Equation 21.9, is normalised to the maximum mode values for these modes, it is likely that the modal stiffness for the fundamental mode would be correct but not the second and third modes in the same direction. This is because all the modes should be normalised to unity at the top of the building where the measurements were taken.

21.7 SUMMARY

The key points presented in this chapter are summarised as follows:

- The relationship between the static stiffness and the modal stiffness of a structure is given in Equation 21.17. The physical meaning of the relationship in the equation is that the deflection of a single spring with static stiffness of K_S is equal to that of a series of springs with stiffness of $K_{m,i} / \phi_{cr,i}^2$ ($i = 1, 2, 3, \ldots, r$) when the same loads are applied on the two spring systems. Thus, the static displacement of a structure can be represented by the summation of the response from each mode.

- The modal stiffness of the fundamental mode is always larger than the static stiffness of a linear system (Equation 21.19) when the critical point and the maximum mode value point are at the same position.
- Equations 21.17 and 21.19 have been verified using analytical, numerical and experimental examples, and three applications of static and modal stiffnesses in structural engineering have been presented.
- Using the measured fundamental natural frequency to calculate the modal stiffness can provide much more reliable results than pure calculations, as the measured fundamental frequency reflects the actual behaviour of the structure and includes the factors that affect the stiffness, such as boundary conditions, elasticity modulus, dimensions and possible minor cracks (Section 21.4.2).
- Whilst the measured fundamental natural frequency is helpful, it may not always be sufficient and measurement of the modal stiffness will provide more useful information (Section 21.5.1).
- An intuitive understanding gained from an analysis of a number of examples is that the more complicated the structure, the larger the difference between the static stiffness and the modal stiffness of the fundamental mode. The implication of this understanding is that a comparison between the static stiffness and the modal stiffness of the fundamental mode can provide an indication of the complexity of the structure.
- The modal stiffness of the fundamental mode is not always smaller than the modal stiffness of any other mode, although, by definition, the natural frequency of the fundamental mode is smaller than that of the other modes (Section 21.4.3).

The modal stiffness, based on Equation 21.9, can have many different values depending on which point a unit value in the mode is selected. When a modal stiffness is mentioned, it should be clearly stated which particular point has been used to determine the modal stiffness rather than quoting a general modal stiffness (Section 21.5).

REFERENCES

1. Zheng, T., Ji, T. and Ellis, B. R. The significance of continuity in a multi-panel composite floor, *Engineering Structure*, 32, 184–194, 2010.
2. Wahyuni, E. and Ji, T. Relationship between static stiffness and modal stiffness of a structure, *Journal for Technology and Science*, 21, 1–5, 2010.
3. Yu, X. *Improving the Efficiency of Structures Using Mechanics Concepts*, PhD Thesis, The University of Manchester, UK, 2012.
4. Clough, R. W. and Penzien J, *Dynamics of Structures*, London: McGraw-Hill, 1993.
5. Grossman, S. I. and Derrick, W. R. *Advanced Engineering Mathematics*, Harper & Row, 1988.
6. Timoshenko, S. *Theory of Plates and Shells*, London: McGraw-Hill, 1970.
7. El-Dardiry, E., Wahyuni, E., Ji, T. and Ellis, B. R. (2002). Improving FE models of a long-span flat concrete floor using natural frequency measurements, *Journal of Computers and Structures*, 80, 2145–2156.
8. Wahyuni, E. *Static Stiffness and Modal Stiffness of a Structure*, PhD Thesis, The University of Manchester, UK, 2007.
9. Ellis, B. R. and Ji, T. *BRE Digest 426: Response of Structures Subject to Dynamic Crowd Loads*, 2nd edn, 2004.
10. Ellis, B. R. and Ji, T. *BRE IP 4/02: Loads Generated by Jumping Crowds: Experimental Assessment*, 2002.
11. Dillon, P. Human loading on barriers, *Proceedings of the Institution of Civil Engineers Structures & Buildings*, 152, 381–393, 2002.

12. Bougard, A. Human loading on staircases, *Proceedings of the Institution of Civil Engineers Structures & Buildings*, 152, 371–380, 2002.
13. Ellis, B. R. and Littler, J. D. The response of cantilever grandstands to crowd loads, Part 2 Load estimation, *Proceedings of the Institution of Civil Engineers Structure & Buildings*, 157, 297–307, 2004.
14. Littler, J. D. and Ellis, B. R. Full-scale measurements to determine the response of Hume Point to wind loading, *Journal of Wind Engineering and Industrial Aerodynamics*, 42, 1085–1190, 1992.

CHAPTER 22

CONTENTS

Static and Dynamic Problems

22

22.1 PRELIMINARY COMMENTS

When a structure is subjected to both static and dynamic loads, static and dynamic analyses are required. However, they are normally conducted independently, because the governing equations for static and dynamic problems are different.

For static problems, the governing equations, in a discrete form, are [1]
Equilibrium problems:

$$[K]\{U\} = \{P\} \tag{22.1}$$

Eigenvalue problems for buckling:

$$([K] - \lambda[K_G])\{\phi_b\} = \{0\} \tag{22.2}$$

For dynamic problems, the governing equations, in a discrete form, are [1]
Equations of motion:

$$[M]\{\ddot{U}(t)\} + [C]\{\dot{U}(t)\} + [K]\{U(t)\} = \{P(t)\} \tag{22.3}$$

Eigenvalue problems for free vibration:

$$([K] - \omega^2[M])\{\phi_v\} = \{0\} \tag{22.4}$$

where:

$\{P\}$ and $\{P(t)\}$	are the static and dynamic load vectors
$[K]$, $[C]$ and $[M]$	are the stiffness, damping and mass matrices
$\{\ddot{U}(t)\}$ and $\{\dot{U}(t)\}$	are the acceleration and velocity vectors
$\{U\}$ and $\{U(t)\}$	are the displacement vectors for static and dynamic problems
$[K_G]$	is the geometric or stress stiffness matrix [2]
λ	is the buckling load factor
ω	is the natural frequency
$\{\phi_v\}$ and $\{\phi_b\}$	are the vibration mode and buckling mode, respectively

From Equations 22.1 through 22.4 it can be observed that

- All four equations include the stiffness matrix $[K]$. Hence, any errors in defining $[K]$ will result in errors in the answers to all of the four problems.
- When the load is dynamic, a function of time, the effects of the mass and the damping of the structure need to be considered.
- Free vibration and buckling can be described mathematically using eigenvalue equations, although they are two different physical phenomena.

These observed mathematical characteristics, of the four basic equations, indicate that the static and dynamic problems are related. Two particular cases can be demonstrated qualitatively.

1. If the static deflection $\{U\}$ and the vibration mode $\{\phi_v\}$ have the same shape, that is,

$$\{U\} = \eta_1\{\phi_v\} \qquad (22.5)$$

where η_1 is a constant. Substituting Equations 22.4 and 22.5 into Equation 22.1, leads to

$$\omega^2[M]\{U\} = \{P\} \qquad (22.6)$$

Equation 22.6 provides a relationship between the static displacement and the fundamental natural frequency of a structure. It indicates that the displacement of a structure can be determined by the known external loads $\{P\}$, and the natural frequency ω; alternatively, the natural frequency of the structure can be determined by the known displacements and loads. In this calculation, the stiffness matrix $[K]$ is not needed.

2. If the buckling mode $\{\phi_b\}$, and the vibration mode $\{\phi_v\}$, have the same shape, that is,

$$\{\phi_b\} = \{\phi_v\} = \{\phi\} \qquad (22.7)$$

substituting Equations 22.4 and 22.7 into Equation 22.2, gives

$$(\omega^2[M] - \lambda[K_G])\{\phi\} = 0 \qquad (22.8)$$

Equation 22.8 provides a relationship between the natural frequency and the buckling load factor of a structure. It indicates that the buckling load factor can be determined from the known natural frequency ω, which extends the application of vibration measurements.

These relationships trigger an examination of some relationships between static and dynamic problems and an exploration of different applications for simplifying analysis, for improving the accuracy of calculations and for solving problems in an effective way. In Chapter 21, the relationships between static and modal stiffnesses were studied; the next four sections consider other relationships between static and dynamic problems.

22.2 MAXIMUM DISPLACEMENT AND FUNDAMENTAL NATURAL FREQUENCY

22.2.1 Relationship Equations

When a beam structure is subjected to distributed self-weight loading, $m(x)g$, the deformation of the structure, $v_s(x)$ can be expressed as

$$v_s(x) = \Delta \cdot \phi_d(x) \qquad (22.9)$$

where:
   Δ is the maximum displacement of the structure
   $\phi_d(x)$ is the deformation shape

The strain energy and the work done by the load are [2]

$$V = \frac{\Delta^2}{2}\int_0^L EI(x)[\phi_d''(x)]^2 dx \qquad (22.10)$$

$$W = -\Delta\int_0^L m(x)g\,\phi_d(x)dx \qquad (22.11)$$

The maximum displacement Δ can be obtained by minimising the total energy in respect of the displacement, that is,

$$\frac{\partial V}{\partial \Delta} + \frac{\partial W}{\partial \Delta} = 0 \qquad (22.12)$$

Substituting Equations 22.10 and 22.11 into Equation 22.12 gives the maximum displacement:

$$\Delta = \frac{g \int_0^L m(x)\phi_d(x)dx}{\int_0^L EI(x)[\phi_d''(x)]^2 dx} \qquad (22.13)$$

The accuracy of Equation 22.13 depends on the correctness of the selected deformation shape $\phi_d(x)$, that is, how close the assumed deformation shape is to the true deformation shape. In this equation, the denominator is likely to experience a larger error due to the double differentiation of the assumed deformation shape, as has been demonstrated in Example 16.2.

When this structure experiences free vibration, the motion $v(x,t)$ can be expressed as

$$v(x,t) = A_v \phi_v(x) \sin \omega t \qquad (22.14)$$

where:

- x is a coordinate defining position along the beam
- t is time
- A_v is the magnitude of the vibration
- $\phi_v(x)$ is the mode shape
- ω is the natural frequency of the structure

The strain energy of the system is [3]

$$V = \frac{1}{2}\int_0^L EI(x)\left(\frac{\partial^2 v}{\partial x^2}\right)dx \qquad (22.15)$$

Substituting Equation 22.14 into Equation 22.15 and letting the displacement reach its maximum value, leads to

$$V_{max} = \frac{A_v^2}{2}\int_0^L EI(x)[\phi_v''(x)]^2 dx \qquad (22.16)$$

The kinetic energy of the system, with mass distribution of $m(x)$, is

$$T = \frac{1}{2}\int_0^L m(x)\dot{v}^2(x,t)dx \qquad (22.17)$$

Differentiating Equation 22.14 with respect to time to obtain velocity, substituting it into Equation 22.17 and letting the velocity reach its maximum, gives the maximum kinetic energy as

$$T_{max} = \frac{1}{2}A_v^2\omega^2\int_0^L m(x)[\phi_v(x)]^2 dx \qquad (22.18)$$

The following equation holds for a conservative system:

$$V_{max} = T_{max} \qquad (22.19)$$

The expression for determining the natural frequency can then be obtained by substituting Equations 22.16 and 22.18 into Equation 22.19, that is,

$$\omega^2 = \frac{\int_0^L EI(x)[\phi_v''(x)]^2 dx}{\int_0^L m(x)\phi_v^2(x)dx} \tag{22.20}$$

where the numerator and denominator are normally termed the *modal stiffness* and the *modal mass*. Similar to Equation 22.13, the accuracy of Equation 22.20 depends on the correctness of the assumed mode shape $\phi_v(x)$, and the numerator of Equation 22.20 is likely to have a larger error.

If it is assumed that the static deformation shape is the same as the vibration mode shape, that is,

$$\phi_d(x) = \phi_v(x) = \phi(x) \tag{22.21}$$

A relationship between the maximum static displacement, in Equation 22.13, and the natural frequency, in Equation 22.20, can be established by removing the modal stiffness as follows:

$$f = \frac{1}{2\pi}\sqrt{\frac{g\int_0^L m(x)\phi(x)dx}{\int_0^L m(x)\phi^2(x)dx}}\sqrt{\frac{1}{\Delta}} = A_d \frac{1}{\sqrt{\Delta}} \tag{22.22}$$

$$\Delta = \frac{g\int_0^L m(x)\phi(x)dx}{4\pi^2 \int_0^L m(x)\phi^2(x)dx}\frac{1}{f^2} = A_f \frac{1}{f^2} \tag{22.23}$$

where

$$A_d = \frac{1}{2\pi}\sqrt{\frac{g\int_0^L m(x)\phi(x)dx}{\int_0^L m(x)\phi^2(x)dx}} \tag{22.24}$$

$$A_f = \frac{g\int_0^L m(x)\phi(x)dx}{4\pi^2 \int_0^L m(x)\phi^2(x)dx} \tag{22.25}$$

A_d and A_f are the factors for deformation transformation and for natural frequency transformation, respectively. Equations 22.22 and 22.23 are based on a flexural beam for simplicity, but they can easily be extended to plates and shells by altering the shape function from $\phi(x)$ to $\phi(x,y)$.

The significance of Equations 22.22 and 22.23 is that

- If the maximum static displacement of a structure is known, then the fundamental natural frequency of the structure can be estimated using Equation 22.22.
- If the fundamental natural frequency of a structure is known by calculation, or by measurement, the maximum static displacement of the structure subjected to the uniformly distributed load can be estimated from Equation 22.23.
- The difficulties and errors in calculating the modal stiffness are removed.
- The accuracy of the estimations depends on the quality of the selected shape function, $\phi(x)$, which should be similar in both the vibration mode and the deformation shape. Hence, if the vibration mode shape and the static deformation shape are dissimilar, this relationship is not valid or it involves a large error.
- The units for g and Δ in the two equations should be consistent, using either metres or millimetres. For example, when the displacement is measured in millimetres, the gravitational acceleration should be $g = 9810$ mm s^{-2}.

22.2.2 Examples

EXAMPLE 22.1

For a simply supported uniform beam with a length L and carrying self-weight only, determine the factor for displacement transformation, A_d, in Equation 22.24 using two different shape functions.

SOLUTION

Choose the function $\phi(x)$ to be the shape of the fundamental mode and the shape of static deformation as follows:

Mode shape:

$$\phi(x) = \sin\left(\frac{\pi x}{L}\right) \tag{22.26}$$

Static deformation shape:

$$\phi(x) = \frac{16x}{5L^4}(L^3 - 2Lx^2 + x^3) \tag{22.27}$$

Substituting Equation 22.26 into Equation 22.24 with the uniform mass distribution:

$$A_d = \frac{1}{2\pi}\sqrt{\frac{9810\int_0^L \sin(x)dx}{\int_0^L \sin^2(x)dx}} = \frac{1}{2\pi}\sqrt{\frac{9810\times(2L/\pi)}{L/2}} = 17.79$$

Substituting Equation 22.27 into Equation 22.24 gives the factor 17.77. These two values are very close and indicate that the two shapes (Equations 22.26 and 22.27) are similar. Table 16.2 provides a value of 17.75 that is based on the existing equations for the maximum displacement and for the fundamental natural frequency of a simply supported beam. For simple beams, the factors in Table 16.2 can be used. Equations 22.22 and 22.23 would be useful when a structure is more complicated than a beam, as long as the mode shape and static deformed shape are similar.

EXAMPLE 22.2

Consider a simply supported uniform plate that has a width of a, a length of b and a mass density of $\bar{m}$. The maximum displacement of the plate, due to its self-weight, is 25 mm. Calculate the fundamental natural frequency of the plate.

SOLUTION

Choose the fundamental mode shape of the plate as the shape function and use Equation 22.22 for the solution. The selected shape function is thus, $\phi(x,y) = \sin(\pi x/a)\sin(\pi y/b)$. The natural frequency is then

$$f = \frac{1}{2\pi}\sqrt{\frac{g\int_0^a\int_0^b \phi(x,y)dxdy}{\int_0^a\int_0^b \phi^2(x,y)dxdy}}\sqrt{\frac{1}{\Delta}} = \frac{1}{2\pi}\sqrt{\frac{9810\int_0^a\int_0^b \sin(\pi x/a)\sin(\pi y/b)dxdy}{\int_0^a\int_0^b [\sin(\pi x/a)\sin(\pi y/b)]^2 dxdy}}\frac{1}{\sqrt{25}}$$

$$= \frac{1}{2\pi}\sqrt{\frac{9810(4ab/\pi^2)}{ab/4}}\frac{1}{\sqrt{25}} = 20.07\times0.2 = 4.01\,\text{Hz}$$

It can be noted that

- The coefficient A_d for a simply supported uniform plate is 20.07, which is about 12% bigger than 18, that is normally used for simple structures [4].
- The dimensions and density of the plate are not required in this calculation because the effects of these data are imbedded in the maximum displacement, which is 25 mm.

EXAMPLE 22.3

A simply supported, flat, reinforced concrete floor has a length $a=8.0$ m, a width $b=6.0$ m and a thickness $t=0.2$ m. The modulus of elasticity and the density of concrete are $E = 30 \times 10^9$ Nm^{-2} and $\rho = 2400$ kgm^{-3}. The fundamental natural frequency of the plate is 14.2 Hz, which is calculated using Equations 21.32 and 21.10. Calculate the maximum displacement of the floor when a uniformly distributed load of 2 kN m^{-2} is applied in the middle part of the floor ($a/4 \leq x \leq 3a/4$ and $b/4 \leq x \leq 3b/4$).

SOLUTION

The fundamental mode shape of the floor is selected as the shape function: $\phi(x,y) = \sin(\pi x/a)\sin(\pi y/b)$. The maximum displacement of the floor can then be estimated, using Equation 22.23 as follows:

$$\Delta = \frac{\int_0^a \int_0^b q(x,y)\phi(x,y)dxdy}{4\pi^2 \int_0^a \int_0^b \overline{m}\phi^2(x,y)dxdy}\frac{1}{f^2} = \frac{q\int_{a/4}^{3a/4}\int_{b/4}^{3b/4}\sin\left(\frac{\pi x}{a}\right)\sin\left(\frac{\pi y}{b}\right)dxdy}{4\pi^2 \times \rho \times t\int_0^a \int_0^b \left[\sin\left(\frac{\pi x}{a}\right)\sin\left(\frac{\pi y}{b}\right)\right]^2 dxdy}\frac{1}{f^2}$$

$$= \frac{2q}{\rho t\pi^4}\frac{1}{f^2} = \frac{2\times 2000}{2400\times 0.2\pi^4}\frac{1}{14.2^2} = 0.0004243\,\text{m} = 0.4243\,\text{mm}$$

The integration in the numerator is conducted in the ranges where the load is applied. The correctness of the analytical solution can be checked by a finite element analysis using 4800 quadrilateral plate elements which gives the fundamental natural frequency and the maximum displacement of the floor as 14.16 Hz and 0.4382 mm, respectively.

22.3 BUCKLING LOAD AND FUNDAMENTAL NATURAL FREQUENCY

22.3.1 Relationship Equations

Consider the free vibration of an axially loaded column with any boundary conditions (Figure 22.1). The motion of the column $v(x,t)$ is a function of time t and the position x, and can be expressed using the variable separation method as

$$v(x,t) = z(t)\phi(x) \tag{22.28}$$

FIGURE 22.1
Vibration of a loaded column.

where:

$z(t)$ is a function of time and defines the magnitude of the motion

$\phi(x)$ is the mode shape

The strain energy of the column, due to bending, is

$$V_1 = \frac{1}{2}\int_0^L EI(x)\left(\frac{\partial^2 v(x,t)}{\partial x^2}\right)^2 dx = \frac{z^2(t)}{2}\int_0^L EI\left(\frac{d^2\phi(x)}{dx^2}\right)^2 dx \qquad (22.29)$$

The loss of the potential energy, due to the reduced distance between the two ends, or due to the bending of the column, is the product of the force and the shortening, that is,

$$V_2 = -P\times\frac{1}{2}\int_0^L\left(\frac{\partial v(x,t)}{\partial x}\right)^2 dx = -\frac{Pz^2(t)}{2}\int_0^L\left(\frac{d\phi(x)}{dx}\right)^2 dx \qquad (22.30)$$

Thus, the total strain and potential energies of the column is $V = V_1 + V_2$.
The kinetic energy of the column is

$$T = \frac{1}{2}\int_0^L m(x)[\dot v(x,t)]^2 dx = \frac{\dot z^2(t)}{2}\int_0^L m(x)\phi^2(x)dx \qquad (22.31)$$

For free vibrations, the Lagrange equation [5] is

$$\frac{d}{dt}\left(\frac{\partial T}{\partial \dot z}\right)+\frac{\partial U}{\partial z}=0 \qquad (22.32)$$

Substituting Equations 22.29 through 22.31 into Equation 22.32 gives

$$M_m\ddot z(t)+K_m z(t)-P\delta z(t)=0 \qquad (22.33)$$

where

$$M_m = \int_0^L m(x)\phi^2(x)dx \qquad (22.34)$$

$$K_m = \int_0^L EI(x)\left(\frac{d^2\phi(x)}{dx^2}\right)^2 dx \qquad (22.35)$$

$$\delta = \int_0^L \left(\frac{d\phi(x)}{dx} \right)^2 dx \tag{22.36}$$

where M_m and K_m are the modal mass and modal stiffness. Equation 22.33 can be written in a common form as follows:

$$\ddot{z}(t) + \frac{K_m - P\delta}{M_m} z(t) = \ddot{z}(t) + \omega_P^2 z(t) = 0 \tag{22.37}$$

where

$$\omega_p = \sqrt{\frac{K_m - P\delta}{M_m}} \tag{22.38}$$

where ω_p is the natural frequency of the column subjected to a compressive force P. Two special cases can be derived from Equation 22.38.

(a) When $P = 0$ then

$$\omega_p = \omega = \sqrt{\frac{K_m}{M_m}} \tag{22.39}$$

This is the natural frequency of an unloaded column.

(b) When $K_m = P_{cr}\delta$ or

$$P_{cr} = \frac{K_m}{\delta} \tag{22.40a}$$

then

$$\omega_p = 0 \tag{22.40b}$$

Equation 22.40a gives the buckling formula for a column. Equation 22.40 indicates that

- The buckling phenomenon can be interpreted as occurring when the structure loses its stiffness and its natural frequency becomes zero.
- The lower the natural frequency of a loaded column, the less the remaining buckling capacity. This conclusion may lead to a nondestructive test method using the measured natural frequencies to detect the remaining buckling capacity of a column.
- Although the derivation is based on a loaded column, the previous two statements and Equation 22.38 may be applicable to more complicated structures.

The accuracy of the prediction (Equation 22.38) depends on the correctness of the shape $\phi(x)$, which is used to describe both the vibration mode and the buckling mode. For a simply supported column with compressive loads at its two ends, the mode shapes of free vibration and buckling of the column are identical. Thus, this mode shape will give an exact solution. For other situations, when the vibration mode shape and the buckling shape are slightly different, Equation 22.38 will lead to an estimate of the natural frequency.

Taking the square of the two sides of Equation 22.38 and then dividing it by $\omega^2 = K_m/M_m$ gives, using the definition in Equation 22.40:

$$\frac{\omega_p^2}{\omega^2} = 1 - \frac{P}{P_{cr}} \quad \text{or} \quad \frac{\omega_p^2}{\omega^2} + \frac{P}{P_{cr}} = 1 \tag{22.41}$$

Equation 22.41 can be rewritten as follows:

$$P_{cr} = \frac{P}{1 - \dfrac{\omega_p^2}{\omega^2}} = \frac{\omega^2 P}{\omega^2 - \omega_p^2} \qquad (22.42)$$

Equation 22.42 indicates that *the critical load can be predicted by taking two measurements of the fundamental natural frequency of a column, one without a load and one with a load of P.* Equation 22.42 is likely to give a more accurate prediction of the buckling load using two natural frequency measurements.

When a column has been loaded and the natural frequency of the unloaded column is not available, the natural frequencies of two loaded cases can be used to predict the critical load. Then, Equation 22.42 becomes

$$P_{cr} = \frac{P_2 \omega_{p1}^2 - P_1 \omega_{p2}^2}{\omega_{p1}^2 - \omega_{p2}^2} \qquad (22.43)$$

where ω_{p1} and ω_{p2} are the fundamental natural frequencies of the column subjected to compressive loads P_1 and P_2, respectively.

EXAMPLE 22.4

A pin-ended column has a length of L and a rigidity of EI. A compressive load $P = EI/L^2$ is applied to the top end of the column. The selected mode shape is $\phi(x) = \sin(\pi x/L)$. Predict the critical load of the column using Equation 22.42.

SOLUTION

Use Equations 22.34 through 22.36 to calculate the following parameters:

$$M_m = \int_0^L m\phi^2(x)\,dx = \int_0^L m\sin^2\left(\frac{\pi x}{L}\right)dx = \frac{1}{2}mL$$

$$K_m = \int_0^L EI\left(\frac{d^2\phi(x)}{dx^2}\right)^2 dx = \int_0^L EI\left(\frac{\pi^2}{L^2}\sin\left(\frac{\pi x}{L}\right)\right)^2 dx = \frac{\pi^4 EI}{2L^3}$$

$$\delta = \int_0^L \left(\frac{d\phi(x)}{dx}\right)^2 dx = \int_0^L \left(\frac{\pi}{L}\cos\left(\frac{\pi x}{L}\right)\right)^2 dx = \frac{\pi^2}{2L}$$

Thus, the natural frequencies of the column without, and with, the load can be calculated using Equations 22.39 and 22.38, respectively,

$$\omega^2 = \frac{K_m}{M_m} = \frac{\pi^4 EI}{mL^4}$$

$$\omega_L^2 = \frac{K_m - P\delta}{M_m} = \left(\frac{\pi^4 EI}{2L^3} - \frac{EI}{L^2}\frac{\pi^2}{2L}\right)\Big/ (mL/2) = \frac{\pi^2(\pi^2 - 1)EI}{mL^4}$$

Substituting the two natural frequencies and the known load into Equation 22.42 gives

$$P_{cr} = \frac{P}{1 - \dfrac{\omega_p^2}{\omega^2}} = \frac{P}{1 - \dfrac{\pi^2 - 1}{\pi^2}} = \pi^2 P = \frac{\pi^2 EI}{L^2}$$

The answer is the same as that given in Equation 11.2. There is little benefit from calculating the critical load of a simply supported column using Equation 22.42. The usefulness of Equation 22.42 lies in practical applications taking measured natural frequencies to predict the critical load.

22.3.2 Vibration-Buckling Test of a Strut

This experiment aims to *verify Equation 22.42 and demonstrate that the buckling load of a strut can be predicted more accurately using natural frequency measurements.*

Figure 22.2 shows the rig, equipment and test specimen used in the vibration-buckling test. The straight steel strut had a width of 35.4 mm, a thickness of 3.20 mm and a length of 711 mm. It was placed in the test rig, with the two ends of the strut having pinned supports. When in position, the strut experienced a load of 30.4 N from the top metal platform. Further weights were gradually placed on the platform until the strut buckled. In parallel with the buckling test, the natural frequency of the loaded strut was measured at each loading stage using a small accelerometer placed at the centre of the strut and linked to a vibration analyser. At each loading stage, a gentle lateral impact was applied to the strut (a tap from a finger) to generate lateral vibrations of the strut which were recorded, from which the fundamental natural frequency of the loaded strut was determined. Table 22.1 shows the applied forces and measured fundamental natural frequencies for 11 loading stages.

The theoretical prediction using Equation 11.2 is

$$P_{cr} = \frac{\pi^2 EI}{L^2} = \frac{\pi^2\, 210 \times 10^3 \times 3.2^3 \times 25.4}{12 \times 711^2} = 284.4 \text{ N}$$

FIGURE 22.2
Vibration-buckling test of a loaded strut.

TABLE 22.1	Natural Frequency and Compression Force of the Test Strut	
No	Total Load (N)	Natural Frequency, f (Hz)
1	30.4	12.75
2	52.7	12.22
3	74.9	11.84
4	97.2	10.84
5	119.5	10.00
6	141.7	9.00
7	164.0	7.875
8	186.3	6.375
9	208.5	4.57
10	230.8	2.52
11	253.1	Buckled

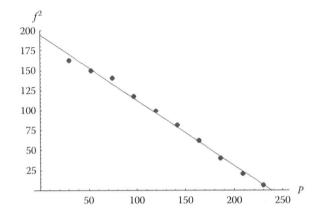

FIGURE 22.3
Relationship between natural frequency and compression force.

Comparing the predicted critical load of 284.4 N and the measured critical load of 253.1 N, it can be seen that the prediction overestimates the critical load by 12%. A more accurate prediction of the critical load of the strut can be made using all of the natural frequency measurements.

Figure 22.3 shows the relationship between the measured natural frequency squared, on the vertical axis, and the applied force, on the horizontal axis. The points show the measurements and the straight line is fitted to the points. It can be seen that the natural frequency squared has a linear relationship with the compressive force, as predicted by Equation 22.41.

The straight line in Figure 22.3 is defined by $f^2 = 194 - 0.815\,P$. The buckling load can be obtained from this equation by letting $f = 0$, that is, $P_{cr} = 194/0.815 = 238$ N, which is 16% smaller than the theoretically predicted value of 284.4 N. The predicted critical load based on the frequency measurements is more accurate than that from Equation 11.2. The measurements do not include the effects of inaccurate data input, such as measurement errors of specimen dimensions and nonideal pinned boundary conditions.

Equation 22.43 can also be used for predicting the critical load, through selecting any 2 of the first 10 sets of results in Table 22.1. The predicted critical load is the intersection point of the horizontal axis and the line defined by any pair of points in Figure 22.3. For example, if the data in the second and seventh rows in Table 22.1 are used, Equation 22.43 gives a critical load of 243 N; if the data in the first two rows are taken, it predicts a critical load of 299 N.

The large difference between the predicted buckling loads of 243 and 299 N, based on the frequency measurements, may indicate the limitation of this nondestructive test method. However, it should also be recognised that the theoretical prediction using Equation 11.2 is also likely to contain errors due to inaccurate input data.

22.4 PERIODIC DYNAMIC LOADS AND CORRESPONDING STATIC LOADS

22.4.1 Relationship Equation

Consider a periodic function, $f(t)$, that has a period of T; the integration of the function over its period is equal to the product of the corresponding constant part, a_0, of the function and the period, that is,

$$\int_0^T f(t)dt = a_0 T \tag{22.44}$$

Equation 22.44 can be proved as follows.
Any periodic function $f(t)$ can be expressed using Fourier series [3]:

$$f(t) = a_0 + \sum_{n=1}^{\infty} \left(a_n \cos \frac{2n\pi x}{T} + b_n \sin \frac{2n\pi x}{T} \right) \tag{22.45}$$

Integrating $f(t)$ from 0 to T gives

$$\int_0^T f(t)dt = \int_0^T a_0 dt + \sum_{n=1}^{\infty} \left[\int_0^T a_n \cos \frac{2n\pi x}{T} dt + \int_0^T b_n \sin \frac{2n\pi x}{T} dt \right] = a_0 T \tag{22.46}$$

since

$$\int_0^T a_n \cos \frac{2n\pi x}{T} dt = 0 \quad \text{and} \quad \int_0^T b_n \sin \frac{2n\pi x}{T} dt = 0$$

This theorem is straightforward mathematically but it links to several engineering applications. When $f(t)$ is a periodic force, such as that induced by machines, human jumping or bouncing in response to music. Equation 22.44 can be interpreted as

The impulse of a periodic forces f(t) over its period of T is equal to the product of the static component of the force, a_0, and the period, T.

Equation 22.44 provides a relationship between a periodic dynamic load and its static load component.

22.4.2 Human Jumping Loads

When people dance following a musical beat, the human-induced load is periodic. The load time history, where jumping is included, can be described by a high contact force for a certain time t_p (contact duration) followed by zero force when the feet leave the floor. The load function for one period for a single person jumping is given by [6]

$$F_s(t) = \begin{cases} K_p G_s \sin\left(\dfrac{\pi t}{t_p}\right) & 0 \le t \le t_p \\ \\ 0 & t_p \le t \le T_p \end{cases} \tag{22.47}$$

where:
  K_p is the impact factor, that is the ratio of the peak value of the dynamic force F_{max} to the body weight of the dancer G_s
  t_p is the contact duration, when feet are on the floor
  T_p is the period of the periodic dynamic load

The contact duration t_p can vary from 0 to T_p, corresponding to different movements and activities. The contact ratio is defined as follows:

$$\alpha = \frac{t_p}{T_p} \le 1.0 \tag{22.48}$$

Thus, Equation 22.47 becomes

$$F_s(t) = \begin{cases} K_p G_s \sin\left(\dfrac{\pi t}{\alpha T_p}\right) & 0 \le t \le \alpha T_p \\[2ex] 0 & \alpha T_p \le t \le T_p \end{cases} \tag{22.49}$$

The weight of the dancer, G_s, and the period of the music beat, T_p, are known and there are two parameters, K_p and α, to be determined to completely describe the load. These two parameters are not independent, because the higher the jumping, the smaller the contact ratio, and the bigger the impact factor. Substituting Equation 22.49 into Equation 22.44 yields the relationship between the contact ratio and the impact factor as follows [7]:

$$K_p = \frac{\pi}{2\alpha} \tag{22.50}$$

Equation 22.50 removes one unknown, so only one parameter, either K_p or α, needs to be determined to describe the human jumping loads. Equation 22.50 is simple and useful. In experiments, the impact factor is easier to measure, and more accurate, than the contact ratio.

It is interesting to note that it has been observed [8] that the mean value of the time history of a vertical load, corresponding to bouncing to music or rhythmic jumping, was always equal to the weight of the performer. This observation can be expressed using Equations 22.44 and 22.49 as follows:

$$\frac{1}{T_p} \int_0^{\alpha T_p} K_p G_s \sin\left(\frac{\pi t}{\alpha T_p}\right) dt = G \tag{22.51}$$

Alternatively, Equations 22.44 and 22.51 can be interpreted as *the average of the integration of a periodic load over its period is equal to the static component of the periodic load*. The use of Equation 22.50 has led to expressions for jumping loads using Fourier series in which α is used to describe different dance activities [7,9].

22.5 TENSION FORCE AND FUNDAMENTAL NATURAL FREQUENCY

22.5.1 Relationship Equation

The vibration of a straight string in tension can be described by the following equation [10]:

$$\bar{m}\frac{\partial^2 v}{\partial t^2} = F\frac{\partial^2 v}{\partial x^2} \tag{22.52}$$

where:
  v is the transverse vibration of the string, and is a function of time t and position x
  F is the tension force in the string
  $\bar{m}$ is the distributed mass of the string

Considering the vibration of the first mode:

$$v(x,t) = A \sin \frac{\pi x}{L} \sin(2\pi f) t \qquad (22.53)$$

where:

    L is the length of the string

    f is the fundamental natural frequency of the string in the transverse direction

Substituting Equation 22.53 into Equation 22.52 gives

$$F = 4\bar{m}L^2 f^2 \qquad (22.54)$$

Equation 22.54 indicates that

- The tension force in the string can be calculated using the fundamental natural frequency obtained from either measurement or analysis.
- The tension force in the string is proportional to the total mass of the string $\bar{m}L$, its span L and its natural frequency squared, f^2.

Similar formulae to Equation 22.54 are available for predicting the tension force in a string or a cable [11].

22.5.2 Tension Force and Natural Frequency of a Straight Tension Bar

This experiment *verifies Equation 22.54 and demonstrates that the force in a tension bar can be predicted using natural frequency measurements* [11].

Two steel bars, with different diameters (6 mm and 8 mm) and lengths (2.55 m and 2.73 m), were fixed to the steel frame shown in Figure 22.4a. Two strain gauges were glued to each of the bars on opposite sides at the midpoints of the bars. Tension forces were applied through turnbuckles at the ends of the bars, and the values of the forces were determined through strain measurements. A small accelerometer was fixed to one of the two bars, as shown in Figure 22.4b, to record vibrations.

Vibration tests of the tension bars were conducted, and accelerations were recorded when the bars were tensioned to different levels. Fundamental natural frequencies were determined from spectral analysis of the acceleration responses. Table 22.2 lists a range of measured fundamental natural frequencies and comparisons of the measured tension forces, in the 6 mm tension bar,

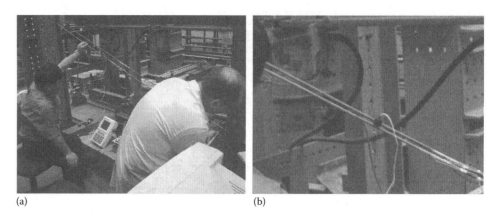

(a)   (b)

FIGURE 22.4
(a,b) Verification test on a tension bar.

TABLE 22.2 Comparison of the Measured and Predicted
Tension Forces for the 6 mm Bar

Experimental Results		Prediction Using Equation 22.54	
Natural Frequency (Hz)	Tension Force (kN)	Tension Force (kN)	Relative Error (%)
13.9	1.16	1.16	0.00
15.5	1.42	1.39	2.11
16.9	1.70	1.65	2.94
18.3	1.98	1.93	2.53
19.6	2.25	2.22	1.33
20.5	2.50	2.43	2.80
21.9	2.77	2.77	0.00
22.8	3.05	3.00	1.67

Source: Tzima, K. Predicting tension of a stayed cable using frequency measurements, MSc dissertation, UMIST, UK, 2003.

and the forces predicted using Equation 22.54. For predictions, the following values were used: $L = 2.55$ m, $\rho = 7850$ kg m^{-3} and $A = \pi r^2 = \pi(0.003)^2$ m^2.

From Table 22.2, it can be seen that the tension forces in the bar can be predicted using Equation 22.54 and the natural frequency measurements.

22.5.3 Tension Forces in the Cables in the London Eye

The wheel of the London Eye is stiffened by cables, as shown in Figure 22.5. There are 16 rim rotation cables, each around 60 mm thick. In addition, there are 64 spoke cables, which are all 70 mm thick and are spun from 121 individual strands in layers. After a period of time, part of the tension in the cables will be lost due to normal operations. Thus, the cables need to be retensioned to maintain their design values. It is not convenient to measure the tension force directly in each of the cables. Instead, transverse impact loads are applied to each cable to generate free vibrations, from

FIGURE 22.5
The London Eye.

which the fundamental natural frequency of the cable in the transverse direction can be determined. Using the measured fundamental natural frequency, the tension in each cable can be predicted using Equation 22.54 or similar equations. Tension losses can then be identified and appropriate action taken to rectify the situation.

22.6 SUMMARY

Following Chapter 21, this chapter further studies some relationships between static and dynamic problems. Such relationships help to simplify analysis, improve the accuracy of analysis and solve problems effectively. The key points presented in this chapter are summarised as follows:

1. Maximum displacement and natural frequency
   a. Equation 22.22 can be used to determine the fundamental natural frequency of a structure based on the known maximum static deflection that can normally be obtained from a static analysis. The accuracy of the prediction depends on the correctness of the selected shape function that should be close to both deformation shape and vibration mode shape.
   b. When the fundamental natural frequency is known, in particular from a measurement, Equation 22.23 can be used to predict the maximum displacement of the structure, which is often more accurate than that obtained without using measurements. A further example is given in Chapter 23.3.
2. Buckling load and natural frequency
   a. An alternative explanation of the buckling phenomenon is that when a member or a structure reaches its critical load, the stiffness of the member/structure is lost and its fundamental natural frequency becomes zero (Equation 22.38).
   b. The relationship between the buckling load and the natural frequency provides a nondestructive test method using measured natural frequencies to predict the buckling capacity of a structure. However, the application is limited as the selected shape function should be close to both the vibration mode and the buckling mode shapes.
3. Periodic load and static load
   a. The average of the integration of a periodic load over its period is equal to the static component of the periodic load. This provides an additional equation for determining one unknown in the definition of a periodic load.
4. Tension force and natural frequency
   a. The tension force in a cable can be predicted using fundamental natural frequency measurements on the cable (Equation 22.54).

REFERENCES

1. Cook, R. D., Malkus, D. S. and Plesha, M. E. *Concepts and Applications of Finite Element Analysis*, New York: John Wiley, 1989.
2. Clough, R. W. and Penzien, J. *Dynamics of Structures*, New York: McGraw-Hill, 1993.
3. Grossman, S. I. and Derrick, W. R. *Advanced Engineering Mathematics*, New York: Harper & Row, 1988.
4. Smith, A. L., Hicks, S. J. and Devine, P. J. *Design of Floors for Vibration: A New Approach*, Ascot: The Steel Construction Institute, P354, 2007.
5. Thomson, W. T. *Theory of Vibration and Applications*, London: Allen & Unwin, 1966.
6. Bachmann, H. and Ammann, W. *Vibration in Structures: Induced by Man and Machines*: Structural Engineering Document No. 3e, IABSE-AIPC-IVBH, Zürich, 1987.
7. Ji, T. and Ellis, B. R. Floor vibration induced by dance-type loads: Theory, *The Structural Engineer*, 72, 37–44, 1994.

8. Tuan, C. Y. and Saul, W. E. Loads due to spectator movements, *Journal of Structural Engineering*, 111, 418–434, 1985.

9. Ellis, B. R. and Ji, T. Floor vibration induced by dance-type loads: Verification. *The Structural Engineer*, 72, 44–50, 1994.

10. Morse, P. M. *Vibration and Sound*, New York: McGraw-Hill, 1948.

11. Tzima, K. Predicting tension of a stayed cable using frequency measurements, MSc dissertation, UMIST, UK, 2003.

CHAPTER 23

CONTENTS

Experimental and Theoretical Studies

23

23.1 CHARACTERISTICS OF EXPERIMENTAL AND THEORETICAL STUDIES

Experimental and theoretical studies are two types of investigation used in science and engineering. They can be conducted independently in structural engineering, but the two types of study have complementary characteristics which provide a basis for their combined use.

23.1.1 Experimental Studies

Experimental studies can provide accurate results, in the sense that measurements are directly taken on a structure, a structural element or a model of a structure, and no further assumptions or simplifications are needed. However, measurements are usually incomplete. For example, it can prove difficult to measure high-order natural frequencies and mode shapes on a structure, and it is normal to take any measurements at a limited number of accessible locations.

Experimental studies often provide direct information to help understand reality. However, it is possible that experiments can be conducted inappropriately, and may involve significant errors if they are undertaken without a good understanding of structural fundamentals. Figure 23.1a shows the test of a full-size beam–column connection which could form part of a real structure. In the test, the only loads involved are an axial load on the column and a vertical load applied to one of the horizontal members. In a joint, such as this, in a real frame structure, the continuity of members would lead to internal axial forces, shear forces and bending moments at what are the free ends in the test specimen. Therefore, the behaviour of this test connection will not fully reflect the actual behaviour of such a joint in a real structure. The experimental results are only truly valid when the boundary conditions and loading conditions are as shown in Figure 23.1a, and then they can be used to validate the results of numerical or other analysis based on the test model.

Figure 23.1b shows a simply supported, reinforced concrete beam subjected to a concentrated load acting at the centre of the beam, which was to be used to measure the fundamental natural frequency of the loaded beam. However, this would not be a suitable model for such a vibration test because the vertical load is applied to the test beam through a stiff steel beam that is supported by two frames. The measured natural frequency would not be that of the loaded beam but would be that of the whole system, consisting of the test beam and the stiff steel beam with its supporting system.

It is assumed for the experimental examples in this chapter that (a) experimental studies are conducted correctly and appropriately, (b) measurement errors are neglected and (c) the measurements may be considered to be correct and accurate. Hence, detailed experimental work will not be described and discussed.

23.1.2 Theoretical Studies

In contrast, theoretical studies, either analytical or numerical, can provide a complete solution to a problem. For instance, the stress and displacement at any point on a structure, or a natural frequency and vibration mode of any order can be calculated. However, the calculated results are normally, or at least sometimes, not accurate due to assumptions and simplifications adopted in the theoretical

(a)   (b)

FIGURE 23.1
Appropriateness of experimental studies. (a) Beam–column connection. (b) A reinforced concrete beam.

models, with errors being introduced due to differences between the actual structure and the theoretical model of the structure.

Theoretical studies can be divided into analytical studies and numerical studies.

Analytical solutions are normally explicit but limited. For example, the analytical solution of the buckling load of a simply supported strut is $P_{cr} = \pi^2 EI/L^2$. This solution states that the buckling load is proportional to the rigidity, EI, of the cross section of the strut and is inversely proportional to its length, L, squared. The solution is applicable to a strut of any length, cross section and material, and provides an understanding of the loading capacity of a strut which is useful for design. Such analytical solutions are normally limited to solving relatively simple problems. For example, it is difficult to obtain analytical solutions for the buckling loads of a multibay and multistorey frame.

Numerical solutions are normally unlimited but implicit. For example, the finite element (FE) method can be used to solve a wide range of problems, from beams to whole structures, with linear or nonlinear material behaviour, under static or dynamic loads. This type of modelling is so specific that if one aspect of the model is altered, such as the number of elements, a material property, a boundary condition, a dimension or the loading, then the whole model needs to be revised and the calculations have to be repeated.

It can be noted that analytical and numerical studies have complementary features and they can be combined where appropriate for reaching better solutions. These two types of studies are considered as theoretical studies in the following sections.

23.1.3 Basis for Combining Experimental and Theoretical Studies

In summary, theoretical solutions tend to be complete but may be inaccurate, whilst experimental solutions are accurate but incomplete. The two types of solutions have complementary characteristics.

When a theoretical study is conducted, the following questions should be answered.

1. Are the assumptions made correct?
2. Has the theory or method been verified experimentally?
3. How accurate are the results?

Similarly, when an experimental study is performed, similar questions need to be answered.

1. Are the measurements sufficient to achieve the desired objectives?
2. Can the measurements be explained theoretically?
3. Has the test been conducted appropriately?

All these questions should be answered if both theoretical and experimental studies are conducted, taking advantage of their complementary characteristics. If possible, it is best to combine the advantages of the two approaches to tackle practical and research problems more effectively. This leads to the further question of how to conduct complementary experimental and theoretical studies; more specifically, what are the relationships between the two types of study and how can these relationships be modelled and used?

23.2 MODELLING RELATIONSHIPS BETWEEN EXPERIMENTAL AND THEORETICAL STUDIES

When talking about modelling in engineering, one would often consider FE modelling. This is a quantitative type of modelling which requires details of a structure, such as its dimensions, material properties, loads and boundary conditions, to be defined before an analysis. Accordingly, the computed results relate to the particular set of input data provided. Any changes of the input data will lead to a different set of results.

Some ideas from economics can be borrowed to examine the relationship between experimental and theoretical studies [1]. The demand-and-supply analysis in economics has been used to explain many economic phenomena in competitive markets for many sectors. Figure 23.2 shows the demand-and-supply model that describes how prices vary as a result of a balance between product availability and demand. If the characteristics of the curves are presented in the format of structural concepts, it can be considered to state:

Demand: *The higher the price of a product, the less of it people would like to buy; or the lower the price, the more of it people would buy.*
Supply: *The higher the price of a product, the more of it would be supplied; or the lower the price, the less of it would be supplied.*
Demand and supply: *When demand becomes higher, as demand curve D_1 moves towards curve D_2, both price and quantity will increase to reach a new equilibrium point A on the supply curve (S), or when demand becomes lower, as curve D_2 moves towards curve D_1, both price and quantity will reduce to reach a new equilibrium point B on S.*

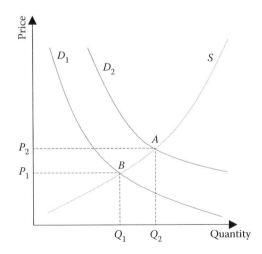

FIGURE 23.2
Model of relationship between demand and supply.

The relationships between the four observable attributes – *demand, supply, price* and *quantity* – are illustrated using simple plots rather than equations. The foregoing statements are an intuitive interpretation of Figure 23.2 and somewhere between those for structural concepts and intuitive understanding. This demonstrates a qualitative modelling based on market observations and the model can be used to explain and explore many economic phenomena in the market.

Similar to this modelling in economics, it is possible to qualitatively model the relationships between experimental and theoretical studies using diagrams based on observations and experience. This chapter explores such models and summarises six relationships between the two studies. Each relationship model is illustrated using a simple diagram and two real examples from engineering practice and research.

23.3 COMPARISON MODEL

23.3.1 Model and Features

The qualitative description of the first relationship between experimental and theoretical studies is shown in Figure 23.3, in which the subject represents a physical subject that may be a structure or a physical model of a structure, when both studies are applied.

Figure 23.3 indicates that both experimental and theoretical studies are applied to the same subject independently and the outcome from the two routes is compared. The features of this model are:

1. The experiment and analysis can be conducted independently without affecting each other.
2. Through a comparison of the results from the experiment and the analysis, the two independent studies benefit from each other and the results from the two studies are complementary.
3. Both experiment and analysis have equal importance.

The comparison between the experimental and theoretical results provides several benefits. If the two independent sets of results are sufficiently close, the results would be mutually supportive. In other words, it helps to answer the questions whether the theoretical results are correct and whether the measurements can be explained. The theoretical study produces data that may be difficult to obtain through experiment, and the experimental study provides evidence to support the theoretical results. If the two sets of results do not match well, the theory or method of application could be revised, based on the feedback from the experiment. Of the six relationship models considered here, the comparison model is the most frequently used in research and design.

23.3.2 Steel-Framed Building

This example shows that

1. *A finite element model of a building was improved, based on the feedback from experiment.*
2. *The finite element model predicted a vertical vibration transmission phenomenon that was not detected in the experiment.*
3. *The calculations were verified by the experiment.*

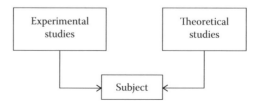

FIGURE 23.3
Comparison model.

Dynamic tests were undertaken on an eight-storey, full-sized steel-framed test building with dimensions: length 45.0 m, width 21.0 m and height 33.5 m. Numerical modelling of the building was carried out in parallel for each of its five distinct construction stages: (1) the bare steel frame; (2) the frame plus steel floor decks; (3) the frame plus composite floors; (4) the frame plus floors and walls; and (5) the frame plus floors, walls and static loads. Figure 23.4 shows the building at Stages 2 and 5 [2].

At Stage 4, when the internal walls were erected, the increases in the calculated natural frequencies were significantly larger than the increases in the measured values. Also, the increases in the natural frequencies for the shorter direction of the building and torsion mode were far larger than that for the longer direction. As both measured and calculated natural frequencies at Stage 3 matched well, it indicated that the walls in the calculations were modelled as being significantly stiffer than they should have been; particularly for the full height walls on the shorter faces. This comparison provided clear and useful feedback for indentifying errors in the modelling of the walls. The walls at the two ends of the building were divided into several areas by beams and columns, as shown in Figure 23.5; and in the construction of the walls, gaps between the walls and columns and beams were unavoidable. However, in the numerical modelling it is difficult to represent such real situations accurately and the assumption of continuity between walls led to significant differences between measurements and calculations. The wall in Figure 23.5 is actually divided into several small wall units by columns and beams, which is very different from a large continuous wall. The model of the walls was subsequently revised to attempt to reflect the actual situations leading to an improvement in the numerical model, based on the feedback from measurements.

A potential dynamic problem in the structure was identified through the FE modelling, as it was discovered that vibration transmission in the vertical direction could occur. Figure 23.6 shows the mode shape, with the left-hand drawing showing the shape on the sectional elevation and the right-hand drawing providing a three-dimensional view of a section of the building. This mode shape indicates that if vibration occurs on any floor between the second storey and the roof, a similar level of vibration will occur in the other floors in the range. This was due to the columns on the inner side of the entrance area being offset by 2 m from the columns on the upper floors. This led to an indirect load path and a relatively low frequency mode of vibration involving the floors from the second storey upwards, resulting in a potential serviceability problem. The FE modelling and analyses identified this phenomenon, which would have been difficult to detect experimentally unless this mode had been expected.

The theoretical prediction of vertical vibration transmission was confirmed by observation. When a loaded trolley was moved on the unfinished concrete floor in the fifth storey of the building

(a)   (b)

FIGURE 23.4
An eight-storey steel-framed test building (copyright of Building Research Establishment Ltd, reproduced by kind permission). (a) At construction Stage 2. (b) At construction Stage 5.

FIGURE 23.5
Shorter face of the building.

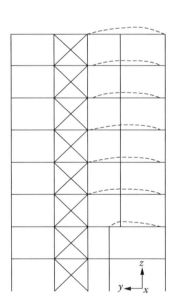 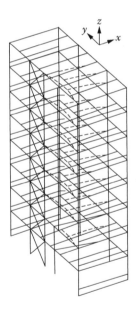

FIGURE 23.6
Vibration transmission in the vertical direction.

passing over the entrance area, several people on the roof were able to feel the vibration generated by the trolley three storeys below.

23.3.3 Appropriate Floor Model

This example shows *how an appropriate floor model was identified for engineering practice by comparing 11 measured and calculated natural frequencies.*

Long-span flat slabs, supported by columns, are widely used for floors in offices, shopping centres and airport terminals. The use of this type of structure can lead to a serviceability problem due to vibrations being produced by people walking, and this should be considered in design. The objective of this study was to identify an appropriate FE model for use in the design and analysis of such floors [3].

The seven-storey in situ concrete building shown in Figure 23.7a was used for this study. The building has seven 3.75 m storeys, giving a total height of 26.25 m. It has three 7.50 m bays, constituting a width of 22.50 m; four 7.50 m bays, making a length of 30.00 m; thus, forming 12 floor areas (panels) between columns (Figure 21.4). The building floors were designed as reinforced concrete flat slabs with a thickness of 0.25 m.

Dynamic tests were conducted and the accelerations at the centres of each of 11 floor panels were monitored in response to a heel-drop test at the centre of each area. The other floor panel had a limited area due to the openings for the staircase and elevator and no measurements were taken on this panel. The responses from the tests were converted to autospectra using a fast Fourier transform (FFT) procedure and the dominant natural frequencies identified, (many spectra identified a number of modes although one mode would dominate the response of each panel). The measurements provided a basis for evaluating the quality of different FE models.

Four commonly used FE models were evaluated by comparing 11 calculated and measured natural frequencies. They were: (1) A floor model with pinned supports from the columns. (2) A floor model with fixed supports from the columns. (3) A floor–column model: The columns linked with the floor are included in the model and the remote ends of the columns are considered to be fixed (Figure 23.7b). (4) A floor–column model considering the effect of reinforcement in the columns.

Table 23.1 compares 11 measured natural frequencies with 12 calculated natural frequencies for the 4 FE models and Table 23.2 summarises the ranges of the 11 ratios of calculated to measured natural frequencies with their averages in brackets. It can be seen that the model with the pinned supports (model 1) is less stiff than the real floor structure, while that with the fixed supports is stiffer. However, the floor–column models gave a much improved representation of the real structure, not only because the errors become smaller but also because the ranges of errors become smaller. This comparison was based on 11 measured natural frequencies of different floor areas, which gives confidence in the identification of the appropriate overall floor model showing that the floor–column model is the most appropriate. Nowadays, it is common to use floor–column models in practice and research.

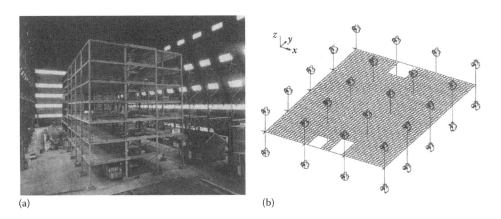

(a)   (b)

FIGURE 23.7
A seven-storey concrete building and an FE model of a typical floor in the building. (a) The concrete building (copyright of Building Research Establishment Ltd, reproduced by kind permission). (b) The column-floor model.

TABLE 23.1 Comparison between Calculated and Measured Natural Frequencies

Mode No.	Pinned Floor Model (1) Freq. (Hz)	Fixed Floor Model (2) Freq. (Hz)	Floor–Column Model (3) Freq. (Hz)	Floor–Column Model (4) Freq. (Hz)	Measured Natural Freq. (Hz)
1	7.12	9.26	8.25	8.38	8.54
2	7.20	9.31	8.30	8.42	8.54
3	7.60	9.42	8.53	8.64	8.54
4	7.70	9.57	8.62	8.73	8.54
5	7.98	10.05	8.96	9.09	9.28
6	8.10	10.19	9.18	9.30	9.28
7	8.75	10.37	9.41	9.53	9.53
8	8.82	10.56	9.73	9.84	10.01
9	9.76	11.54	10.72	10.83	–
10	10.21	11.57	10.76	10.89	10.74
11	10.50	11.90	11.04	11.13	11.72
12	10.94	12.44	11.56	11.67	11.96

TABLE 23.2 Ranges of the Ratios of Calculated to Measured Frequencies

Model No.	Ratio% (Average%) FE/Measurements	Maximum Difference (%)
Model 1	75–91 (87)	16
Model 2	95–112 (106)	17
Model 3	94–101 (98)	7
Model 4	95–102 (99)	7

23.4 INTEGRATION MODEL

23.4.1 Model and Features

As a theoretical solution may be complete, but inaccurate, and an experimental solution may be accurate, but incomplete, it is sensible to integrate the two methods. If a problem cannot be solved using either experimental or theoretical methods, it might be solvable when the two methods are integrated.

A qualitative description of the second relationship between experimental and theoretical studies is shown in Figure 23.8.

Figure 23.8 indicates that both experimental and theoretical studies are integrated before being applied to the subject.

The features of this model are

1. Integration of the two studies will lead to an outcome that is more accurate than the theoretical solution and more complete than the experimental solution.

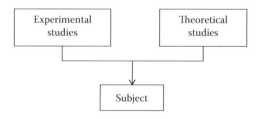

FIGURE 23.8
Integration model.

2. Due to their interaction, theoretical and experimental solutions cannot be performed independently.

3. The experimental and theoretical studies are equally important.

23.4.2 Dynamic Response of a Reinforced Concrete Beam

This example shows that *the accuracy of a response prediction is significantly improved when a measured natural frequency is integrated into the calculation.*

Modelling errors are often introduced in analysis due to differences between the model and the actual structure. One type of modelling error comes from inaccurate input data, such as the dimensions of a structure, or assumed material properties or boundary conditions. An example of a simply supported beam is given in Section 21.4.2 which shows a significant difference between the calculated and the measured fundamental natural frequencies, even for this simple structure.

An analytical method has been developed to predict the response of a structure subjected to individual jumping loads. For examining possible errors in this method using experiments conducted on a reinforced concrete beam, the possible errors in modelling the beam were minimised. This was achieved by substituting the measured fundamental natural frequency of the beam into the analysis [4].

A dynamic test was conducted on the reinforced concrete beam shown in Figure 20.6a. A person jumping up and down at the centre of the beam followed a clear musical beat. Displacements at the centre of the beam were recorded. In parallel, calculations were conducted in which the measured natural frequency was used. Figure 23.9 compares the measured and calculated dynamic displacements. The close match between the two sets of data indicates that the modelling errors for the beam were reduced by using measured data and the prediction method was reasonably accurate.

23.4.3 Floor in a Sports Centre

This example shows *how an integration model was used to solve a practical vibration problem that was difficult to deal with using either experimental or theoretical methods alone.*

A floor in a sports centre, normally used for bowling, experienced noticeable vibration when dance activities were held. This led to a requirement for a dynamic assessment to be carried out to assess the suitability of the existing structure for rhythmic exercises, including dancing to music, for up to 120 people. The sports centre had been built over 40 years prior to the time of assessment and Figure 23.10 shows the floor in question.

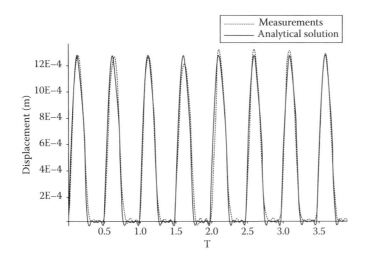

FIGURE 23.9

Comparison between measured and predicted dynamic displacements at the centre of a reinforced concrete beam when an individual was jumping following a musical beat.

FIGURE 23.10
A sports centre floor that experienced human-induced vibration.

It was not feasible to perform a test with many dancers since the safety of holding such activities was in question. Also, analyses could not be carried out as no drawings for the floor structure were available.

However, it was possible and easy to carry out dynamic tests on the unloaded floor to determine the fundamental natural frequency and the associated damping ratio of the floor. The measured natural frequency and damping ratio were then used in an analytical model to predict the response of the floor subjected to rhythmic crowd loads. The mode shape of the floor was not measured, hence it had to be assumed in the analysis. However, such an assumption would not usually lead to unacceptable modelling errors. The accuracy of the integration model has been demonstrated, as shown in Section 23.4.2. Thus, the integrated use of experimental and theoretical methods allowed the assessment of the suitability of the floor for dance activities.

23.5 VERIFICATION MODEL

23.5.1 Model and Features

When a new method, or model, is developed from theoretical studies, it is desirable that experimental studies are conducted to verify its correctness and accuracy. The experimental verification may be conducted in the laboratory or on-site. With appropriate verification, the method, or model, can be used with confidence when the prediction matches the measurements sufficiently well, or it can be improved if there are obvious differences. In this instance, experimental studies play a supporting role, while theoretical studies benefit from the experimental studies. The role of verification based on experimental studies includes

1. Ratification of the authenticity of the assumptions. Assumptions are normally used when developing a new method or a new model in theoretical studies. Different assumptions lead to different outcomes and incorrect ones give incorrect results. Some assumptions can be examined experimentally, which helps to provide a solid basis for theoretical studies.
2. Reinforcement of the certainty of predictions, since theoretical predictions have been verified by experimental results.
3. Justification of the feasibility of the proposed method.
4. Validation of the suggested model.
5. Checking the accuracy of the method.

The qualitative description of the third relationship, between experimental and theoretical studies, is shown in Figure 23.11.

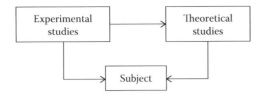

FIGURE 23.11
Verification model.

Figure 23.11 shows that both experimental and theoretical studies are applied to the subject, with the experimental studies supporting the theoretical studies. The features of this model can be summarised as follows:

1. Theoretical studies could be conducted independently of the experimental studies.
2. The theoretical outcome could be strengthened by experimental verification.
3. The theoretical studies play the main role while the experimental studies play a supporting role.

23.5.2 Ratification of the Authenticity of an Assumption

This example shows that *assumptions used for developing a theoretical study can be ratified using experimental studies. Without such ratification, the theoretical study might not be developed.*

Assumptions are often needed when developing a model or a method. For instance, when modelling human rhythmic jumping action on structures, which is encountered at pop concerts and in aerobics classes, it was assumed that human jumping was simply a dynamic load. Three scenarios were considered to model this human action:

1. Human rhythmic jumping acting as a load
2. Human rhythmic jumping acting as both loading and as a mass
3. Human rhythmic jumping acting as loading and as a mass-spring-damper

The three possible assumptions are qualitatively different and lead to different predicted structural responses caused by human rhythmic jumping. Theoretical studies could not resolve which assumption correctly reflects the actual situation, but experimental studies were ideal for examining which was correct. Two sets of experimental studies were conducted.

The differences between the three possible assumptions are that one is without a 'human body' mass, one has a mass and the other has a mass-spring-damper. Thus, these could be detected through measuring the fundamental natural frequency of a simply supported, reinforced concrete beam when a person jumped up and down on it. In one test case, the ratio of the body mass of the jumper to the modal mass of the beam was 0.52. The measurements showed that the fundamental natural frequency of the beam did not change when the person jumped on the beam and indicated that the person jumping acted solely as a load on the structure.

Another test was to compare the measured response of the beam subjected to an individual jumping with the theoretical prediction in which human rhythmic jumping was modelled as a dynamic load only. The comparison of the two sets of data is given in Figure 23.9.

The tests confirmed that human rhythmic jumping acts as a dynamic load only and the mass of the person does not contribute to the vibration of the structure. This scenario has been adopted in UK design guides [4–6]; however, the human mass is considered in U.S. and Canadian design guides [7].

23.5.3 Verification of Two Predictions

This example shows that *two theoretical predictions were enhanced, to different degrees, by experiments that confirmed the correctness of the theories.*

Individual inducing dance-type loads, where jumping is included, can be described as follows [8]:

$$F(t) = \begin{cases} \dfrac{G\pi}{2\alpha}\sin\left(\dfrac{f_p\pi}{\alpha}t\right) & 0 \le t \le \alpha T_p \\[2em] 0 & \alpha T_p < t \le T_p \end{cases} \tag{23.1}$$

where:

f_p is the load frequency, $T_p = 1/f_p$
α is the contact ratio, corresponding to different activities
t is time
G is the human body weight

Alternatively, Equation 23.1 can be represented in the following form through an analytical derivation [9]:

$$F(t) = G\left[1.0 + \sum_{n=1}^{\infty} r_n \sin(2nf_p t + \phi_n)\right] \tag{23.2}$$

where r_n and ϕ_n are Fourier coefficient and phase lag. Equation 23.2 indicates that the human loads can be represented as a summation of a series of harmonic loads. In other words, if one person jumps at a frequency of f_p, the dynamic load generated would have frequencies of f_p, $2f_p$, $3f_p$ and so on. Is this prediction correct?

Figure 23.12 shows an autospectrum based on acceleration measurements taken on a grand-stand with many spectators during a pop concert. It clearly shows that the structure responded to human action at 2.03 , 4.07 , 6.10 Hz and so on. This confirmed the correctness of the format of Equation 23.2.

A theoretical study of floor vibrations induced by dance-type loads found that resonance might occur on a relatively stiff dance floor with a fundamental frequency over 10 Hz. This was a prediction of possible resonance that had not been reported in research and practice, and one which might affect the serviceability of floors.

This resonance was simulated experimentally [10]. The natural frequency of a reinforced concrete beam was adjusted to 14.4 Hz. According to the theoretical prediction, resonance would occur at the sixth load component for jumping at exactly 2.4 Hz. This proved to be true. One of the difficulties with this experiment was providing the correct dance frequency, which is especially

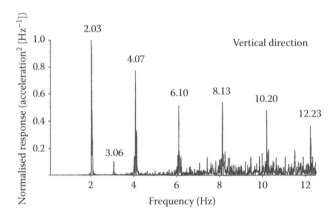

FIGURE 23.12
Autospectrum from vibration measurement at a pop concert.

important when dealing with a system with a very low damping value (in this case, 0.35% critical). If the calculation was repeated using a dancing frequency of 2.37 Hz (instead of 2.4 Hz), the evaluated maximum acceleration was one-third of the resonance value. This experiment not only validated the prediction but also verified the prediction method.

23.6 EXPLANATION MODEL

23.6.1 Model and Features

When an experimental study is undertaken, it is desirable that a corresponding theoretical study is conducted to explain any new observations. In this case, theoretical studies play a supporting role while experimental studies benefit from theoretical explanations. The explanation role based on theoretical studies includes the following.

1. Explaining measurements: Measurements may bring new or unexpected observations, which require explanation. The explanation should have a sound theoretical basis, to avoid any misinterpretation.
2. Exploring the nature of observation: This is more than simply providing an explanation, as it examines the causes of the observation based on a theoretical understanding possibly leading to new findings.
3. Identifying areas for further testing: Theoretical understanding can also help to identify further useful tests.

The qualitative description of the fourth relationship, between experimental and theoretical studies, is shown in Figure 23.13.

Figure 23.13 indicates that both experimental and theoretical studies are applied to the same subject, and the theoretical studies support the experimental studies. The features of the models are

1. The experimental studies could be conducted independently of the theoretical studies.
2. The experimental outcome could be strengthened by theoretical explanations.
3. The experimental studies play the main role, while theoretical studies play a supporting role.

This model is a mirror reflection of the verification model.

23.6.2 Effect of Stationary People in Structural Vibration

This example shows that *a theoretical study helped to explain experimental observations and led to the design of new tests to clarify the initial findings.*

In Chapter 20, stationary human body models in structural vibration were discussed and it was mentioned that a stationary human body acts as, at least, a mass-spring-damper system. Thus,

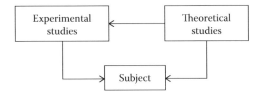

FIGURE 23.13
Explanation model.

a person (or a crowd) and a structure form a two-degree-of-freedom (TDOF) system in which a human body has a large damping ratio, of the order of 30%–50% [11].

From experimental studies it has been observed that

- The two resonance frequencies, or natural frequencies, of the human–structure system are rarely both observable.
- When the natural frequency of a human body is larger than the natural frequency of the structure, the first natural frequency of the human–structure system is often observed to be lower than the natural frequency of the structure.
- When the natural frequency of a human body is less than the natural frequency of the structure, the second natural frequency of the human–structure system is often observed to be higher than the natural frequency of the structure.

To help explain the experimental observation, a theoretical study was conducted to identify the conditions necessary for the presence, or experimental identification, of the two resonance frequencies of a highly damped TDOF system that included the human–structure system [12]. The study revealed that the presence of the two resonance frequencies of such a system depends on the damping ratio of the human body, the mass ratio of the human body to the structure and the frequency ratio of the human body to the structure. This provided a guide to the design of new tests for observing the presence of the two resonance frequencies.

A new test was designed for increasing the mass ratio while keeping other parameters unchanged. Figure 23.14 shows a single-degree-of-freedom (SDOF) test rig for studying people experiencing vertical vibration. Figure 23.15 provides the spectral responses for acceleration in the vertical direction for: (a) the empty test rig, (b) the rig with one standing person and (c) the rig with four standing people. It clearly shows the following:

- When one person stood on the rig, only one frequency was observed (Figure 23.15b). As the natural frequency of the test rig is lower than that of a standing human body, the first natural frequency of the human–structure system was measured, which is smaller than the natural frequency of the test rig (Figure 23.15a).
- When four persons stood on the same rig, there were clearly two frequencies present (Figure 23.15c), with one smaller and the other larger than the natural frequency of the rig (Figure 23.15a).

(a)   (b)

FIGURE 23.14
Human–structure interaction test. (a) Vertical SDOF test rig. (b) Four people standing on the rig.

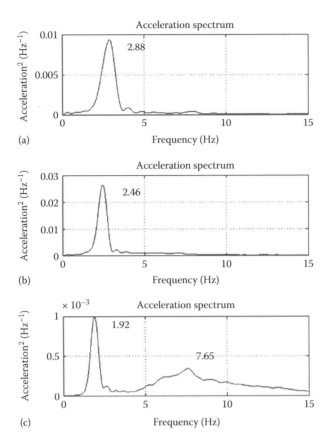

FIGURE 23.15
Spectral response for acceleration in the vertical direction. (a) Test rig. (b) One person standing on the rig. (c) Four people standing on the rig.

When the damping ratio of the upper SDOF system reduces, the two frequencies of the TDOF system would be present. The damping ratio of a human body in the lateral direction is much smaller than that in the vertical direction; hence, if the free vibration of a human–structure system in the lateral direction is tested, the two frequencies can be observed. This is evident in Figure 20.9d.

23.6.3 Lateral Stiffness of Temporary Grandstands

This example shows that *a theoretical study demonstrates the significant effect of the bracing pattern on the lateral stiffness of frame structures and explains why temporary grandstands have low fundamental natural frequencies in their horizontal directions.*

The structural safety of temporary grandstands had been considered to be an important issue following several incidents, the most serious being the collapse of the rear part of a temporary grandstand in Corsica in May 1992.

Fifty demountable stands, of 15 different types, were tested by the Building Research Establishment over several years [13]. The seating capacities of the grandstands varied from 243 to 3500. Only one stand had a vertical natural frequency below 8.4 Hz (at 7.9 Hz), indicating that there was no concern for human-induced vibration in the vertical direction. However, the natural frequencies in the two horizontal directions were low. Table 23.3 summarises the distribution of the natural frequencies in the sway and front-to-back directions.

The relatively low natural frequencies indicated that the structures had relatively low stiffness in the horizontal directions. Temporary grandstands are assembled using slender circular steel tubes, usually with the same cross section with a small second moment of area, and the links between the vertical and horizontal members are closer to an idealised pin connection than to a rigid

TABLE 23.3	Principal Horizontal Natural Frequencies of Temporary Grandstands	
	No. of Stands	
Frequency (Hz)	**Sway**	**Front to Back**
Under 3.0	15	10
3.0–3.9	17	13
4.0–4.9	13	9
5.0 or over	5	18

connection. Therefore, frames formed by horizontal and vertical members only are mechanisms and have little lateral stiffness. When enough bracing members are added, the structure becomes stable and it possesses a lateral stiffness. In other words, the low horizontal stiffnesses of temporary grandstands is due to their bracing arrangements. This point can be demonstrated by examining a frame structure with different bracing arrangements.

A four-bay and four-storey pin-connected frame structure is shown in Figure 23.16a. A pair of concentrated lateral forces of 0.5 kN are applied antisymmetrically at the two top corner nodes. The frame is stabilised and stiffened using eight bracing members, following the rules: (a) on each storey, two of the four panels should be braced (Figure 23.17a); and (b) in each of the braced panels, there are two possible bracing orientations (Figure 23.17b).

For bracing two panels from four in any storey, there are six options. For each braced panel, there are two possible bracing orientations for one bracing member. Thus, there are 24 bracing options in each storey, that is, $6 \times 2 \times 2 = 24$. Arranging bracing members for all four storeys gives a total of $24^4 = 331776$ options [14]. The best and worst bracing arrangements, in terms of the lateral stiffness of the frame, are identified from the all possible options and shown in

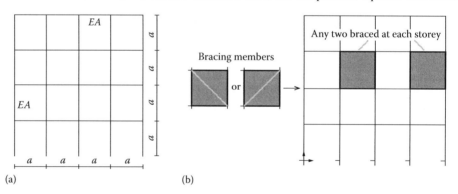

(a)   (b)

FIGURE 23.16
A four-bay, four-storey frame. (a) Geometry. (b) Bracing panels and orientations.

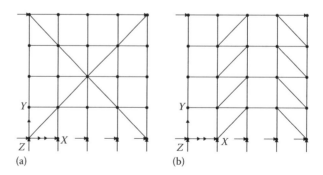

(a)   (b)

FIGURE 23.17
The effect on the lateral stiffness of a frame due to the action of bracings. (a) The stiffest frame. (b) The least stiff frame.

Figure 23.17. The maximum displacement of the frame in Figure 23.17b is 3.91 times that of the frame in Figure 23.17a.

This study shows the significant effect of the bracing pattern on the lateral stiffness of a simple frame and illustrates which bracing pattern should be used, or avoided, in practice.

23.7 CREATION MODEL

23.7.1 Model and Features

In the verification model, theoretical studies are generated from experimental studies by explaining the experimental results and identifying further experimental studies. However, there may be opportunities to develop theoretical studies further based on the experimental evidence, which may lead to the creation of a new theory or new methods.

The qualitative description of the fifth relationship between experimental and theoretical studies is shown in Figure 23.18.

Figure 23.18 indicates that theoretical studies are developed based on experimental evidence and the outcome from theoretical studies can be applied to practice independently of the experimental studies.

The features of the models are summarised as follows:

1. The creation of the new theory and method is based on measurements, that is, measurements provide a foundation for the development of theoretical studies.
2. The new theory, or method, can be applied independently to an application.
3. Without measurements, the new theory and method may not be advanced.

23.7.2 Human Body Models in Structural Vibration

This example shows that *a correct human body model was developed, based on the observation of human–structure interaction, which helps to advance understanding.*

The experimental studies of human–structural vibration on-site and in the laboratory have demonstrated that a stationary person does not act as an inert mass, but acts, at least, as an SDOF system in structural vibration. This has been discussed in Chapter 20. The experimental evidence has suggested a few questions for theoretical studies. How should a human body be modelled in structural vibration? How should the interaction between a structure and a person, or a crowd, on the structure be modelled?

It is normal practice to first model a standing human body (Figure 23.19a) as an SDOF system on a fixed support (Figure 23.19b), and simplify the support structure to be another SDOF system (Figure 23.19c). The body model is then combined with the structure model to form a human–structure interaction model (Figure 23.19d). This methodology for modelling human–structure interaction implies that the fixed base body model is valid when placed on a vibrating structure. This assumption needs to be verified.

A standing body is now placed directly on an SDOF structure system before modelling, as shown in Figure 23.20. This correctly considers the human body vibration in a vibrating structure [15].

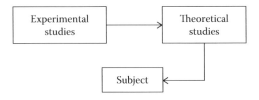

FIGURE 23.18
Creation model.

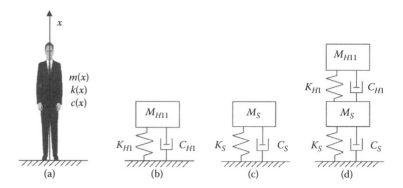

FIGURE 23.19
A standing body is modelled on a fixed base and then added to a vibrating structure to form a human–structure system. (a) Standing body (Hemera Technology 2001). (b) Body model. (c) Structural model. (d) Human–structure model.

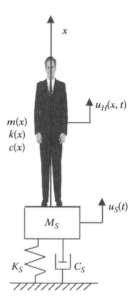

FIGURE 23.20
An individual standing on an SDOF structure system.

If only the first mode of vibration of the human body in the vertical direction is considered, the human–structure interaction model is shown in Figure 23.21a, in which the upper mass-spring-damper is the SDOF human whole-body model (Figure 23.21b) in structural vibration.

The features of the human–structure models are

1. The whole-body mass contributes to vibration
2. There is a mass coupling between the human and structure masses
3. There is a support mass in the body model
4. The mass matrix in the equations of motion is not a diagonal matrix

This theoretical study has led to an improved human body model in structural vibration, as shown in Figure 23.21b, and it has initiated further research, both experimental and theoretical.

23.7.3 Effective Bracing Systems for Structures

This example illustrates that *a new theory was developed to design stiffer structures which was based on the experimental work on the low horizontal stiffnesses of temporary grandstands.*

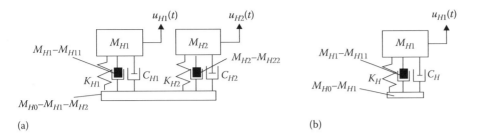

FIGURE 23.21
Human–structure interaction model and human body model. (a) Human–structure interaction model. (b) Human body model in structural vibration.

As many temporary grandstands can be considered to be pin-jointed truss structures, that is, their members are subjected primarily to axial forces, tension or compression, the maximum displacement of any such structure subjected to a unit force at its critical point is (Chapter 9)

$$\Delta = \sum_{i=1}^{s} \frac{N_i^2 L_i}{E_i A_i}$$

(23.3)

This equation has been in use for over 150 years [16] but it tends now not to be a major component of undergraduate courses in analysis of structures and mechanics of materials. This is because it requires a calculation of the internal forces N_i of all members of a structure which are difficult to determine for a statically indeterminate structure. Although this equation is taught at universities, students are normally only asked to use it to calculate deflections of simple statically determinate trusses.

Equation 23.3 is valid for a whole structure, with any layout and any number of members. The low lateral stiffness of temporary grandstands stimulated a close examination of Equation 23.3 and the intuitive interpretation of this equation led to the following three important structural concepts:

- The more direct the internal force paths, the stiffer the structure
- The more uniform the distribution of the internal forces, the stiffer the structure
- The smaller the internal forces, the stiffer the structure

These concepts have wide application in the design of structures, and some of them have been demonstrated in Chapters 9 and 10.

23.8 EXTENSION MODEL

23.8.1 Model and Features

In the verification model, experimental studies are used to support theoretical studies by providing experimental evidence and checking the accuracy of the theoretical results. However, there are also opportunities to develop experimental studies further to take the leading role, based on new or existing theories.

The qualitative description of the sixth relationship between experimental and theoretical studies is shown in Figure 23.22.

Figure 23.22 indicates that experimental studies or methods are developed based on a theory, and the outcome from experimental studies can be applied to practice independently of theoretical studies. The features of the models are summarised as follows:

1. The extension of experimental ability is based on theory, that is, the theory provides a basis for the new use of the experimental results.

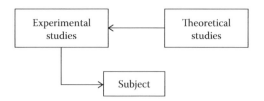

FIGURE 23.22
Extension model.

2. The experimental method can be applied to solving practical problems independently.
3. Without the theory, the experimental method or technique may not be proposed.

This model is a mirror reflection of the creation model.

23.8.2 Measurement of Human Whole-Body Frequency

This example shows *how a structural dynamics method was developed to measure the fundamental natural frequency of a stationary human body, based on the model of human–structure interaction. The experimental method can then be used independently.*

To study human–structure vibrations, it is necessary to know the fundamental natural frequency of a human whole-body. To measure the fundamental natural frequency of a human whole body, the technique usually adopted requires the use of a shaking table with an electrodynamic or hydraulic vibrator and a person sitting in a chair fixed to a stiff plate connected to the shaking table [11].

An alternative experimental method has been developed using the test rig shown in Figure 23.14a and based on the human–structure interaction models shown in Figure 23.21a. A person is asked to stand on the test rig, a shaker generates a harmonic input at a given frequency range and the acceleration–time history of the rig for 20 sec is recorded. In parallel to the measured acceleration processed in the frequency domain, the theoretical equation of the human–structure system subjected to a harmonic load can be presented in the frequency domain. The unknown parameters of a standing person, including the natural frequency and damping ratio, in the theoretical equation can then be determined from the best fit between the measured and calculated spectral curves.

This measurement method is based on the theory of human–structure interaction, but the experiment is now conducted independently to determine the fundamental natural frequency of a standing human body.

23.8.3 Identification of Possible Cracks in the Pinnacles at Westminster

This example shows *an experimental method that was developed, based on theoretical understanding, to identify cracks in the pinnacles at Westminster.*

A problem was encountered with the integrity of the decorative stone pinnacles on the Palace of Westminster in London. The pinnacles, which are made of limestone and contain a central vertical metal dowel, can be quite substantial structures with a height of several metres and, like all structures, the pinnacles will deteriorate with time. Deterioration can lead to a reduction in strength, and over many years several pinnacles have failed, with consequential damage to the structure below the pinnacle. Theoretically, if a pinnacle is cracked, or suffers other forms of deterioration, its stiffness will be reduced and this can be identified by measuring its fundamental frequency. As there are many similar pinnacles, the characteristics of any one pinnacle can be compared with those within the same group, thus identifying pinnacles which are significantly different to the norm.

FIGURE 23.23
Laser equipment used to measure natural frequencies of large pinnacles on the Palace of Westminster.

Proof tests were undertaken to establish the viability of a test method aimed at measuring the fundamental natural frequency, and hence stiffness, of individual pinnacles. Eventually, two types of test were established: an impact test for the smaller pinnacles, which involved attaching a transducer to a pinnacle and monitoring the pinnacle response to a gentle impact load; and a remote laser test for the larger pinnacles, which involved monitoring their ambient (wind) response. The full survey of the 534 pinnacles proved useful for identifying five suspect pinnacles and provided control data for comparison for future surveys [14]. Figure 23.23 shows the laser equipment, on the bank of the River Thames, aimed at a pinnacle on the Palace of Westminster.

23.9 LINKS BETWEEN THE RELATIONSHIP MODELS

The six relationship models and their respective features have been discussed individually but it is also useful to examine the logical links between the models.

For combined theoretical and experimental studies, theoretical studies may be undertaken before, at the same time or after experimental studies; the theoretical solutions may support or be supported by the experiment studies; and the two types of studies may act either independently or dependently. If the theoretical studies relate to the experimental studies in terms of time order and roles, the links between the six models may be represented as shown in Figure 23.24.

In Figure 23.24, a single line with an arrow indicates that one model could be developed from another model and a single line with two arrows indicates a symmetric/reflective relationship between two models. For the extension and explanation models, theoretical studies play a supporting role while experimental studies are being supported, or assisted, and take a leading role. Symmetrically, experimental studies play a supporting role while theoretical studies are being supported and take a leading role in the verification and creation of models.

Viewing the symmetric features shown in Figure 23.24 indicates that, as a whole, experimental and theoretical studies are equally important in research when the two types of studies are related.

It is recognised that benefits can be achieved through the combination of theoretical and experimental studies. In other words, the work performed using their combined capabilities is more useful, powerful and fruitful than the sum of the work produced by independently undertaken theoretical and experimental studies. Their combined capabilities strengthen the advantages and reduce the weaknesses of independent studies.

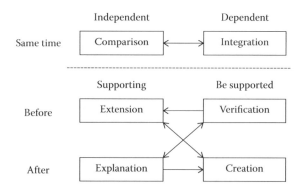

FIGURE 23.24
Relationships between the six models. (Theoretical studies relate to experimental studies in terms of time and role.)

Stimulated by modelling work in economics, the relationships between experimental and theoretical studies have been modelled qualitatively and have been illustrated using examples. The models can be used to explain many combined experimental and theoretical studies, but more importantly, it is hoped that the models could help to generate further higher-quality studies based on existing knowledge.

REFERENCES

1. Sloman, J., Wride, A. and Garratt, D. *Economy*, 8th edn, Harlow: Pearson Education, 2012.
2. Ellis, B. R. and Ji, T. Dynamic testing and numerical modelling of the Cardington steel framed building from construction to completion, *The Structural Engineer*, 74, 86–192, 1996.
3. El-Dadiry, E., Wahyuni, E., Ji, T. and Ellis, B. R. Improving FE models of a long-span flat concrete floor using frequency measurements, *Computers and Structures*, 80, 2145–2156, 2002.
4. Ellis, B. R. and Ji, T. *BRE Digest 426: The Response of Structures to Dynamic Crowd Loads*, Watford: Building Research Establishment, 2004.
5. Smith, A. L., Hicks, S. J. and Devine, P. J. *P354: Design of Floors for Vibration: A New Approach*, Ascot: The Steel Construction Institute, 2007.
6. Institution of Structural Engineers. *Dynamic Performance Requirements for Permanent Grandstands Subject to Crowd Action: Recommendations for Management, Design and Assessment*, London: Institution of Structural Engineers, 2008.
7. Murray, T.M., Allan, D. E. and Ungar, E. E. *Floor Vibration due to Human Activity, Steel Design Guide 11*, Chicago, IL: American Institute of Steel Construction, 2003.
8. Bachmann, H. and Ammann, W. *Vibration in Structures: Induced by Man and Machines*, Zurich: IARSE-AIPC-IVBH, 1987.
9. Ji, T. and Ellis, B. R. Floor vibration induced by dance type loads: Theory, *Journal of Structural Engineer*, 72, 37–44, 1994.
10. Ellis, B. R. and Ji, T. Floor vibration induced by dance type loads: Verification, *Journal of Structural Engineer*, 72, 45–50, 1994.
11. Griffin, M. *Handbook of Human Vibration*, New York: Academic Press, 1990.
12. Wang, D., Ji, T., Zhang, Q. and Dueate, E. The presence of resonance frequencies in a two degree-of-freedom system, *Journal of Engineering Mechanics*, 140, 406–417, 2014.

13. Ellis, B. R., Ji, T. and Littler, J. The response of grandstands to dynamic crowd loads, *Proceedings of the Institution of Civil Engineers, Structures and Buildings*, 140, 355–365, 2000.
14. Yu, X., Ji, T. and Zheng, T. Relationships between internal forces, bracing patterns and lateral stiffnesses of a simple frame. *Engineering Structures*, 89, 147–161, 2015.
15. Ji, T., Zhou, D. and Zhang, Q. Models of a standing human body in vertical vibration, *Proceedings of Institution of Civil Engineers, Structures and Buildings*, 166, 367–378, 2013.
16. Timoshenko, S. P. *History of Strength of Materials*, New York: McGraw-Hill, 1953.
17. Ellis, B. R. Non-destructive dynamic testing of stone pinnacles on the Palace of Westminster, *Proceedings of the Institution of Civil Engineers, Structures and Buildings*, 128, 300–307, 1998.

CHAPTER 24

CONTENTS

Theory and Practice

24

24.1 PRELIMINARY COMMENTS

It is often recognised that there are gaps between theory and practice. This chapter addresses the questions what are the gaps between theory and practice in structural engineering and how can such gaps be narrowed or bridged or both?

It is appropriate and more productive to examine the relationships between theory, concepts and practice than to study the relationships between theory and practice. This is because structural concepts give a clear interpretation of the mathematical relationships between physical quantities relating to structures and provide a basis for application in practice. In other words, concepts lie between theory and practice and provide a link between them. The study of structural concepts is unavoidably related to both theory and practice, to solving challenging engineering problems and improving structural efficiency.

Currently, there are no established methods for studying structural concepts. One of the reasons for this is that structural concepts are not physical objects. For example, a simple steel frame can be studied experimentally, or theoretically. Its behaviour can be examined from linearity and elasticity through to nonlinearity and nonelasticity. In addition, the effects of different loading schemes and different connections can be examined. This method of research can be used to investigate other types of structure, and structural components, but such research methods do not appear to be useful for studying structural concepts.

This chapter explores the relationships between theory, concepts and practice, and suggests several ways to study structural concepts based on the relationships. Once again, examples will be provided for illustration. It will demonstrate that the study of structural concepts helps to bridge the gap between theory and practice, and can draw together engineering practice, research and teaching. It also shows that an integrated study leads to any one of them contributing to, and benefiting from, the other two. However, some of the methods suggested in this chapter are neither mature nor complete, and future development and refinements are to be expected.

24.2 THEORETICAL AND PRACTICAL SOURCES FOR STRUCTURAL CONCEPTS

Structural engineering is considered to be a mature discipline. It is therefore questionable whether there are any new concepts to discover and whether all the structural concepts embedded in engineering structures and textbooks are fully recognised. To answer this question, it is necessary to examine the sources of the structural concepts, and where they can be accessed.

There are three independent sources of reference.

1. *Textbooks*: This is perhaps the major source for learning structural concepts. There are many structural concepts in textbooks, but not all of them are clearly identified and explored, or even understood. For example, Equation 9.6 in Section 9.2 can be seen in many textbooks which consider structural and stress analysis [1,2]. For many years, students have been taught to use this equation to calculate the displacements of simple trusses. However, insufficient effort has been given to identifying and presenting the important structural concepts embedded in the equation. Structural concepts in textbooks have helped engineers tackle many challenging problems, but some structural concepts are still not fully understood. Examples will be given to show how the concepts embedded in textbooks can be better presented for assisting research, practice, teaching and learning.

2. *Well-known structures and structural failures*: Many engineers and architects are quite creative when dealing with practical engineering projects. Studying their projects can show what particular measures have been used in solving challenging problems and what structural concepts are embedded in the solutions. Fortunately, many well-known structures are documented and published in either technical papers or books for the general public [3–5]. Important lessons can also be learnt from failures, some of which have been due to a misunderstanding or an ignorance of structural concepts. For example, Figure 9.9 shows the collapse of a scaffolding structure due to wind loads. It can be seen that no diagonal (bracing) members were present in the structure, that is, no direct internal force paths were provided. A study of this accident could have resulted in a study of the force paths in the structure, producing results which could in turn be applicable to many other structures.

3. *Newly emerged engineering problems*: As these are new problems that have not been fully investigated, it may be possible to identify new concepts and develop a new understanding. An example of this is the observation and identification of the phenomenon of human–structure interaction through dynamic tests on a grandstand in the United Kingdom [6]. This has led to both theoretical and experimental studies, a new understanding of stationary human body models in structural vibration and to the theory of human–structure interaction [7,8]. Some aspects of this problem are discussed in Chapter 20. Another example was the 'wobbling' of the London Millennium Footbridge [9]. Excessive lateral vibrations of the footbridge occurred when several thousand people walked across the bridge on its opening day, but such a phenomenon had not been thoroughly studied before. Experimental studies of structural problems in the laboratory, or on-site, provide opportunities to identify new structural concepts and gain an improved understanding of them.

Examples from these three sources will be illustrated in the next two sections.

24.3 RELATIONSHIP BETWEEN THEORY AND PRACTICE

24.3.1 Structural Concepts and Intuitive Understanding

When designing a structure, the first ideas for conceptual designs often come from the designer's intuitive knowledge and experience, and theory and analysis are only applied after conceptual design schemes have been produced. Pier Luigi Nervi, the Italian architect, engineer, constructor and educator, said 'The mastering of structural knowledge … is the result of a physical understanding of the complex behaviour of a building, coupled with an intuitive interpretation of theoretical calculation' [10]. Intuitive understanding of structures implies some direct knowledge of structures and structural concepts, without conscious learning and derivation. This can be gained from experience, and can be applied in practice. The physical models and practical examples provided in this book can help the reader to gain an intuitive understanding of structures.

Figure 19.8 shows that tyres were used to isolate the vibration of a generator in a rural area. People in the area did not have a university education, did not know the theory of vibration and probably had never heard of resonance; however, they knew how to reduce vibration from their own experience, or from that of others. The solution of the vibration problem was found using intuitive knowledge rather than through theory. An everyday extension of this solution is the use of rubber studs between washing machines and the floor, to reduce vibrations. Whilst the tyre solution worked, it would not have been possible to determine by how much the vibration would be reduced, or how the best level of vibration reduction could be achieved. This would have required a knowledge of the theory of vibration.

University students may not have had the opportunity to gain practical experience, but they can still gain an intuitive knowledge from theoretical studies and from examining concepts. Figure 24.1 illustrates the sources of intuitive knowledge, in which theory can be converted to intuitive knowledge through structural concepts, or how intuitive knowledge may be obtained from experience.

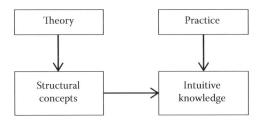

FIGURE 24.1
Sources of intuitive knowledge.

The following example will help to illustrate how intuitive knowledge is obtained from theory. Chapter 1 has shown that the maximum displacement of a simply supported uniform beam subjected to a uniformly distributed load is

Theory:

$$\Delta = \frac{5}{384}\frac{qL^4}{EI} = \frac{5}{6EIq}M_{max}^2 \tag{24.1}$$

where:

    q is the uniform loading
   EI is flexural rigidity
  $M_{max} = qL^2/8$

Equation 24.1 can be interpreted in terms of a structural concept as follows:

Structural concept: Deflection is proportional to the span to the power four, or to the maximum bending moment squared, and to the inverse of the rigidity of the cross section of the beam.
    The following intuitive knowledge or intuitive understanding can be developed from the concept:
    Intuitive understanding: Reducing span, internal forces and/or increasing the rigidity of the cross section will reduce deflection. Alternatively, a smaller span, smaller internal forces or larger cross-sectional rigidity or both will lead to a smaller deflection.

Equation 24.1 defines the deflection with specific properties and yields a precise value of the deflection. However, this equation is strictly for a simply supported uniform beam subjected to a uniformly distributed load, otherwise it is not valid.
    This structural concept is a qualitative and concise interpretation of Equation 24.1, formed by abstracting the essence of the equation and removing the specific value of 5/384 and the load q. The concept extends the application of Equation 24.1 in a qualitative manner. It is noted that the concept is straightforward and is a logic deduction from Equation 24.1, which effectively helps to understand the equation.
    Intuitive knowledge is an extension of the structural concept, and removes the specific descriptions of span, internal forces and rigidity in the concept. It is then applicable to many structures subjected to different types of loading. Due to the extension, the rigorous presentation of a theory is lost, but the physical essence of the theory remains. This is the knowledge that is often first applied in structural design by engineers and architects.
    Textbooks do not normally highlight concepts, or refer to intuitive knowledge, such as that shown here. However, readers can try to suggest intuitive interpretations of related equations which not only helps to gain a better understanding, but also prepares the readers to use the concepts behind the equations.
    A further example concerns the second moment of area of a cross section, which is defined as

Theory:

$$I = \int_A y^2 dA \tag{24.2}$$

where y is the distance between area dA and the neutral axis of the cross section. Equation 24.2 is used to calculate the second moment of area, I, of a cross section with any shape. The intuitive interpretation of Equation 24.2 is

Intuitive understanding: The farther away the material is from the neutral axis of a cross section, the larger the contribution of this material to the second moment of area of the section.

Readers can try to state the concept embedded in Equation 24.2 based on the definition of a structural concept given in Chapter 1. This example shows that an intuitive interpretation of Equation 24.2 is more straightforward than the presentation of the structural concept embodied in the same equation.

Intuitive knowledge can also be gained from practice, through either observation or experience. For example, a vibration investigation was conducted on a grandstand when it was empty and when it was full of spectators. Measurements showed that an additional natural frequency was observed in the occupied stand and the damping increased significantly when people were present (Section 20.4.1). The intuitive understanding from the measurements was that *a crowd of stationary people did not act as an inert mass but acted like a mass-spring-damper in structural vibration,* which is different to what has been illustrated in some textbooks. This intuitive understanding has led to further research to verify the observations, to develop human body models, to study human–structure interaction and to explore new applications (Chapter 20).

24.3.2 Structural Concepts and Physical Measures

If a theory, or a relationship, provides a basis for an application, then concepts or intuitive understanding suggest the direction in which physical measures should be developed for particular solutions.

The concept and intuitive understanding abstracted from Equation 24.1 indicate that reducing span, reducing internal forces or increasing second moment of area or both, can reduce deflections. However, particular physical measures have to be developed to realise the concept which should then address their use in meeting design requirements. Four practical examples are given in Figure 24.2, to illustrate how different measures are used to reduce the deflections or bending moments of real structures.

1. A support was placed at the centre of a footbridge which effectively increased the stiffness and reduced the internal forces in the footbridge. This was achieved through using a mast and cables as the physical means to realise the concept (Figure 24.2a).
2. A short-span deck was supported by scaffolding members. To increase the second moment of area of the support, several members were used to form a simple truss structure that had a significantly increased second moment of area (Figure 24.2b).
3. Cables were used as elastic supports to reduce the deflections and bending moments of the bridge deck (Figure 24.2c). The internal forces in the cables are balanced by the curved mast and a thick stayed cable.
4. In an airport waiting lounge (Figure 24.2d), roof trusses with overhangs were used to reduce the span between two supporting columns. In addition, the loads acting on the overhangs would reduce the bending moments in the midspans of the trusses. Example 6.2 examines the effect of a simply supported beam with overhangs.

These examples demonstrate that even though a concept, or intuitive understanding, may be relatively simple, different forms of physical measures can be developed to link the concept and design requirements. Such physical measures were used to solve practical problems in the structures illustrated in Figure 24.2a and b, and to improve the efficiency of the structures illustrated in Figure 24.2c and d.

The process of using structural concepts in the foregoing examples can be summarised using the diagram in Figure 24.3.

This process can also be conducted inversely, identifying a particular measure from a practical case and then abstracting a concept or intuitive understanding from the measure. Examples of this are provided in Section 24.4.

(a)

(b)

(c)

(d)

FIGURE 24.2
Different measures used to reduce deflections of structures. (a) An additional support is used.
(b) Forming a truss to increase the I value. (c) Cables are used to provide elastic supports.
(d) Overhangs are used to reduce the span and internal forces.

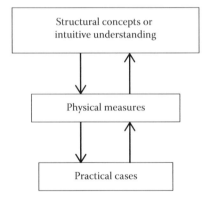

FIGURE 24.3
Bridging the gap between concepts and practice.

24.3.3 Theory and Practice

It is often noted that there are gaps between theory and practice which are reflected in two aspects relating to structural concepts. First, theories have not been, or cannot be, used in practice to solve engineering problems effectively. Second, many innovative measures in practical designs, which have provided effective solutions to challenging problems, have not been, or cannot be, developed into theory with wider applications.

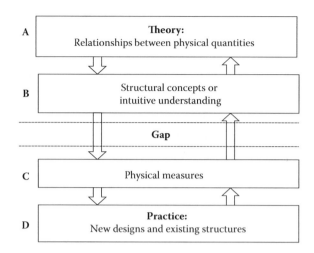

FIGURE 24.4
A relationship between theory and practice.

If the two diagrams in Figures 24.1 and 24.3 are combined, Figure 24.4 results. Here, the structural concepts and intuitive understanding are placed together, suggesting that intuitive understanding can be thought of as an extended expression of structural concepts:
    Figure 24.4 illustrates that

1. Through the study of structural concepts, the gap between theory and practice is effectively reduced to the gap between structural concepts (or intuitive understanding) and physical measures.
2. The gap can be bridged by either developing appropriate physical measures for practical structures based on concepts (or intuitive understanding), or identifying intuitive understanding (or concepts) from available measures used in existing structures.

24.4 BRIDGING GAPS BETWEEN THEORY AND PRACTICE

Figure 24.4 illustrates two approaches for studies involving structural concepts, one starting from theory and then moving downwards to end with practice and the other starting from practice and moving upwards to end with theory. The examples given in Section 24.3 follow the first approach.

24.4.1 Downward Approach: From Theory to Practice

The downward approach starts from the basic theoretical equations that provide relationships between physical quantities (A in Figure 24.4), then the concepts, or intuitive knowledge, are abstracted from the equations (B); the next step is to develop appropriate measures (C), based on the concepts for applications (D). The procedure can be represented as A→B→C→D. One example has been given in Section 24.3. Two well-known examples are now illustrated following the downward approach:

(A) Three *relationships of physical quantities* in textbooks:

$$\Delta = \frac{5}{384} \frac{qL^4}{EI} \tag{24.1}$$

$$I = \int_A y^2 dA \tag{24.2}$$

$$\sigma = \frac{My}{I} \tag{24.3}$$

(B) The concepts and intuitive understanding can be abstracted from the three equations as
  1. Stress and deflection are proportional to the inverse of the I value.
  2. The farther away the material from the neutral axis of a cross section, the larger its contribution to the second moment of area of the section.
  3. The material around the neutral axis has only a small contribution to the I value.
  4. Normal stress close to the neutral axis of the section is small.

(C) Two measures based on (B) have been developed and used in practice:
  1. Based on the first two statements in (B), the measure is to place as much material as possible away from the neutral axis of a cross section of a beam. For an I-shape cross section, two flanges are placed at the two ends of the web that links the flanges to form the section, which creates a section with a larger second moment of area than is the case with other cross sections using a similar amount of material. This is one of the reasons why I cross section beams, and columns, are widely used in building structures.
  2. Based on the third and fourth statements in (B), the measure is to take away material close to the neutral axis of I-section beams, thus forming cellular beams and columns. This measure reduces the I value and the loading capacity of the beams insignificantly. However, it does make more effective use of material (and also offers other nonstructural benefits).
      The second and third statements come from Equation 24.2 and highlight the development of different measures for the effective use of material. One involves placing material as far away as possible from the neutral axis of a section and the other involves removing material around the neutral axis of a section. A further example of using the second statement is illustrated in Figure 24.5.
      A newly built floor, with supporting beams, had a low fundamental natural frequency which might have led to excessive vibration when dance-type activities were held on the floor. To increase the fundamental natural frequency, or stiffness, of the floor, the possible solutions were, as indicated by Equation 24.1, to reduce the span using additional supports, or to increase the second moment of area of the supporting beams, or introduce measures to reduce internal forces. Increasing the second moment of area of the supporting beam was selected for this design. Additional materials were placed at the bottom side of the lower flanges of the I-sectioned beams, effectively increasing their second moments of area. The diagram to the right in Figure 24.5 shows an innovative use of the second statement: two vertical thick plates were used to increase the distance between the added bottom plate and the neutral axis of the beam to create an even larger second moment of area.

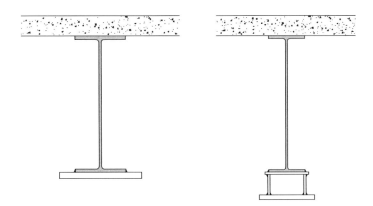

FIGURE 24.5
Measures for increasing the second moment of area of beams.

The downward approach in Figure 24.4 also indicates four other ways to study:

A→B: Work out structural concepts based on available equations, that is, relationships between physical quantities.

B→C: Develop physical measures based on existing concepts which make structures more effective.

C→D: Examine possible application of the measures.

B→C→D: Develop appropriate measures for dealing with particular practical problems.

A→B, B→C, C→D and B→C→D appear as a separation of A→B→C→D; however, the difference is that the four subactivities can be conducted independently and cover a wider range of topics related to structural concepts.

24.4.2 Upward Approach: From Practice to Theory

The upward approach requires identifying effective measures (C in Figure 24.4) used in practice (D), then abstracting the concepts, or intuitive knowledge (B), embedded in the measures; and finally, conducting a theoretical study to work out the relationships between the physical quantities (A). The procedure can be summarised as D→C→B→A. Once again, an example is used for illustration.

(D) The existing structure

The well-known Raleigh Arena is used in this case study. Section 10.4.1 provides a description of the structure and its internal force paths. Figure 24.6a shows a finite element model of the arena [11].

(C) The measure

Tendons (underground ties) were used between the ends of two pairs of inclined arches as a physical measure to balance a significant portion of the horizontal forces transmitted from the arches, thereby effectively reducing the forces on the foundations. In order to understand the effect of the tendons (the measure), a physical model was abstracted from the structure. The lower part of the two arches and a tendon can be simplified as a rigid frame, with a tendon linking the two ends of the frame, as shown in Figure 24.6b. The vertical force resulted from the upper part of the arches is applied on the top of the frame. Using symmetry, the

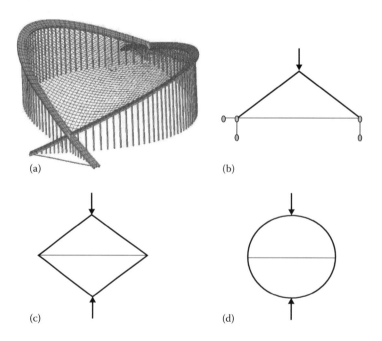

(a)   (b)   (c)   (d)

FIGURE 24.6

Identification of the measure (using a tendon) in the Rayleigh Arena. (a) A finite element model of the Rayleigh Arena. (b) A simplified model. (c) Equivalent representation of the model in (b). (d) A simplification of the model in (c).

open triangular frame in Figure 24.6b can be represented by the equivalent closed frame shown in Figure 24.6c, which allows removal of the supports of the model in Figure 24.6b. For easy construction of a physical model, the model in Figure 24.6c is converted to a ring with a tie at its middle (Figure 24.6d), which captures the physical essence of what is shown in Figure 24.6c and also the lower part of the structure.

(B) Concept and intuitive understanding

Figure 10.8 shows the deformations of two rubber rings, one with a horizontal wire tie across the diameter of the ring and one without the tie, loaded with the same vertical weight. It is apparent that the ring with the wire tie is much stiffer than the ring without the wire tie. The bending moment induced by the wire tie is in the opposite direction to that induced by the vertical forces, leading to smaller internal forces in the ring with the wire tie. This leads to the concept, or intuitive understanding, that the smaller the internal forces, the stiffer the structure. It also gives rise to the idea for creating self-balance systems for reducing internal forces in other structures.

(A) Theory

The theoretical study of the rings with and without a wire tie has been conducted in Section 10.2 and several equations or relationships between physical quantities have been provided. This is a study of the idealised models for identifying its fundamental characteristics. A further numerical study of the Raleigh Arena has been conducted including the finite element model shown in Figure 24.6a [11]. The theoretical background of the concept of smaller internal forces is presented in Chapter 10.

This example demonstrates how a concept embedded in a measure (the tendons) in a practical design was identified, studied and eventually developed to become a general concept that has wider applications in practice. The tendons were used to balance part of the horizontal forces from the arches in the Rayleigh Arena, and to reduce their action on the foundations. However, the research using the upward approach demonstrates that this measure creates a partly self-balanced system and makes a structure stiffer.

This example shows that it is beneficial to conduct research based on engineering practice. Engineering practice provides a source of ideas and appropriate measures for making structures efficient. Research is required to identify such ideas which may lead to new concepts and further applications.

Similar to the downward approach, the upward approach indicates other ways of study:

D→C: Identify the measures used in existing structures which may have positive effects on the performance of the structures.

C→B: Gain an intuitive understanding, or identify a concept embedded in the measures and their actions.

B→A: Develop a theoretical basis that can support the intuitive understanding and explain the action of the measures.

Similar to the downward approach, D→C, C→B and B→A are separations of D→C→B→A but the three can be conducted independently, without following the full sequence.

24.4.3 Interdisciplinary Approach: Combining Practice, Research and Teaching

Interdisciplinary research is effective for providing better solutions to problems. This involves studying a subject from different angles and using various methods, sometimes cutting across different disciplines to form new methods for understanding a subject. A feature of interdisciplinary research is that there are two or more disciplines and a common subject. For example, structural dynamics and body biomechanics are two different disciplines and human–structure interaction is a subject involving both disciplines. Therefore, different methods used with the two disciplines can be used to examine the subject and the methods developed for one discipline may be useful to the other. In addition, the outcome from interdisciplinary research is likely to be more significant than that produced from the independent research in either of the two disciplines.

Based on the definition of structural concepts and the diagram shown in Figure 24.4, structural concepts link the fundamental theories taught to students with real structures in engineering

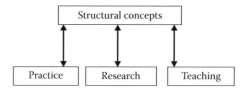

FIGURE 24.7
Interdisciplinary research.

practice. Research can be conducted between any two adjacent blocks in Figure 24.4. If engineering practice, research and teaching are treated as three different disciplines or areas, and structural concepts are considered to be the common subject across the three disciplines/areas, different methods can then be applied to the study of structural concepts, as shown in Figure 24.7.

The rationale for developing and studying structural concepts across practice, research and teaching is that

- Engineering practice provides a source for identifying new concepts and for creative use of existing concepts.
- Research and teaching help to identify new concepts and improve current understanding. The outcome then serves practice more effectively.
- Teaching and learning structural concepts requires input from engineering practice and research.
- The integrated research allows any of the three to contribute to and to benefit from the other two, and may lead to a more significant outcome than that from any one of the activities.

The contents of Chapters 9 and 13 are now briefly examined in the context of the integrated study of structural concepts through practice, research and teaching.

24.4.3.1 Designing for stiffer structures

The structural safety of temporary grandstands has been considered to be an important issue. Attention has been drawn to the problem by a number of incidents that have occurred, the most serious being the collapse of the rear part of a temporary grandstand in Corsica in May 1992. Insufficient bracing of the grandstand was the main cause of the incident [12].

Consider a pin-jointed structure, such as a truss, containing s bar members and n pinned joints, with a unit load applied at the critical point of the structure where the maximum displacement, Δ, of the structure is likely to occur. The displacement at the critical point can be expressed as a function of the internal forces N_i in the members as follows:

$$\Delta = \sum_{i=1}^{s} \frac{N_i^2 L_i}{E_i A_i} \tag{24.4}$$

A detailed explanation of this equation has been given in Chapter 9. Equation 24.4 contains internal forces and it can be a demanding task to determine the internal forces for a large truss or for a statically indeterminate truss. Therefore, this equation has limited use in practice and in the courses of *Structural Analysis and Mechanics of Materials* [1,2]. Equation 24.4 has been known for over 150 years [13], but the essence of the equation had not been explored until recently [14].

Triggered by the fact that many temporary grandstands in the United Kingdom had low natural frequencies in their horizontal directions, Equation 24.4 was studied for improving the stiffness of these grandstands, many of which can be considered as truss-type structures. The essence of Equation 24.4, through the links between the displacement and the internal forces, was identified and presented as three structural concepts [15]:

1. The more direct the internal force paths, the stiffer the structure
2. The more uniform the distribution of the internal forces, the stiffer the structure
3. The smaller the internal forces, the stiffer the structure

These concepts are important for practice, as they are applicable to whole structures due to the nature of Equation 24.4. Appropriate measures based on any of the three concepts can be developed. For example, effective bracing systems were developed based on the first concept, to make temporary grandstands significantly stiffer without using extra material. The new concepts and related material have also been used in teaching to enrich the contents of existing textbooks.

The downward approach was used in this example. It was a practical problem that triggered the research that identified Equation 24.4, as the theoretical basis. The research outcome not only provided an effective way to improve the lateral stiffness of temporary grandstands and other structures, but it also developed new concepts. This example also demonstrates that it is the measures developed, based on the concepts, that bridge the gap between theory and practice, and indicates that there may be other equations in textbooks that have not yet been fully explored.

24.4.3.2 Horizontal resonance of a frame structure due to vertical dynamic loading

It is normally considered that horizontal deflections of a structure are induced by horizontal loads. In textbooks, there are few examples of the lateral displacement of a frame due to a vertical load. Whilst this might not be an important issue in many situations when a structure, such as a temporary grandstand, is subjected to dominant vertical loads, nominal horizontal loads still need to be taken into account in design (often as a percentage of the vertical load) [16], implying that the lateral deformation of the structure is due to the horizontal loads. This is different from the concept of the horizontal response of a frame due to vertical loads, which is related to the structural form, including the height and width of a structure, and cross section properties of members.

The vibration of a permanent grandstand was monitored during a pop concert when spectators responded to the music. Several accelerometers were placed at the end of cantilever tiers as it was expected that this part of the structure would experience the largest vibration due to the vertical rhythmic crowd loads. However, the vibration of the whole grandstand in the front-to-back direction was far larger than that of the tiers and could be seen by observers at the event. This phenomenon had not been reported and discussed before in practice, research or textbooks. However, this observation led to an intuitive understanding that *the structure experienced resonance when its natural frequency in the front-to-back direction matched one the frequencies of the vertical rhythmic loads.*

Simple symmetric, antisymmetric and asymmetric frames suitable for teaching and learning can be used for deriving relationships between physical quantities (equivalent horizontal load, ratio of the height to width of the frame, ratio of the second moment of area of the beam to column and location of the loads) when vertical, static and harmonic loads are applied. Several concepts have been identified from this research and the phenomenon has been reproduced from numerical modelling and experiment in Chapter 13.

This research was also generated by an observation of a practical problem that a structure responded in the front-to-back direction when it was subjected to vertical rhythmic loads. The theory was established based on simple frame structures from which structural concepts were abstracted. The integrated research not only explained the vibration phenomenon and raised awareness for similar practical problems, but also enriched structural theories and led to new teaching and learning material for undergraduates.

There are several other examples in which practice, research and teaching are closely related in this book, such as pendulums in Chapter 15, human body models in structural vibration in Chapter 20 and the relationship between the static and modal stiffnesses of a structure in Chapter 21.

24.5 SUMMARY

This chapter provides a philosophical consideration of the relationships between theory and practice and the study of structural concepts for bridging the gap between theory and practice. The key points presented in this chapter are summarised as follows:

- There are gaps between theory and practice. It has been clarified that one of the gaps between theory and practice can be narrowed to that between structural concepts based on theory and physical measures used in practical cases (Figure 24.4). It is the measures developed from the concepts that bridge the gap. The measures can also be observed and identified from existing structures.

- As structural concepts link theory, physical understanding and practice, it is logical to develop several possible ways of studying structural concepts, as illustrated in Figures 24.4 and 24.7, which are supported by a number of real examples.
- Examples used in this chapter demonstrate that it is the practical problems that stimulate the study of structural concepts and research. The outcome from the studies contributes to the identification of new structural concepts, solutions for challenging practical problems, improvement in the efficiency of structures and enrichment of existing structural theories.
- Intuitive understanding of structures is close to structural concepts and can be obtained from practice and from structural concepts.
- Studying structural concepts brings practice, research and teaching together, allowing any one of the three to contribute to and benefit from the two other.

REFERENCES

1. Hibbeler, R. C. *Structural Analysis*, 8th edn, Singapore: Pearson Education South Asia, 2012.
2. Gere, J. M. and Timoshenko, S. P. *Mechanics of Materials*, Boston: PWS-KENT, 1990.
3. Schlaich, J. and Bergermann, R. *Light Structures*, London: Prestel, 2005.
4. Parkyn, N. *SuperStructures: The World's Greatest Modern Structures*, London: Merrell, 2004.
5. Jadidio, P. *100 Contemporary Architects*, Cologne: Taschen, 2008.
6. Ellis, B. R. and Ji, T. Human–structure interaction in vertical vibrations, *Structures and Buildings. The Proceedings of Civil Engineers*, 122, 1–9, 1997.
7. Wang, D., Ji, T., Zhang, Q. and Dueate, E. The presence of resonance frequencies in a two degree-of-freedom system, *Journal of Engineering Mechanics*, 140, 406–417, 2014.
8. Ji, T., Zhou, D. and Zhang, Q. Models of a standing human body in vertical vibration, *Proceedings of ICE, Structures and Buildings*, 166, 367–378, 2013.
9. Dallard, P., Fitzpatrick, A. J., Flint, A., Le Bourva, S., Low, A., Ridsdill Smith, R. M. and Willford, M. The London Millennium Footbridge, *The Structural Engineer*, 79, 17–33, 2001.
10. Nervi, P. L. *Structures*, New York: F.W. Dodge Corporation, 1956.
11. Yu, X. Improving the efficiency of structures using structural concepts, PhD thesis, University of Manchester, 2012.
12. Damiani, B., Poineau, D. and Millan, A. L. A technical report on the collapse of the north stand at Armand Cesari Stadium, France 1992 (in French), 1992.
13. Timoshenko, S. P. *History of Strength of Materials*, New York: McGraw-Hill, 1953.
14. Ji, T. and Ellis, B. R. Effective bracing systems for temporary grandstands, *The Structural Engineer*, 75, 95–100, 1997.
15. Ji, T. Concepts for designing stiffer structures, *The Structural Engineer*, 81, 36–42, 2003.
16. Institution of Structural Engineers. *Temporary Demountable Structures. Guidance on Procurement, Design and Use*, 3rd edn, London: Institution of Structural Engineers, 2007.

Index